Intelligent Systems and Simulation

This book covers a wide range of topics such as mathematical modeling, computational methods, simulation techniques in engineering, and their applications in environmental sciences. It begins with an introduction to the fundamental concepts and principles of mathematical modeling with subsequent sections delving into various simulation techniques, including numerical methods, statistical modeling, and bio-mathematical modeling. It also explores the application of these techniques in addressing complex environmental problems.

Features:

- Covers mathematical and computational modeling techniques for simulation in engineering, mathematical, and environmental domains.
- Explores intelligent techniques and simulation methods to enhance accuracy and efficiency in modeling and simulation.
- Includes practical applications and studies that demonstrate the effectiveness of techniques in real-world scenarios.
- Incorporates pertinent newly developed methodologies and applications.
- Discusses CFD software tools like Fortran, MATLAB®, and Mathematics.

This book is aimed at researchers and graduate students in computer engineering, applied mathematics, and applied ecology.

Intelligent Data-Driven Technology for Sustainability
Series Editors:
Siddhartha Bhattacharyya, Mario Koeppen, Sergey Gorbachev, Khan Muhammad, and Naba Kumar Mondal

The proposed series aims to describe evolution, design, and development of intelligent technology and models for analysis of data involved in environmental engineering with an overall objective of ensuring sustainability. This series explains intelligent and cognitive models to address issues related to effective monitoring of environmental pollution and ushering in a sustainable environmental design, thereby contributing to the overall well-being of the global environment for better sustenance and livelihood. It centres on evolving novel intelligent/cognitive models and algorithms to evolve sustainable solutions for the mitigation of environmental pollution. Sufficient coverage of novel cognitive models for the purpose of effective environmental pollution data management at par with the standards laid down by World Health Organization, is also envisaged.

Technological Advancement in E-waste Management: Towards Smart, Sustainable, and Intelligent Systems
Edited by Biswajit Debnath, Abhijit Das, Potluri Anil Chowdary, and Siddhartha Bhattacharyya

Intelligent Systems and Simulation: Mathematical and Environmental Modeling
Edited by Sharadia Dey, Sabyasachi Mondal, Abhijit Paul, Jyoti Sekhar Banerjee, and Ahmed J. Obaid

For more information about this series, please visit: www.routledge.com/Intelligent-Data-Driven-Technology-for-Sustainability/book-series/CRCIDTS

Intelligent Systems and Simulation

Mathematical and Environmental Modeling

Edited by
Sharadia Dey, Sabyasachi Mondal,
Abhijit Paul, Jyoti Sekhar Banerjee,
and Ahmed J. Obaid

CRC Press
Taylor & Francis Group
Boca Raton London New York

CRC Press is an imprint of the
Taylor & Francis Group, an **informa** business

Designed cover image: Shutterstock

First edition published 2025
by CRC Press
2385 NW Executive Center Drive, Suite 320, Boca Raton FL 33431

and by CRC Press
4 Park Square, Milton Park, Abingdon, Oxon, OX14 4RN

CRC Press is an imprint of Taylor & Francis Group, LLC

ISBN: 9781032799810 (hbk)
ISBN: 9781041070092 (pbk)
ISBN: 9781003638414 (ebk)

DOI: 10.1201/9781003638414

Typeset in Times
by codeMantra

Sabyasachi Mondal and Sharadia Dey would like to dedicate this book to their dear parents and loving children.

Abhijit Paul would like to dedicate this book to his parents, wife Sonali, and beloved son Soham.

Jyoti Sekhar Banerjee would like to dedicate this volume to his beloved parents, Late. Chandra Sekhar Banerjee and Mrs Rekha Banerjee, wife Arpita, and loving son Rishik.

Ahmed J. Obaid would like to dedicate this book to his family, many friends, and colleagues whose unwavering support and encouragement have been the guiding light in his life. A special feeling of gratitude to his loving wife Zainab R. Raheem; son Hamzah; and daughters Fatema, Malak, Teba, and Basmalah. He also dedicates this work to the University of Kufa, Faculty of Computer Science and Mathematics.

Contents

Preface

Mathematical modeling has undergone remarkable transformations in recent years, driven by advancements in computational power, data availability, and algorithmic techniques. These developments have opened new avenues for researchers and practitioners, enabling the construction of more accurate and sophisticated models that capture the complexities of real-world systems. Moreover, the growing concerns about the environment and the pressing need for sustainable solutions have propelled the application of modeling techniques to address environmental issues effectively.

In line with these advancements and the importance of modeling in enhancing our understanding of complex problems, this book, titled *Intelligent Systems and Simulation: Mathematical and Environmental Modeling*, sheds light on the role of modeling and simulation in science, technology, and engineering. It explores various modeling and simulation methods and tools that can be employed to tackle real-life complex problems. The contents of this book are thoughtfully organized into four parts, each focusing on specific aspects of modeling techniques for addressing research problems. The chapters in each section demonstrate the application of diverse modeling techniques and their efficacy across different domains.

The first part focuses on modeling in 'Pure Mathematics' and encompasses topics related to fixed point theory and its applications, fuzzy sets and soft sets with applications, and mathematical methods in functional analysis. The chapters in this part cover stability in iterative approximations, the Facility Location Problem (FLP) model using triangular membership functions, and fixed point models based on interval-valued fuzzy metrics.

Chapter 1: This chapter introduces groundbreaking methodologies aimed at revolutionizing the approach to tackling the Facility Location Problem (FLP) by harnessing the innovative power of triangular fuzzy numbers. Unlike traditional methods, this novel approach employs a meticulously crafted membership function to derive optimal solutions, significantly enhancing computational efficiency. Beyond its immediate application in FLP, the versatility of this approach extends across various domains, addressing numerous challenges in operations research (OR) and engineering sciences. This broad applicability suggests its potential to transform decision-making processes across diverse industries and sectors. Moreover, the proposed method not only offers a novel solution to FLP but also presents a systematic framework for decision-making under uncertainty, providing decision-makers with robust tools to tackle complex real-world problems. This chapter demonstrates the efficacy of the proposed techniques through a comprehensive set of mathematical examples, which validate the theoretical foundations of the approach and offer tangible insights into its practicality and effectiveness. Through this thorough exploration, the manuscript paves the way for the widespread adoption and implementation of these innovative methodologies, advancing problem-solving and decision-making in complex, uncertain environments.

Chapter 2: This chapter delves deeply into the intricate development and conceptualization of the continuous interval-valued t-norm and a pivotal construct built upon the foundational principles of continuous t-norm and interval numbers. It also introduces and rigorously defines the innovative notion of interval-valued fuzzy metric space (IVFMS). The extension of IVFMS is firmly grounded in the amalgamation of IVF(G) and continuous interval-valued t-norm, enriching the theoretical framework and expanding the horizons of fuzzy metric space theory. The exploration reveals that subject to specific conditions, each metric within this framework engenders an interval-valued fuzzy metric, elucidating novel insights into the interplay between metrics and fuzzy logic. This chapter's focus also meticulously elucidates various fixed-point models for expansion mappings on IVFMS and practical applications within this domain.

Chapter 3: This chapter explores the approximation of common fixed points within iterative sequences, focusing on modified iterations designed for generalized Lipschitzian mappings in Hilbert spaces. It presents an analysis of stability within these iterations, providing theoretical proofs and numerical examples to illustrate the concepts. By examining nonexpansive semigroups, the study delves into the intricacies of convergence and stability in iterative approximation methods. The modified iterative processes introduced offer novel insights into the dynamics of fixed point approximation, particularly in the context of generalized Lipschitzian mappings. Theoretical results establish stability criteria, enhancing understanding and applicability in various mathematical settings. Numerical demonstrations further validate the theoretical constructs, offering practical insights into the efficacy of the proposed methods. Overall, this chapter contributes to the advancement of iterative approximation techniques and their stability analysis, with implications for diverse fields reliant on fixed-point computation. The study considers the approximation of common fixed points in iterative sequences and demonstrates the modified versions of iterative processes for generalized Lipschitzian mappings in Hilbert spaces. Additionally, several results on the stability of these iterations are proved, and some numerical examples are presented.

The second part of this book bridges pure mathematics and applied mathematics modeling that mainly focuses on qualitative structure-property relationship (QSPR) modeling and the process of intermetallic compounds, which are the most critical man-made materials.

Chapter 4: This chapter demonstrates the power of qualitative structure-property relationship (QSPR) modeling through two examples. Topological indices—numerical values associated with graphs encoding structural information—play a pivotal role in QSPR modeling. To facilitate the computation of these indices, the M-polynomial, a mathematical expression representing molecular structure, is first calculated for the graph of silicon dioxide nanoparticles. Next, relevant topological indices, including the Zagreb index, modified Zagreb index, Randić index, forgotten index, harmonic index, and symmetric division degree index, are computed from the M-polynomial expression. The chapter then explores the fundamentals of QSPR analysis, focusing on simple and multiple linear regression models for silicon dioxide nanostructures. Models are created, analyzed, and assessed, revealing that the multiple regression model provides a more accurate prediction. Finally, the chapter concludes with a discussion on the importance of model validation, future outlooks, and the integration of QSPR modeling with machine learning techniques.

Chapter 5: This chapter focuses on intermetallic compounds, which are among the most critical man-made materials of interest to inorganic chemists, physicists, and materials scientists. Understanding systematic trends related to the phase stability of intermetallic compounds is a crucial aspect of chemistry. These compounds exhibit superior properties compared to ordinary metals. However, certain characteristics, such as brittleness, low room-temperature ductility, and poor fracture toughness, strictly limit their applications. The modified three-body interaction potential model (MTBIM), which incorporates the role of temperature, is used to calculate the electronic properties of americium telluride (AmTe), including bulk modulus, lattice constants, and structural phase transitions. The phase transition of AmTe occurs at a high pressure of 15 GPa. At ambient temperature, AmTe exhibits an NaCl structure, but under high pressure, its structure transforms from B1 (NaCl) to B2 (CsCl). This model effectively explains the phase transition of americium telluride and its other properties. Furthermore, the results obtained using this theory are consistent with those from other experimental and theoretical studies.

The third part of this book explores the methods and techniques of modeling in 'Applied Mathematics.' It aims to deepen the understanding of real-world challenges by advancing mathematical modeling, introducing novel applications, or combining both. The chapters in this part cover a wide range of topics, including heat transfer analysis in a ferrofluid, interactions with multiple fluid types in cosmology, phase transitions, heat capacity calculations, rumor propagation, the Hirota bilinear method, and the multi-domain local linearization method.

Chapter 6: This chapter examines the hydroelastic properties of a thin elastic disc submerged in deep water under wave impacts. The hydroelastic behavior of the elastic structure and its effects on physical quantities are investigated using modal analysis. A numerical solution to this one-dimensional Fredholm integral equation is obtained using the Nyström technique in conjunction with the Gauss-quadrature rule. The computed numerical findings are verified by comparing them with existing results for rigid discs submerged horizontally in deep water. This study presents the results of various parametric-dependent surface elevations and hydrodynamic forces. Additionally, it explores the impact of structural vibration on these physical quantities. The findings reveal that the peak magnitude of the hydrodynamic force increases with greater flexibility and disc size, while it decreases with increasing submergence depth. Furthermore, as the submergence depth decreases, the peaks of the surface elevation become sharper.

Chapter 7: This chapter presents a comprehensive exploration of multiple fluid cosmological models within interaction scenarios, offering a nuanced understanding of the universe's dynamics. This chapter primarily focuses on a binary interaction model, where three distinct types of fluids engage in intricate interactions that shape the cosmic landscape. Notably, the authors introduce variations in both the gravitational constant and the cosmological constant, examining the implications of modified gravity geometry. In this context, the modification of Einstein's gravity takes the form of a variable cosmological constant, adding layers of complexity to the analysis. Within this framework, the authors carefully examine the interplay between linear dark matter, an inhomogeneous fluid, and a novel nonlinear component, shedding light on their collective influence on cosmic evolution. Throughout the chapter, a fluid dynamical approach is adopted to analyze cosmic phases by varying essential parameters such as the scale factor, effective energy density, pressure, equation of state parameter, and gravitational constant. This meticulous examination reveals intricate relationships between different cosmological elements, offering valuable insights into the universe's evolution mechanisms.

Chapter 8: This chapter explores soliton theory, which efficiently describes nonlinear features such as regular, singular, and lump soliton structures embedded within nonlinear evolution equations (NEEs). Various collisionless plasma systems, expressed in their canonical forms, give rise to distinct NEEs, including Korteweg–de Vries-type equations and the Kadomtsev–Petviashvili equation, both of which are discussed in detail. This chapter analytically derives key results, including the phase shift in the interaction of regular and singular solitons. Additionally, it examines the influence of parameters such as the number density of hot electrons, the number density of cold electrons, and the spectral index on the formation of lump solitons.

Chapter 9: As we know, the oceans play a crucial role in climate variability, influencing changes in oceanic processes and circulation patterns over timescales ranging from seasons to centuries. This chapter presents a modeling framework for circulation and surface and subsurface dipole patterns over the Indian Ocean (IO) using a regional ocean modeling system (ROMS). The study incorporates air-sea fluxes from TropFlux products and boundary conditions from Simple Ocean Data Assimilation (SODA). The analysis demonstrates that ROMS can reproduce sea surface temperatures, with a warm bias of approximately 0.2°C–0.6°C. The model effectively simulates observed sea surface height variations, capturing anticyclonic eddies during spring and summer and cyclonic eddies during the fall season in the Bay of Bengal (BoB). Additionally, the regional model framework successfully represents seasonal currents in the BoB, the equatorial current, and the subtropical gyres in the southern IO. The findings indicate that the subsurface layer (30–120m) is significantly warmer (cooler) by about 2°C compared to the surface at the equator during positive (negative) Indian Ocean Dipole events. This study highlights the effectiveness of the proposed modeling framework in analyzing seasonal and climate modes in the Indian Ocean.

Chapter 10: This chapter illustrates how a drug-eluting stent (DES) with partially embedded struts, having a circular cross-section, distributes and retains the drug after implantation. A jump condition is imposed at the interfaces between the three layers of the artery wall: the intima, media, and adventitia. The drug is transported through the artery wall in two distinct forms: free and bound. Additionally, the drug is initially released from the strut surface through two mechanisms—time-dependent and time-independent methods. This chapter also provides a thorough analysis of the drug delivery mechanism in three different states: unsteady, quasi-steady, and steady-state. Furthermore, it highlights the effects of various clinically significant factors on the drug-binding mechanism.

Chapter 11: Nowadays, the spread of rumors has a profound impact on people's lives and their social circles. Many factors, such as important news, the religious or political significance of that news, the sudden outbreak of an epidemic, and the influence of media associations, affect the spreading process. In this chapter, we recommend that the credibility of rumor and user maturity have a significant impact on the spread of rumor. Considering the different attitudes of users toward rumors, we divide the total population into five groups: mature ignorant, immature ignorant, latent, spreader, and aware. We then derive the threshold value of influence, which outlines the requirements for rumors to be eradicated from social media or to continue spreading. Next, we investigate the local stability of the system around the steady states using the Routh-Hurwitz criterion and the global stability around the rumor-prevailing equilibrium using the Lyapunov stability theorem. Moreover, we present numerical simulations to support our theoretical study.

Chapter 12: This chapter examines a numerical study of heat transfer due to free convection of ferrofluid (Fe_3O_4/H_2O) inside a tilted enclosure under the influence of a magnetic field and subjected to heat generation or absorption. The enclosure has a sinusoidally heated left wall, a cold right wall, and thermally insulated curved walls. The dimensionless conservation equations of mass, momentum, and energy are transformed into a stream function-velocity formulation and solved using the FORTRAN code. The Bi-Conjugate Gradient Stabilized (BiCGStab) method is applied to solve the coupled equations, followed by the Tri-Diagonal Matrix Algorithm (TDMA) method. Results are obtained for different effective parameters: Rayleigh number ($10^3 \leq Ra \leq 10^5$), Hartmann number ($0 \leq Ha \leq 40$), concentration of Cu-water nanoparticles ($0 \leq \phi \leq 0.04$), heat generation or absorption parameter ($-10 \leq Q_0 \leq 10$), and inclination angle of the enclosure ($0° \leq \delta \leq 180°$). The computations show that the Nusselt number increases with the Rayleigh number and decreases with the Hartmann number. Additionally, at a high Rayleigh number, an increase in the heat generation parameter leads to a decrease in the heat transfer rate.

Chapter 13: This chapter presents a numerical investigation of natural convection nanofluid flow over the vertical frustum of a cone under the influence of thermal radiation, Soret, and Dufour effects. The equations are solved using a novel numerical approach—the bivariate multi-domain local linearization method (MD–SLLM). This method employs local linearization to linearize and decouple the equations. The time domain of the resulting linear system is decomposed into non-overlapping subdomains to ensure efficient implementation over a very large time domain. The equations are then solved within each subdomain using the spectral collocation method to discretize the independent variables. The solution and residual error norms are analyzed to verify the convergence and accuracy of the numerical method. Computational efficiency and the number of iterations required to achieve convergence are also investigated. The results suggest that MD–SLLM is a highly accurate method that requires few grid points to produce convergent solutions in minimal time. Furthermore, the findings indicate that regardless of the size of the time domain, MD–SLLM consistently yields highly accurate results.

Chapter 14: This chapter introduces an improved multi-derivative hybrid block method (IMDHBM) of one-step number and order $M+6$. The approach employs multi-step collocation and interpolation methods using a power series polynomial as the basis function, while the intra-step or off-step points are obtained from the derivative of a shifted Legendre (S-L) polynomial of degree four. The accuracy, consistency, and stability properties of the method are analyzed. The proposed method, combined with a Gauss-Seidel-type iterative method, is used for the numerical solution of systems of nonlinear initial value problems (IVPs), as well as IVPs exhibiting general chaotic behavior, namely the Rössler system (RS) and the Arneodo–Coullet system (ACS). The approximate numerical outcomes demonstrate that the method is accurate and highly effective in solving IVPs over large time intervals. IMDHBM emerges as a powerful tool for tackling complex and chaotic dynamical systems, promising significant contributions to the fields of numerical analysis and nonlinear dynamics.

The third part covers chapters on various modeling techniques with applications in different engineering fields. It specifically emphasizes mathematical methods in computer engineering and information technology, such as a book recommendation system using machine learning-based collaborative filtering and the assessment and prediction of water quality using artificial neural network techniques. The section on environmental modeling focuses on simplifying complex environmental systems and problems through appropriate modeling approaches. These chapters explore computational methods for addressing contemporary water and air pollution issues. Topics covered include a computational model for estimating human inhalation particle dosimetry based on the physical and physiological parameters governing transport within the respiratory tract, a state-of-the-art modeling framework with multiscale validations of a regional ocean model, a chemical kinetics-based model for understanding the formation of tropospheric ozone, the HYSPLIT model for back trajectory analysis of horizontal pollutant transport, and an ANN model for evaluating water quality.

Chapter 15: In an era of severe water constraints, poor water quality is a significant limitation to ecological functions and the maintenance of a healthy life. To develop effective strategies for proper water management, it is essential to have a clear understanding of an area's water quality. In this regard, this chapter aims to create a neural network model to forecast dissolved oxygen (DO) levels based on biochemical oxygen demand (BOD) and chemical oxygen demand (COD) at seven monitored stations across Kolkata. To evaluate the variability of these parameters, box-whisker plots are drawn for each variable. Different neural network models are developed to predict DO levels at various sampling stations in Kolkata. Three multilayer perceptron (MLP) models are selected for this study. The efficacy of artificial neural network (ANN) models is analyzed using root mean square error and mean absolute error. The results show that the MLP-3 model achieves the lowest forecast error in predicting DO levels at the seven sampling stations. These findings indicate that ANN models can be effective tools for forecasting DO levels in the study area.

Chapter 16: Tropospheric ozone is a secondary pollutant that is formed through a series of complex photochemical reactions in the presence of oxides of nitrogen (NO_x) and volatile organic compounds (VOCs). This chapter identifies the occurrence of elevated ozone episodes over Three Rivers site of Gauteng, South Africa, during January to December 2015. A chemical kinetics-based model is used for understanding the role of basic NO_2-NO-O_3 photochemical cycle in ozone production and tropospheric ultraviolet and visible (TUV) radiation model (Version 4.1) of NCAR is used for computing the rate constant for the photolysis of NO_2 $\left(j_{NO_2}\right)$. Ozone photochemistry is found to be only NO_x driven during morning and evening hours. But some other ozone precursors (benzene, toluene, xylene, and carbon monoxide) also play a vital role in ozone production during the late morning to afternoon hours. Maximum deviation among the observed and model-predicted ozone values is observed around noon. Statistical analysis and computation of Leighton Ratio also support the contribution of other chemical reactions (apart from the basic NO_x cycle) in the photochemical production of tropospheric ozone. Long-range horizontal transport of ozone and its precursors from polluted industrial regions is portrayed by HYSPLIT back trajectory simulations. Meteorological variables (temperature, humidity, solar radiation, precipitation, and cloud cover) influence the concentration of surface ozone over the study site.

Chapter 17: In recent times, remotely sensed satellite data have become indispensable in pollution monitoring studies. As these data sources gain popularity, it is essential to ensure their accuracy by comparing it with ground-based data. This chapter focuses on the correlation between particulate matter representations from satellite data and ground observation stations (WBPCB) in the Lower Gangetic Plain (LGP) of West Bengal. MODIS AOD and CAMS satellite data for PM_{10} are found to overestimate actual values across the LGP, particularly in the eastern and southeastern regions, where anthropogenic activities are minimal (Pearson correlation coefficient ranging between −0.43 and 0.22). However, satellite data effectively depict the high pollution levels in urban areas, exceeding the standard limits, while TROPOMI AI (aerosol index) data successfully identify pollution hotspots in the LGP. Back trajectory analysis of particulate matter using HYSPLIT data traces pollution sources in the LGP to long-range transport from the northwest during the winter season, as well as local sources such as industrial emissions and crop-residue burning. The study concludes that while there is a generally positive correlation between satellite and ground-level data, significant discrepancies exist in actual values.

Chapter 18: The burning of firecrackers during Diwali, combined with prevailing meteorological conditions, leads to poor air quality in various parts of India. Increased $PM_{2.5}$ levels in the atmosphere have a pronounced impact on health, particularly in children. This chapter investigates the ambient concentration of $PM_{2.5}$ and meteorological parameters over Victoria Memorial, Kolkata, during the pre-Diwali, Diwali, and post-Diwali periods from 2018 to 2022, as well as the deposition of $PM_{2.5}$ in different parts of the respiratory tract in children using the Multiple Path Particle Dosimetry (MPPD) model. Findings indicate that the pulmonary region receives the highest deposited mass rate, followed by the head region and the tracheobronchial (TB) region. Due to a very slow clearance mechanism in the pulmonary region, $PM_{2.5}$ accumulates most heavily in this area. Meteorology plays a crucial role in $PM_{2.5}$ deposition in the respiratory tract. The variability of the deposited mass rate of PMis explained by three meteorologica is explained by three meteorological parameters—temperature, relative humidity, and wind speed—accounting for 79.4%, 85.7%, 91.9%, 72.6%, and 71.8% of the variations during the post-Diwali periods of 2018, 2019, 2020, 2021, and 2022, respectively. This study provides valuable insights into the potential health risks associated with exposure to and inhalation of high concentrations of ambient $PM_{2.5}$ in children.

Chapter 19: This chapter explores the development of a book recommender system employing collaborative filtering, a widely used technique in recommendation systems within the broader field of artificial intelligence. The investigation covers fundamental aspects such as the definition, tools, and types of both machine learning and recommendation systems. Recommendation systems have become integral to daily life, with widespread applications. This research specifically focuses on collaborative filtering, utilizing tools such as SciPy and Scikit-learn. A thorough analysis of these tools and algorithms is conducted to optimize precision, ultimately leading to the development of a robust book recommendation system. The implementation process involves key steps, including data gathering, data cleaning, model building, model training, model testing, and model optimization. The chapter concludes with an evaluation of the tools' efficiency, insights gained from the project, and suggestions for future work in similar system domains. This work contributes to a broader understanding of collaborative filtering and provides practical insights into the development of effective recommendation systems.

This book brings together a collection of contributions from leading researchers and experts, offering a multidisciplinary perspective that bridges the gap between mathematics, computer science, and environmental science. Emerging problems are addressed in detail, employing theoretical, computational, and simulation techniques. This book is structured with clear steps and calculations to enhance students' and researchers' comprehension of the subjects.

This book will be an invaluable resource for students, researchers, academicians, and technology developers engaged in research on modeling and simulation in science, technology, and engineering. Its comprehensive coverage and practical approach will aid readers in gaining a thorough understanding of the subject matter and its applications.

Sharadia Dey, Ph.D.
Kolkata, India

Sabyasachi Mondal, Ph.D.
Shillong, India

Abhijit Paul, Ph.D.
Kolkata, India

Jyoti Sekhar Banerjee, Ph.D.
Kolkata, India

Ahmed J. Obaid, Ph.D.
Kufa, Iraq

Acknowledgements

We express our deepest gratitude to all the contributors who have generously shared their expertise and insights. Without their valuable contributions, this book would not have been possible. We also extend our appreciation to the editorial and production teams, whose diligent efforts have brought this book to fruition.

Editor Biographies

Dr. Sharadia Dey is working as an Assistant Professor in the Department of Environmental Studies at St. Xavier's College (Autonomous), Kolkata, West Bengal, India. She completed B.Sc. (Honours) in Chemistry from St. Xavier's College, Kolkata, University of Calcutta, West Bengal, India, and M.Sc. in Atmospheric Science from Ballygunge Science College, University of Calcutta, West Bengal, India. She obtained her Ph.D. from the Department of Environmental Science, University of Burdwan, West Bengal, India, and she completed her postdoctoral research at the University of KwaZulu-Natal, South Africa.

Her research interests include air quality, health, and environmental modeling. Along with 12 years of teaching experience, she is also actively engaged in research . She has published her research work in reputed international journals, conferences, and book chapters. Additionally, she has been successfully working on three national and international research projects. She has research collaborations with scientists in India and abroad.

Dr. Sabyasachi Mondal is an Associate Professor in the Department of Mathematics at North-Eastern Hill University, Shillong, Meghalaya, India. He earned his B.Sc. (H) in Mathematics, M.Sc. in Applied Mathematics, and Ph.D. in Computational Fluid Dynamics from Visva-Bharati Santiniketan, West Bengal, India. He worked as a Postdoctoral Research Fellow at the University of KwaZulu-Natal, South Africa.

His areas of research interest include computational fluid dynamics, heat and mass transport, nanofluid flow, etc. Along with his 13 years of teaching experience, he has more than 90 international publications in different reputed journals, books, and conferences in his credit. Also, he published one Indian patent. He delivered invite talks at different universities and colleges. He is an editorial board member in different international journals and a reviewer of various reputed journals.

He has received various prestigious funds and fellowships like the UKZN PDF fellowship, Claude Leone Postdoctoral fellowship, NRF travel grants, etc. He is a lifetime member of the Calcutta Mathematical Society and the International Association of Engineers (IAENG). He has research collaborations with various eminent scientists in India and abroad, like South Africa, Botswana, Malaysia, Iran, etc. His research has led to him being selected on the 'Top 2% Most Influential Scientists List' published by Stanford University for the year 2022–2023.

Dr. Abhijit Paul is an Associate Professor in the Department of Computer Science at Swami Vivekananda University, Kolkata, India. Prior to this role, he served at Assam Down Town University, Guwahati, and Amity University, Kolkata. He holds a B.Sc. (Hons) in Mathematics from Assam University, Silchar; an M.C.A. from the West Bengal University of Technology; and a Ph.D. in Computer Science from Assam University, Silchar.

His research interests encompass IoT, sensor networks, ad hoc networks, and machine learning. With 8 years of teaching experience, Paul has contributed to over 11 international publications in reputed journals, books, and conferences. He is also the inventor of one Indian patent and serves as a reviewer for prestigious journals. He actively collaborates with prominent researchers both in India and internationally.

Dr. Jyoti Sekhar Banerjee is currently serving as the Head of the Department of Computer Science and Engineering (AI & ML) at the Bengal Institute of Technology (BIT), Kolkata, India. He is also the Professor-in-Charge, R & D and Consultancy Cell & Nodal Officer of the IPR Cell of BIT. Since 2024, he also works as a Remote Researcher in the Internet of THings & AppliCAtions (ITHACA) Lab at the Department of Electrical and Computer Engineering, University of Western Macedonia, Greece. He is the former Remote Research Fellow of the Cognitive Computing and Brain Informatics (CCBI) Research Group at Nottingham Trent University (NTU), UK. He did his Postdoctoral Fellowship at Nottingham Trent University, UK, in the Department of Computer Science. He also completed the Postgraduate Diploma in IPR & TBM from MAKAUT, West Bengal. He has teaching and research experience spanning 20 years and completed one IEI-funded project. He is a member of the CSI, IEEE, ISTE, IEI, ISOC, and IAENG and a fellow of IETE. He is the present Secretary-cum-Treasurer of the ISTE WB Section and Secretary of the IETE, Kolkata Centre. He is the Immediate Past Secretary of the Computer Society of India, Kolkata Chapter. He is also elected as the Vice Chairman Cum Chairman Elect in Computer Society of India, Kolkata Chapter for the year 2025–2027. He has published over 80 papers in various international journals, conference proceedings, and book chapters. So far, he has published 15 edited books and four text books. He is serving as the Series Editor for *Advances in Disruptive Technologies and Generative Artificial Intelligence*", published by CRC Press, USA. He is also the Series Editor for the *Conference Proceedings Series on Futuristic Intelligent and Smart Technologies (FIST)*, CRC Press, USA. Currently, he is serving as the General Chair of ICHCSC 2022, 2023, 2024, and 2025; and General Co-Chair of GCAIA 2021. He is the Program Chair of the Human 2023, 2024, and 2025 and AAIDS 2024. He is also the General Chair of the AIEST-2024, NEIAIS-2025, WIN 6.0 2025, and ICADAC 2025. He is the Organizing Chair of the EAIT 2024, ICNSBT 2023. 2024, and ICACA 2024. His areas of research interests include computational intelligence, cognitive radio, sensor networks, AI/ML, network security, different computing techniques, IoT, WBAN (e-healthcare), and expert systems.

Ahmed J. Obaid is an Assistant Professor in the Department of Computer Science, Faculty of Computer Science and Mathematics, University of Kufa, Iraq. He holds a bachelor's degree in Computer Science and a degree in Information Systems from the College of Computers, University of Anbar, Iraq (2001–2005); a master's degree (M. TECH) in Computer Science Engineering (CSE) from the School of Information Technology, Jawaharlal Nehru Technological University, Hyderabad, India (2010–2013); and a Ph.D. in Web Mining from the College of Information Technology, University of Babylon, Iraq (2013–2017). He is a Certified Web Mining Consultant with over 14 years of experience working as a Faculty Member at the University of Kufa, Iraq. He has taught courses in web designing, web scripting, JavaScript, VB.Net, MATLAB Toolbox, and other courses on PHP, CMC, and DHTML from more than 10 international organizations and institutes in the USA and India. He is a member of the Statistical and Information Consultation Center (SICC), University of Kufa, Iraq. His main line of research is web mining techniques and applications, image processing in web platforms, image processing, genetic algorithm, information theory, and medical health applications. He has supervised several final year projects for bachelor's and master's degrees in his main line of work. He authored and co-authored several scientific publications in journals, books, and conferences with more than 75+ journal research articles, 5+ book chapters, 15+ conference papers, 10+ conference proceedings, 8+ books editing, and 2+ patents. He attended and participated as Keynote Speakers (60+ Conferences), Webinars (10+), and Session Chairs (10+) in many international events in the following countries: India, Turkey, Nepal, Philippines, Vietnam, Thailand, Indonesia, and other countries.

Contributors

Salma Ahmedai
School of Mathematics, Statistics and Computer Science
University of KwaZulu-Natal
Pietermaritzburg, South Africa

Jyoti Sekhar Banerjee
Department of CSE (AL and ML)
Bengal Institute of Technology
Kolkata, India

Ritwick Banerjee
Department of Mathematics
Heritage Institute of Technology
Kolkata, West Bengal, India

Purvee Bhardwaj
Department of Physical Science
Rabindranath Tagore University
Bhopal, India

Ramakant Bhardwaj
Department of Mathematics
Amity University
Kolkata, India

Somangee Chakrabarti
Department of Microbiology
St. Xavier's College (Autonomous)
Kolkata, India

Bidisha Chatterjee
Department of Microbiology
St. Xavier's College (Autonomous)
Kolkata, India

Prasanta Chatterjee
Department of Mathematics
Visva-Bharati (A Central University)
Bolpur, India

Pritha Das
Department of Mathematics
Indian Institute of Engineering Science and Technology
Howrah, India

Vaishali Deshmukh
Shrimant Madhavrao Scindia Govt. Model Science College
Gwalior, India

Sharadia Dey
Department of Environmental Studies
St. Xavier's College (Autonomous) Kolkata
Kolkata, West Bengal, India

Debjani Dutta
Department of Environmental Science
The University of Burdwan
Burdwan, India

Jibitesh Dutta
Mathematics Division, Department of Basic Sciences and Social Sciences
North-Eastern Hill University
Shillong, India

Sanghita Dutta
Department of Mathematics
North-Eastern Hill University
Shillong, India

Moumita Ghosh
Department of Mathematics
Indian Institute of Engineering Science and Technology
Shibpur, India

Uday Narayan Ghosh
Department of Mathematics
KKM College, A Constituent unit of Munger University
Jamui, India

Sicelo Goqo
School of Mathematics, Statistics and Computer Science
University of KwaZulu-Natal
Pietermaritzburg, South Africa

Sourav Gupta
Department of Mathematics
Sardar Vallabhbhai National Institute of Technology
Surat, India

Srimanta Gupta
Department of Environmental Science
The University of Burdwan
Bardhaman, India

Bhanu Pratap Singh Jadon
Shrimant Madhavrao Scindia Govt. Model Science College
Gwalior, India

Alokananda Kar
Department of Physics
Indian Institute of Technology (ISM) Dhanbad
Dhanbad, India

Diamond Kharkongor
Department of Mathematics and Statistics
ICFAI University
Shillong, India

Sukanya Kool
Department of Information Technology
Amity University
Kolkata, India

Souvik Kundu
Department of Mathematics
Heritage Institute of Technology
Kolkata, India

Tapas Ray Mahapatra
Department of Mathematics
Visva-Bharati (A Central University)
Bolpur, India

Tapas Mal
Department of Mathematics
Sardar Vallabhbhai National Institute of Technology
Surat, India

Arijit Malakar
Department of Mathematics
Indian Institute of Engineering Science and Technology
Howrah, India

Sandeep Kumar Malhotra
Dr. Shyama Prasad Mukherjee, Science College Bhopal
Bhopal, India

Akash Pradip Mandal
Department of Mathematics, Science Building
Ananda Chandra College
Jalpaiguri, India

Prashanta Kumar Mandal
Department of Mathematics
Siksha Bhavana, Visva-Bharati
Bolpur, India

Kaushik Mishra
Media Studies Department
Chandigarh University
Chandigarh, Punjab, India

Arup Kumar Mitra
Department of Microbiology
St. Xavier's College (Autonomous)
Kolkata, India

Sabyasachi Mondal
Department of Mathematics
North-Eastern Hill University
Shillong, India

Sayantika Mukherjee
Amity University Kolkata
Kolkata, India

Tamoghna Mukherjee
Department of Computer Science and Engineering
Amity University
Kolkata, India

Snehalata Nasipuri
Department of Mathematics
Visva-Bharati (A Central University)
Bolpur, India

Sumanta Nayek
Amity University
Kolkata, India

Shina Daniel Oloniiju
Department of Mathematics
Rhodes University
Grahamstown, South Africa

Olumuyiwa Otegbeye
University of the Witwatersrand
Johannesburg, South Africa

Ibukun Sarah Oyelakin
University of the Witwatersrand
Johannesburg, South Africa

Özen Özer
Department of Mathematics, Faculty of Science and Arts
Kirklareli University
Kirklareli, Turkey

Bhavna Pandey
Department of Physical Science
Rabindranath Tagore University
Bhopal, India

Rujda Parveen
Department of Basic Science and Humanities
Dream Institute of Technology
Kolkata, India

Abhijit Paul
Department of Computer Science
Swami Vivekananda University
Kolkata, India

Saikat Pramanik
Indian Institute of Tropical Meteorology
Pune, India

Abdolrahman Razani
Department of Pure Mathematics, Faculty of Science
Imam Khomeini International University
Qazvin, Iran

Uthman O. Rufai
School of Mathematics, Statistics and Computer Science
University of KwaZulu-Natal
Pietermaritzburg, South Africa

Shouvik Sadhukhan
Department of Physics
Indian Institute of Space Science and Technology
Thiruvananthapuram, India

Gunajyoti Saharia
Department of Mathematics
North-Eastern Hill University
Shillong, India

Bharti Saxena
Department of Mathematics
Rabindranath Tagore University
Bhopal, India

Arti Shrivastava
Department of Mathematics
Rabindranath Tagore University
Bhopal, India

Precious Sibanda
School of Mathematics, Statistics and Computer Science
University of KwaZulu-Natal
Pietermaritzburg, South Africa

Sonam
Department of Mathematics
Amity University
Kolkata, India

Sourav Sil
School of Earth, Ocean and Climate Sciences
Indian Institute of Technology
Bhubaneswar, India

Ruchi Singh
Pandit S.N. Shukla University
Shahdol, India

Yusuf Olatunji Tijani
Department of Mathematics
Rhodes University
Grahamstown, South Africa

1 Establishment of FLP Model by Applying Triangular Membership Functions

Arti Shrivastava, Bharti Saxena, Sonam, Ramakant Bhardwaj, and Kaushik Mishra

1.1 INTRODUCTION

In the present cost-effective market, there is a burden on industries to discover effective solutions to generate and deliver goods to consumers. How to produce and supply the necessary quantity of goods to the consumers, while ensuring profitability, is a very challenging procedure. Linear programming (LP) provides a strong model to handle this type of problem. It ensures well-organized production and timely transportation of all types of necessary materials and completed items. The first studies of LP were applied to diet problems, which were published between 1950 and 1960. This research was started by Jerry Cornfield, who developed "The Diet Problem" for the Army during World War II (1941–1945); in this research, he developed a model for a minimum-cost diet that would provide the required nutrition to a soldier. The economist George Stigler tried to develop optimization procedures to establish the lowest cost diet providing enough proteins, energy, minerals, and vitamins. After this study, many researchers continued their work in this area. Dantzig [1] studied linear programming as an efficient tool in operations research.

An effective decision maker should utilize linear programming as one of their means. It is a very important tool in operational research to handle real-world problems. Since linear programming is now well established, a broader range of methods, including the simplex method, have been developed [2]. Campos and Verdegay [3] laid the groundwork for fuzzy programming by introducing its foundational principles. In a parallel effort, Bellman and Zadeh [4] elucidated the concept of a fuzzy decision-making framework, adding depth to the understanding of fuzzy systems. Simultaneously, Zimmermann [5] made significant contributions to the field by formulating fuzzy linear programming (FLP), thereby advancing the practical application of fuzzy logic in optimization problems. Subsequently, numerous researchers have delved into various types of FLP problems with proposed multiple methodologies to address these issues [2,6–9]. Fuzzy set theory has found widespread

DOI: 10.1201/9781003638414-1

applications across diverse fields, including control theory, management science, mathematical modeling, and industrial sectors. Particularly noteworthy is the discovery that the most effective approach involves comparing fuzzy numbers using linear ranking functions [3,10–14].

The application of ranking functions in fuzzy programming lacks universally accepted conditions, primarily due to the diverse requirements and objectives across different research studies. Scholars have customized linear ranking functions to align with their specific needs and the characteristics of the problem at hand. This customization often involves adjusting parameters or criteria within the ranking functions to better capture the nuances of the fuzzy sets being compared. Consequently, the flexibility of linear ranking functions allows researchers to adapt them according to the context of their investigation, leading to a diverse array of methodologies and approaches within the field of fuzzy programming. In such scenarios, the prevailing approach often involves establishing a crisp model that corresponds to an FLP problem and subsequently utilizing the optimal solution of the crisp model to address the FLP problem. The authors of Refs. [11,12] extended the duality theory in fuzzy non-linear programming (FNLP) problems, similar to the type of problems that guide the dual simplex algorithm [1] in elucidating such issues. Generally, to handle these types of processors, researchers introduced a crisp model that is equivalent to FNLP. In Ref. [3], a general ranking function is used to consider a fuzzy linear programming problem with triangular numbers and solve it by matrices, simplex with ranking, and the crisp method. Our main contribution here is the establishment of a new method to solve FLPP with the help of a membership function. Moreover, we illustrate our method with an example.

1.2 PRELIMINARIES

This section provides an overview of fundamental definitions and arithmetic operations.

1.2.1 Fuzzy Set

A fuzzy set is defined by a membership function that maps elements from a domain, space, or universe of discourse X to the unit interval [0, 1], denoted as $A=\{(x,\mu_A(x)): x\in \mathrm{X}\}$. Here, $\mu_A : X \rightarrow [0,1]$ represents the membership function of the fuzzy set A, and $\mu_A(x)$ signifies the membership value of x in X within the fuzzy set A. These membership grades are typically denoted by real numbers spanning from 0 to 1.

1.2.2 Triangular Fuzzy Numbers

A triangular fuzzy number, represented by $\tilde{A}=(a_1,a_2,a_3)$, consists of real numbers a_1,a_2,a_3, with its membership function denoted as $\mu_{\tilde{A}}(x)$,

$$\mu_{\tilde{A}}(x) = \begin{cases} 0 & \text{if } a < x_1 \\ \dfrac{x - a_1}{a_2 - a_1} & \text{if } a_1 \le x \le a_2 \\ \dfrac{a_3 - x}{a_3 - a_2} & \text{if } a_2 \le x \le a_3 \\ 0 & \text{if } x > a_3 \end{cases}$$

Applying min and max operators, the expression obtained for the proceeding equation is,

$$\text{Triangle}(x : a, b, c) = \max\left(\min\left(\frac{x-a}{b-a}, 1, \frac{c-x}{c-b}\right)\right).$$

1.2.3 Trapezoidal Fuzzy Number

A trapezoidal fuzzy number, denoted as $\tilde{A} = (a_1, a_2, a_3, a_4)$, comprises real numbers a_1, a_2, a_3, a_4, with its membership function $\mu_{\tilde{A}}(x)$ defined as follows:

$$\mu_{\tilde{A}}(x) = \begin{cases} 0 & \text{if } x < a_1 \\ \dfrac{x - a_1}{a_2 - a_1} & \text{if } a_1 \le x \le a_2 \\ 1 & \text{if } a_2 \le x \le a_3 \\ \dfrac{a_4 - x}{a_4 - a_3} & \text{if } a_3 \le x \le a_4 \\ 0 & \text{if } x > a_4 \end{cases}$$

An alternative concise expression using min and max is:

$$\text{Trapezoidal}(x : a, b, c, d) = \max\left(\min\left(\frac{x-a}{b-a}, 1, \frac{d-x}{d-c}\right)\right).$$

1.2.4 Membership Function for Triangular Fuzzy Number

When a number $\tilde{A}$ is a triangular fuzzy number, it is represented as $\tilde{A} = (a_1, a_2, a_3)$, where a_1, a_2, a_3 are real numbers, and its membership function $\mu_{\tilde{A}}(x)$ is defined as follows:

$$\mu_{\tilde{A}}(x) = \begin{cases} 0 & \text{if } a < x_1 \\ \dfrac{x - a_1}{a_2 - a_1} & \text{if } a_1 \le x \le a_2 \\ \dfrac{a_3 - x}{a_3 - a_2} & \text{if } a_2 \le x \le a_3 \\ 0 & \text{if } x > a_3 \end{cases}$$

To calculate the confidence interval for all the level α, the triangular shapes will be expressed as functions of α in the following manner:

$$\alpha = \frac{x - a_1}{a_2 - a_1}$$

$$\alpha = \frac{a_3 - x}{a_3 - a_2}$$

$$C_{ij} = [a\alpha + b, d - c\alpha]$$

where $a = a_2 - a_1$; $b = a_1$; $c = (a_3 - a_2)$ and $d = a_3$.

1.3 FUZZY LINEAR PROGRAMMING MODEL

Minimize or Maximize: $Z = \sum_{j=1}^{n} \widetilde{c_j}\tilde{x}_j$

Subject to constraint:

$$\sum_{j=1}^{n} \tilde{a}_{ij}\tilde{x}_j \ (\leq,=,\geq)\ \tilde{b}_i \ i = 1,\ 2,\ \ldots\ldots\ldots. m$$
$$\widetilde{x_j} \geq 0 j = 1,\ 2,\ \ldots\ldots\ldots. n \tag{1.1}$$

here,

$\tilde{c}_j$ = Per unit contribution of decision variable $\tilde{x}_j$,
$\tilde{a}_{ij}$ = Input-output coefficient,
$\tilde{b}_i$ = Total availability of the ith resource.

1.4 NUMERICAL ANALYSIS

Consider an FLP problem,

$$\text{Max } Z = (10,15,20)x + (15,17,19)y + (12,14,16)z$$
$$(6,9,12)x + (3,6,9)y + (22,24,26)z \leq (6,7,8)$$
$$(1,5,9)x + (3,7,11)y + (3,9,15)z \leq (4,8,12)$$
$$(7,8,9)x + (8,16,24)y + (3,4,5)z \leq (6,16,26)$$
$$x, > 0, y > 0 \text{ and } z > 0$$

Membership function of triangular fuzzy number (TFN) (10,15,20).

$$\mu_{\tilde{A}}(x) = \begin{cases} 0 & \text{if } 10 < x \\ \dfrac{x-10}{15-10} & \text{if } 10 \leq x \leq 15 \\ \dfrac{20-x}{20-15} & \text{if } 15 \leq x \leq 20 \\ 0 & \text{if } x > 20 \end{cases}$$

Now,

$$\alpha = \frac{x-10}{15-10}$$

$$\alpha = \frac{20-x}{20-15}$$

$$(10,15,20) = [5\alpha+10, 20-5\alpha].$$

Similarly, we can deduce

$$\begin{aligned}
&(10,15,20) = [5\alpha+10, 20-5\alpha], && (15,17,19) = [2\alpha+15, 19-2\alpha],\\
&(12,14,16) = [2\alpha+12, 16-2\alpha], && (6,9,12) = [3\alpha+6, 12-3\alpha],\\
&(3,\ 6,\ 9) = [3\alpha+3, 9-3\alpha], && (22,\ 24,\ 26) = [2\alpha+22, 26-2\alpha],\\
&(6,7,8) = [\alpha+6, 8-\alpha], && (1,5,9) = [4\alpha+1, 9-4\alpha],\\
&(3,7,11) = [4\alpha+3, 11-4\alpha], && (3,9,15) = [6\alpha+3, 15-6\alpha],\\
&(4,8,12) = [4\alpha+4, 12-4\alpha], && (7,8,9) = [\alpha+7, 9-\alpha],\\
&(8,16,24) = [8\alpha+8, 24-8\alpha], && (3,4,5) = [\alpha+3, 5-\alpha],\\
&(6,16,26) = [10\alpha+6, 26-10\alpha].
\end{aligned}$$

Fuzzy linear programming problem after applying membership function,

$$\text{Max } Z = (5\alpha+10, 20-5\alpha)x + (2\alpha+15, 19-2\alpha)y + (2\alpha+12, 16-2\alpha)z$$

subject to the constraints:

$$\begin{gathered}
(3\alpha+6, 12-3\alpha)x + (3\alpha+3, 9-3\alpha)y + (2\alpha+22, 26-2\alpha)z \le (\alpha+6, 8-\alpha)\\
(4\alpha+1, 9-4\alpha)x + (4\alpha+3, 11-4\alpha)y + (6\alpha+3, 15-6\alpha)z \le (4\alpha+4, 12-4\alpha)\\
(\alpha+7, 9-\alpha)x + (8\alpha+8, 24-8\alpha)y + (\alpha+3, 5-\alpha)z \le (10\alpha+6, 26-10\alpha)\\
x, > 0, y > 0 \text{ and } z > 0
\end{gathered}$$

Now, by using the separation method in the above fuzzy linear programming problem and considering the lower bound fuzzy linear programming problem in integer form, we obtain

$$\text{Max } Z = (5\alpha+10)x + (2\alpha+15)y + (2\alpha+12)z$$

subject to the constraints:

$$\begin{gathered}
(3\alpha+6)x + (3\alpha+3)y + (2\alpha+22)z \le (\alpha+6)\\
(4\alpha+1)x + (4\alpha+3)y + (6\alpha+3)z \le (4\alpha+4)\\
(\alpha+7)x + (8\alpha+8)y + (\alpha+3)z \le (10\alpha+6)\\
x, > 0, y > 0 \text{ and } z > 0
\end{gathered}$$

To achieve the optimality of lower bound fuzzy linear programming problem, the simplex method is applied, and the value of the basic variables is deduced,

$$x_1 = \frac{-22\alpha^2 + 8\alpha + 30}{20\alpha^2 + 41\alpha + 27} \text{ and}$$

$$x_2 = \frac{2886\alpha^5 + 16264\alpha^4 + 48990\alpha^3 + 14554\alpha^2 + 63266\alpha - 8016}{(8\alpha + 18)(13\alpha^2 + 178\alpha + 167)(20\alpha^2 + 41\alpha + 27)}$$

Optimal solution for lower bound fuzzy linear programming problem is

$$Z = \frac{(5\alpha + 10)(-22\alpha^2 + 8\alpha + 30)}{(20\alpha^2 + 41\alpha + 27)}$$

$$+ \frac{(2\alpha + 15)(2886\alpha^5 + 16264\alpha^4 + 48990\alpha^3 + 14554\alpha^2 + 63266\alpha - 8016)}{(8\alpha + 18)(13\alpha^2 + 178\alpha + 167)(20\alpha^2 + 41\alpha + 27)}$$

Also, the upper bound fuzzy linear programming problem in integer form is

$$\text{Max } Z = (20 - 5\alpha)x + (19 - 2\alpha)y + (16 - 2\alpha)z$$

subject to the constraints:

$$(12 - 3\alpha)x + (9 - 3\alpha)y + (26 - 2\alpha)z \leq (8 - \alpha)$$
$$(9 - 4\alpha)x + (11 - 4\alpha)y + (15 - 6\alpha)z \leq (12 - 4\alpha)$$
$$(9 - \alpha)x + (24 - 8\alpha)y + (5 - \alpha)z \leq (26 - 10\alpha)$$
$$x, > 0, y > 0 \text{ and } z > 0.$$

In order to obtain the optimality of the upper bound fuzzy linear programming problem, the simplex method is applied, and the basic variable values are obtained,

$$x_1 = \frac{-20\alpha^2 + 80\alpha - 42}{21\alpha^2 - 114\alpha + 207} \text{ and}$$

$$x_2 = \frac{-2990\alpha^5 + 81418\alpha^4 - 664778\alpha^3 + 303285\alpha^2 - 2941016\alpha - 499140}{(24 - 8\alpha)(13\alpha^2 - 244\alpha + 579)(21\alpha^2 - 114\alpha + 207)}$$

Optimal solution of the upper bound fuzzy linear programming problem,

$$Z = \frac{(20 - 5\alpha)(-20\alpha^2 + 80\alpha - 42)}{(21\alpha^2 - 114\alpha + 207)}$$

$$+ \frac{(19 - 2\alpha)(-2990\alpha^5 + 81418\alpha^4 - 664778\alpha^3 + 303285\alpha^2 - 2941016\alpha - 499140)}{(24 - 8\alpha)(13\alpha^2 - 244\alpha + 579)(21\alpha^2 - 114\alpha + 207)}$$

Thus, the optimality of fuzzy linear programming problems is,

$$Z = \begin{bmatrix} \dfrac{(5\alpha+10)\left(-22\alpha^2+8\alpha+30\right)}{\left(20\alpha^2+41\alpha+27\right)} + \\ \dfrac{(2\alpha+15)\left(2886\alpha^5+16264\alpha^4+48990\alpha^3+14554\alpha^2+63266\alpha-8016\right)}{(8\alpha+18)\left(13\alpha^2+178\alpha+167\right)\left(20\alpha^2+41\alpha+27\right)}, \\ \dfrac{(20-5\alpha)\left(-20\alpha^2+80\alpha-42\right)}{\left(21\alpha^2-114\alpha+207\right)} + \\ \dfrac{(19-2\alpha)\left(-2990\alpha^5+81418\alpha^4-664778\alpha^3+303285\alpha^2-2941016\alpha-499140\right)}{(24-8\alpha)\left(13\alpha^2-244\alpha+579\right)\left(21\alpha^2-114\alpha+207\right)} \end{bmatrix}$$

Special Cases:

1. For $\alpha = 0.25$, the optimal solution is [−8.04, 13.36].
2. For $\alpha = 0.50$, the optimal solution is [8.63, −25.629].
3. For $\alpha = 0.75$, the optimal solution is [6.51, −4.15].
4. For $\alpha = 1.00$, the optimal solution is [4.91, −81.32].

1.5 CONCLUSION

In this chapter, the FLP problem of triangular fuzzy numbers has been simplified with the help of a membership function. The membership function is used to obtain both the lower bound and the upper bound of the TFN. This method provides more options and serves as an important tool in decision-making problems. By using the membership function, our problem is divided into two parts: the lower bound FLP problem and the upper bound FLP problem in integer form. We obtained two results as respective breakups of the problems: the lower bound and the upper bound. It is concluded that the optimal solutions are in generalized crisp numbers. The advantage of this method lies in its computational efficiency. This method can be applied to other problems in OR and Engineering Sciences.

REFERENCES

[1] Dantzig, G. B. (1963). *Linear Programming and Extensions*. Princeton University Press.

[2] Ebrahimnejad, A., Nasseri, S. H. & Mansourzadeh, S. M. (2010). Bounded primal simplex algorithm for bounded linear programming with fuzzy cost coefficients. *International Journal of Operations Research Information Systems*, 2(1), 100–124.

[3] Campos, L. & Verdegay, J. L. (1989). Linear programming problems and ranking of fuzzy numbers. *Fuzzy Sets and Systems*, 32, 1–11.

[4] Bellman, R. E. & Zadeh, L. A. (1970). Decision making in a fuzzy environment. *Management Science*, 17, 141–164.
[5] Zimmermann, H. J. (1978). Fuzzy programming and linear programming with several objective functions. *Fuzzy Sets and Systems*, 1, 45–55.
[6] Ebrahimnejad, A. & Nasseri, S. H. (2009). Using complementary slackness property to solve linear programming with fuzzy parameters. *Fuzzy Information and Engineering*, 3, 233–245.
[7] Ebrahimnejad, A., Nasseri, S. H., Lotfi, F. H. & Soltanifar, M. (2010). A primal–dual method for linear programming problems with fuzzy variables. *European Journal of Industrial Engineering*, 4(2), 189–209.
[8] Ebrahimnejad, A., Nasseri, S. H. & Lotfi, F. H. (2010). Bounded linear programs with trapezoidal fuzzy numbers. *International Journal of Uncertainty, Fuzziness and Knowledge-Based Systems*, 18(3), 269–286.
[9] Ebrahimnejad, A. & Nasseri, S. H. (2010). A dual simplex method for bounded linear programmes with fuzzy numbers. *International Journal of Mathematics in Operational Research*, 2(6), 762–779.
[10] Tanaka, H., Okuda, T. & Asai, K. (1974). On fuzzy mathematical programming. *The Journal of Cybernetics*, 3, 37–46.
[11] Mahdavi-Amiri, N. & Nasseri, S. H. (2007). Duality results and a dual simplex method for linear programming problems with trapezoidal fuzzy variables. *Fuzzy Sets and Systems*, 158, 1961–1978.
[12] Mahdavi-Amiri, N., Nasseri, S. H. & Yazdani, A. (2009). Fuzzy primal simplex algorithms for solving fuzzy linear programming problems. *Iranian Journal of Operations Research*, 1, 68–84.
[13] Yager, R. R. (1981). A procedure for ordering fuzzy subsets of the unit interval. *Information Sciences*, 24, 143–161.
[14] Fortemps, P. & Roubens, M. (1996). Ranking and defuzzification methods based on area compensation. *Fuzzy Sets and Systems*, 82, 319–330.

2 Fixed Point Model for Expansion on Interval-Valued Fuzzy Metric Space

Vaishali Deshmukh, Bhanu Pratap Singh Jadon, Ruchi Singh, Sandeep Kumar Malhotra, and Sonam

2.1 INTRODUCTION

Initialization Kramosil and Michálek [6] first proposed the concept of fuzzy metric. Following the introduction of this idea by Kaleva and Seikkala [7], and Osman [9], fuzzy metric space was established independently using several methodologies. Later, a number of authors, including George and Veeramani [4], Chakrabarty et al. [3], have discussed a number of issues pertaining to fuzzy metric space in depth or widely from different perspectives. Zadeh first suggested the idea of an interval-valued-fuzzy set in 1975. A further generalization of the fuzzy-set concept is the interval-valued-fuzzy-set, which is defined by an interval-valued-membership-function. Three different types of distance between two interval-valued-fuzzy-sets (or members) specified on the real line R were introduced by Li [8].

Additionally, he pointed out that each type of distance is a metric on the corresponding sets and that each interval-valued fuzzy metric space (IVFMS) is full. Shen et al. [12] first proposed the idea of IVFMS with the interval-valued-τ-norm (continuous), which was a generalization of the fuzzy metric space concept by George and Veeramani [4]. The expansion mapping was first introduced by Wang [16] and later developed by Popa [10] and Jain [5]. Agrawal and Chouhan [1] worked on specific fixed-point theorems for expansion mappings. Fixed-point results for P-1 compatibility in fuzzy manger space and fixed-point results for fuzzy manger space with common properties were worked out by Singh et al. [13, 14]. Sewani et al. [11] applied the fuzzy iterated contraction abstraction to create some new results in intuitionistic-fuzzy metric space. Finding results for fixed-point theorems for expansion mapping on IVFMS is the primary goal of this paper. Recently, Singh et al. [15] established some common invariant point theorem for multivalued generalized fuzzy mappings in b-Metric space, and Bhardwaj et al. [2] worked on F-Contractive Type Mappings in Soft Matrix Space.

DOI: 10.1201/9781003638414-2

2.2 PRELIMINARIES

In this section, we provide some fundamental definitions for establishing our main results.

Definition 2.2.1

[2] Assuming G is a typical non-empty set, the mapping $R_m : G \to [I_{iv}]$ is referred to as an interval-valued fuzzy set on G. IVF(G) is used to identify all interval-valued fuzzy sets on G. If $R_m \in \text{IVF}(G)$, then let $R_m \in \text{IVF}(G)$, and let $R_m(p) = [R_m^-(p), R_m^+(p)], R_m^-(p) \le R_m^+(p)$ for all $p \in G$. In this case, the conventional fuzzy sets $R_m^- : G \to [I_{iv}]$ and $R_m^+ : G \to [I_{iv}]$ are referred to as the Lower* – fuzzy set and the Upper* – fuzzy set, respectively. In particular, R_m is referred to as a degenerate fuzzy set, if $R_m^-(p) = R_m^+(p)$ for any $p \in G$.

Definition 2.2.2

[2] An interval-valued τ-norm is a binary*operation* of the form $*_{I_{iv}} : [I_{iv}] \times [I_{iv}] \to [I_{iv}]$ on $[I_{iv}]$ such that the following four condition are met for all $\bar{\alpha}, \bar{\beta}, \bar{\gamma} \in [I_{iv}]$:

(T1) Commutativity: $\bar{\alpha} *_{I_{iv}} \bar{\beta} = \bar{\beta} *_{I_{iv}} \bar{\alpha}$;
(T2) Associativity: $\bar{\alpha} *_{I_{iv}} [\bar{\beta} *_{I_{iv}} \bar{\gamma}] = [\bar{\alpha} *_{I_{iv}} \bar{\beta}] *_{I_{iv}} \bar{\gamma}$;
(T3) Monotonicity: $\bar{\alpha} *_{I_{iv}} \bar{\beta} \le \bar{\beta} *_{I_{iv}} \bar{\gamma}$ whenever $\bar{\beta} \le \bar{\gamma}$;
(T4) Boundary condition: $\bar{\alpha} *_{I_{iv}} \bar{1} = \bar{\alpha}, \bar{\alpha} *_{I_{iv}} \bar{0} = [\alpha^-, \alpha^+] *_I [0,1] = [0, \alpha^+]$.

Example 2.2.1

i. $\bar{\alpha} *_{I_{iv}} \bar{\beta} = [\alpha^-.\beta^-, \alpha^+.\beta^+]$
ii. $\bar{\alpha} *_{I_{iv}} \bar{\beta} = [\alpha^- \wedge \beta^-, \alpha^+ \wedge \beta^+]$

Definition 2.2.3

[2] Let $\{\overline{\alpha_n} = [\alpha^-, \alpha^+], n \in \aleph^+$ be a sequence of interval numbers in $[I_{iv}], \bar{\alpha} = [\alpha^-, \alpha^+] \in [I_{iv}]$. If $\lim_{n\to\infty} \alpha_n^- = \alpha^-$ and $\lim_{n\to\infty} \alpha_n^+ = \alpha^+$, then we say that the sequence $\{\overline{\alpha_n}\}$ is convergent to $\bar{\alpha}$, which is denoted by $\lim_{n\to\infty} \overline{\alpha_n} = \bar{\alpha}$.

Definition 2.2.4

[2] An interval-valued τ-norm $*_{I_{iv}}$ is continuous if and only if it is continuous in its first component, i.e., for each $\bar{\beta} \in [I_{iv}]$, if $\lim_{n\to\infty} \overline{\alpha_n} = \bar{\alpha}$, then $\lim_{n\to\infty} (\overline{\alpha_n} *_1 \bar{\beta}) = \lim_{n\to\infty} \overline{\alpha_n} *_{I_{iv}} \bar{\beta} = \bar{\alpha} *_{I_{iv}} \bar{\beta}$, where $\overline{\alpha_n} \subseteq [I_{iv}], \bar{\alpha} \in [I_{iv}]$.

Definition 2.2.5

[2] If G is a temporary set, $*_{I_{iv}}$ is a continuous interval-valued τ-norm on $[I_{iv}]$ and is an IVF(G) on $G^2 \times (0,\infty)$ meeting the following conditions, the triple $(G,\Bbbk,*_I)$ is known as Interval-valued fuzzy metric space (IVFMS) on G : Following requirements:

(C1) $\Bbbk(\omega,\ \rho,\ \tau-\text{norm}) > I$,
(C2) $\Bbbk(\omega,\ \rho,\ \tau-\text{norm}) = I$ if and only if $\omega = \rho$,
(C3) $\Bbbk(\omega,\ \rho,\ \tau-norm) = \Bbbk(\rho,\ \omega,\ \tau)$,
(C4) $\Bbbk(\omega,\ \rho,\ \tau-\text{norm})\ *I\ \Bbbk(\ \omega,r,s) \le \Bbbk(\ \omega,r,\tau-\text{norm}+s)$,
(C5) $\Bbbk(\omega,\ \rho,\ .) : (0,\infty) \to [I]$ is continuous,
(C6) $\lim_{\eta\to\infty} \Bbbk(\omega,\rho,\tau-\text{norm}) = I$, where $\omega,\ \rho,\ r \in G$ and $\tau-\text{norm},\ s > I$.

Example 2.2.2

Assume that $G = R$. Define $\bar{\alpha} *_I \bar{\beta} = [\alpha^-.\beta^-,\alpha^+.\beta^+]$, then the IVFMS is taken as $k(\wp,q,t) = [k^-(\wp,q,t),k^+(\wp,q,t)] = \left[e^{\frac{-k|\wp-q|}{t}},e^{\frac{-|\wp-q|}{t}}\right]$, $k \ge 1$. For any $\wp,q \in G$ and $t \in (0,\infty)$, then $(G,k,*_I)$ is an IVFMS.

Definition 2.2.6

[2] Let $(G,k,*_I)$ is an IVFMS,

1. For all $\wp,q \in G$, if $s > t > 0$ then $k(\wp,q,t) \le k(\wp,q,s)$.
2. A sequence $\{\wp_n\}$ in G is referred to as a Cauchy sequence if for all $\bar{\varepsilon} > \bar{0}$ and $t > 0$, there is an exit a $\eta \in \aleph$ such that $k(\wp_n,\wp_l,t) > \bar{1} - \bar{\varepsilon}$ for all $n,l \ge \eta$.
3. An IVFMS in which every Cauchy sequence is convergent, is said to be a complete IVFMS.
4. $(G,k,*_I)$ is complete IVFMS If every sequence in G has a convergent subsequence $\{\wp_n\}$.

Definition 2.2.7

[2] In interval-valued-metric-space, a function is continuous if $\wp_n \to \wp, q_n \to q \Rightarrow \lim_{n\to\infty} k(\wp_n,q_n,t) \to k(\wp,q,t)$.

Definition 2.2.8

[2] Two mapping $\wedge$ and $\uplus$ on IVFMS G are weakly commuting by the expression if $k(\wedge \uplus u,\uplus \wedge u,t) \ge k(\wedge u,\uplus u,t)$, $u \in G$.

Lemma 2.2.1

Given $\{q_n\}$ be a sequence in an IVFMS $(G,k,*_I)$ with $\lim_{n\to\infty} k(\wp,q,t)=\bar{1}$, if there exit number $x\in(0,1)$ such that $k(q_{n+2},q_{n+1},xt)\ge k(q_{n+1},q_n,t)$ for all $t>0$ and $n=1,2,3....$, then $\{q_n\}$ is a Cauchy sequence in G.

Lemma 2.2.2

If for all $\wp,q\in G$, $t>0$ and for a number $x\in(0,1), k(\wp,q,xt)\ge k(\wp,q,t)$ then $\wp=q$.

2.3 MAIN RESULTS

In this section, we have presented some fixed-point models:

Theorem 2.3.1

Assume that $(G,\ *_I)$ is an entire IVFMS and that (G,D) is a self-mapping of G. Then the Mapping D meets the following requirement:

$$k(Dp,Dq,t)\ge \min\left[\begin{array}{l}\dfrac{k(p,D[p],t)k(q,D[q],t)}{k(p,q,t)},\ \dfrac{k(p,D[q],t)k(q,D[p],t)}{k(p,q,t)},\\ k(p,D[p],t),k(q,D[q],t),k(p,q,t)\end{array}\right]$$

for all p,q in G with $p\ne q$ and D is onto. A fixed point in G is then admitted by D.

Proof: Let $p_0\in G$. Since D is onto, there exists an element p_1 satisfying p_1 in $D^{-1}p_0$. Similarly, if $p_{l-1}=p_l$ for some l, then p_1 is a fixed-point of D. We can choose, p_n in $D^{-1}p_{n-1}$, where $n=2,3,4....$

Without losing generality, we can assume that $p_{n-1}\ne p_n$ for every n, therefore,

$$k(p_{n-1},p_n,t)=k(D[p_n],D[p_{n+1}],t)$$

$$\ge \min\left\{\begin{array}{l}\dfrac{k(p_n,D[p_n],t)k(p_{n+1},D[p_{n+1}],t)}{k(p_n,p_{n+1},t)},\ \dfrac{k(p_n,D[p_{n+1}],t)k(p_{n+1},D[p_n],t)}{k(p_n,p_{n+1},t)},\\ k(p_n,D[p_n],t),k(p_{n+1},D[p_{n+1}],t),k(p_n,p_{n+1},t)\end{array}\right\}$$

$$\Rightarrow k(p_{n-1},p_n,t)\ge k(p_n,p_{n+1},t).$$

A Cauchy sequence in G is $\{p_n\}$. Since G is complete, for some p, p_n converges to p, and for some $p \in G$.
Since, D is onto, there exists $q \in G$.such that q in $D^{-1}p$ and for infinitely many n, $p_n \neq p$, for such n.

$$k(p_n, p, t) = k(D[p_{n+1}], D[q], t)$$

$$\geq \min\left\{ \frac{k(p_n, D[p_n], t)k(p_{n+1}, D[p_{n+1}], t) + k(p_n, D[p_{n+1}], t)k(p_{n+1}, D[p_n], t)}{k(p_n, q, t)}, \right.$$

$$\frac{[k(p_{n+1}, D[p_{n+1}, t)]^2 + k(p_n, D[p_{n+1}], t)k(p_{n+1}, D[p_n], t)\}}{k(p_n, q, t)}, k(p_n, D[p_n], t),$$

$$\left. k(p_{n+1}, D[p_{n+1}], t), k(p_n, p_{n+1}, t) \right\}$$

$$= \min\left\{ \frac{k(p_{n+1}, p_n, t)k(q, p, t)}{k(p_{n+1}, q, t)}, \frac{k(p_{n+1}, p, t)k(q, p_n, t)}{k(p_{n+1}, q, t)}, \right.$$

$$\left. k(p_{n+1}, p_n, t)k(q, p, t)k(p_{n+1}, q, t) \right\}$$

On considering the limit as $n \to \infty$, we have

$$k(p_n, p, t) \geq k(q, p, t)$$

$$\Rightarrow \overline{1} \geq k(q, p, t) \text{ or } \overline{1} \geq \lim_{n\to\infty} k(p_{n+1}, q, t)$$

So, $k(q, p, t) = \overline{1}$ or $\overline{1} = \lim_{n\to\infty} k(p_{n+1}, q, t)$.

So, in both cases, we obtain $p = q$.

Thus, a fixed point in G is then admitted by D.

Theorem 2.3.2

Assume that $(G, k, *_I)$ is an entire IVFMS and that (G, D) is a self-mapping of G. The Mapping D meets the following requirement:

$$k(p, D[p], t), k(q, D[q], t), k(p, q, t)$$

$\forall p,q \in G$ with $p \neq q$. A fixed point in G is then admitted by D.

Proof: Let $p_0 \in G$ since D is onto there is an element p_1 satisfying, $p_1 \in D^{-1}p_0$. Similarly, if $p_{l-1} = p_l$ for-some l, then p_l is a fixed-point of D. We can choose $p_n \in D^{-1}p_{n-1}$, where $n = 2,3,4\ldots$.

Without losing generality, we can assume that $p_{n-1} \neq p_n$ for every n, therefore,

$$k(p_{n-1},p_n,t) = k(D[p_n],D[p_{n+1}],t)$$

$$\geq \min\{ \frac{k(p_n,D[p_n],t)k(p_{n+1},D[p_{n+1}],t)+k(p_n,D[p_{n+1}],t)k(p_{n+1},D[p_n],t)}{k(p_n,q,t)},$$

$$\frac{[k(p_{n+1},D[p_{n+1},t)]^2+k(p_n,D[p_{n+1}],t)k(p_{n+1},D[p_n],t)\}}{k(p_n,q,t)}, k(p_n,D[p_n],t),$$

$$k(p_{n+1},D[p_{n+1}],t), k(p_n,p_{n+1},t)$$

$$\Rightarrow k(p_{n-1},p_n,t) \geq \min k(p_n,p_{n-1},t), k(p_n,p_{n+1},t),$$

$$\Rightarrow k(p_{n-1},p_n,t) \geq k(p_n,p_{n+1},t)$$

A Cauchy sequence in G is p_n. As a result of G being complete, p_n converges to p, for some $p \in G$. Therefore, using D's continuity, we may write

$$D(p_n) = p_{n-1} \text{ as } n \to \infty.$$

Hence, $D(p) = p$. This completes the proof.

Theorem 2.3.3

Assume that $(G,k,*_I)$ is an entire IVFMS and that (G,D) is a self-mapping of G. The Mapping D meets the following requirement:

$$k(Dp,Dq,t) \geq \min\left\{\frac{k(p,D[p],t)k(q,D[q],t)}{k(p,q,t)}, \frac{k(p,D[q],\tau)k(q,D[p],t)}{k(p,q,t)},\right.$$

$$\left. k(p,D[p],t), k(q,D[q],t), k(p,q,t)\right\}$$

$\forall p,\, q \in G,$ with $p \neq q,$ there exists a point c in G such that $T(c) = \sup\{T(p) : T(p) = k(p,D[p],\tau), p \in G\}$. A fixed point in G is then admitted by D.

Proof: Assume that $c \neq D[c]$, otherwise c is a fixed point of D. Put $p = c$ and $q = D[c]$, we have

$$k\left(D[c],D[D(c)],t\right)=k\left(D[c],D^2[c],t\right)$$

$$\geq \min\Bigg\{ \frac{k(c,D[c],t)k\left(D[c],D[D(c)],t\right)}{k(c,D[c],t)},$$

$$\frac{k\left(c,D[D(c)],t\right)k(D[c],D[c],t)}{k(c,D[c],t)},k(c,D[c],t),k\left(D[c],D[D(c)],t\right),$$

$$k(c,D[c],t)$$

$$=k\left(D[c],D[D(c)],t\right) \text{ or } k(c,D[c],t)$$

$$=k\left(D[c],D[D(c)],t\right)$$

$\geq k\left(D[c],D[D(c)],t\right)$ which is not possible.

So,

$$k\left(D[c],D[D(c)],t\right)\geq k(c,D[c],t) \tag{2.1}$$

Similarly, if we set $p=D[c]$ and $q=c$, we obtain

$$k\left(D[D(c)],D[c],t\right)=k\left(D^2[c],D[c],t\right)$$

$$\geq \min\left\{ \frac{k\left(D[c],D[D(c)],t\right)k(c,D[c],t)}{k(D[c],c,t)},\frac{k(D[c],D[c],t)k\left(c,D[D(c)],t\right)}{k(D[c],c,t)},\right.$$

$$\left. k\left(D[c],D[D(c)],t\right),k(c,D[c],t),k(D[c],c,t)\right\} \tag{2.2}$$

$$=k\left(D[c],D[D(c)],t\right) \text{or } k(c,D[c],t)$$

$$=k\left(D^2[c],D[c],t\right)\geq k(c,D[c],t)$$

From equations (2.1) and (2.2), as a result, we have

$$k\left(D[c],D^2[c],t\right)\geq k(c,D[c],t)$$

$$\text{So, } k\left(D[c],D^2[c],t\right)>k(c,D[c],t)=T(c).$$

This follows that $T\left(D[c]\right)>T(c)$, which leads to a contradiction.

As a result, we must have $D(c)=c$, meaning that c admits a fixed point for D in G.

Theorem 2.3.4

Assume that $(G,k,*_I)$ is an entire IVFMS and that (G,D) is a self-mapping of G. The Mapping D meets the following requirement:

$$k(Dp,Dq,t) \leq \max\left\{\begin{matrix}\dfrac{k(p,D[p],t)k(q,D[q],t)}{k(p,q,t)}, \\ k(p,D[q],t)k(q,D[p],t), k(p,D[p],t), k(q,D[q],t), k(p,q,t)\end{matrix}\right\}$$

$\forall p,q \in G$ with $p \neq q$ and D is onto. A fixed point in G is then admitted by D.

Proof: Let c be an element of G. Since D is onto there is an element p_1 satisfying $p_1 \in D^{-1}p_0$. Similarly, if $p_{l-1} = p_l$ for-some c, then p_l is a fixed-point of D. We can choose $p_n \in D^{-1}p_{n-1}$, where $n = 2,3,4\ldots.$

Without losing generality, we can assume that $p_{n-1} \neq p_n$ for every n.

Therefore,

$$k(p_{n-1},p_n,t) = k(D[p_n],D[p_{n+1}],t)$$

$$\leq \max\left\{\frac{k(p_n,D[p_n],t)k(p_{n+1},D[p_{n+1}],t)}{k(p_n,p_{n+1},t)},\right.$$

$$\left.k(p_n,D[p_{n+1}],t)k(p_{n+1},D[p_n],t), k(p_n,D[p_n],t), k(p_{n+1},D[p_{n+1}],t), k(p_n,p_{n+1},t)\right\}$$

$$= \max\left\{\frac{k(p_n,p_{n-1},t)k(p_{n+1},p_n,t)}{k(p_n,p_{n+1},t)}, k(p_n,p_n,t)k(p_{n+1},p_{n-1},t),\right.$$

$$\left.(p_n,p_{n-1},t), k(p_{n+1},p_n,t), k(p_n,p_{n+1},t)\right\}$$

$$\Rightarrow k(p_{n-1},p_n,t) \leq k(p_n,p_{n+1},t).$$

A Cauchy sequence in G. is p_n. As a result of G being complete, p_n converges to p, for some $p \in G$.

Since, D is onto, there exits $q \in G$ such that $q \in D^{-1}(p)$ and for infinitely many n, $p_n \neq q$, we have n.

$$k(p_n, p, t) = k(D[p_{n+1}], D[q], t)$$

$$\leq \max \left\{ \frac{k(p_{n+1}, D[p_{n+1}], t) k(q, D[q], t)}{k(p_{n+1}, q, t)}, k(p_{n+1}, D[q], t) k(q, D[p_{n+1}], t), \right.$$

$$\left. k(p_{n+1}, D[p_{n+1}], t), k(q, D[q], t), k(p_{n+1}, q, t) \right\}$$

$$= \max \left\{ \frac{k(p_{n+1}, p_n, t) k(q, p, t)}{k(p_{n+1}, q, t)}, \right.$$

$$\left. k(p_{n+1}, p, t) k(q, p_n, t), k(p_{n+1}, p_n, t), k(q, p, t), k(p_{n+1}, q, t) \right\}$$

taking limit as $n \to \infty$, we have,

$$k(p_n, p, t) \leq k(q, \rho^*, t),$$

$$k(q, p, t) = \overline{1} \text{ or } \overline{1} = \lim_{n \to \infty} k(p_{n+1}, q, t).$$

So, in cases we get $p = q$. Thus, a fixed point in G is then admitted by D.

Theorem 2.3.5

Assume that $(G, k, *_I)$ is an entire IVFMS and that (G, D) is a self-mapping of G, then the Mapping D meets the following requirement:

$$k(Dp, Dq, t) \leq \max \left\{ k(p, D[p], t) k(q, D[q], t) \right\},$$

$$\left. \frac{[k(q, D[q], t)]^2 + k(p, D[q], t) k(q, D[p], t)}{k(p, q, t)}, \right\}$$

$$\left. k(p, F[p], t), k(q, D[q], t), k(p, q, t) \right]$$

$\forall p, q \in G$ with $p \neq q$. A fixed point in G is then admitted by D.

Proof: *Let* $p_0 \in G$. Since, D is onto, there is an element p_{1} satisfying

p_{1}\in D^{−1}p_{0}. Similarly, if $p_{l-1} = p_l$ for-some l, then p_l is a fixed-point of D. We can choose, $p_n \in D^{-1} p_{n-1}$, where $n = 2, 3, 4 \ldots.$

Without losing generality, we can assume that $p_{n-1} \neq p_n$ for every n,

Therefore,

$$k(p_{n-1}, p_n, t) = k(D[p_n], D[p_{n+1}], t)$$

$$\leq \max\Big\{k(p_n, D[p_n], t)k(p_{n+1}, D[p_{n+1}], t),$$

$$\frac{k(p_{n+1}, D[p_{n+1}, t)]^2 + k(p_n, D[p_{n+1}], t)k(p_{n+1}, D[p_n], t)}{k(p_n, q, t)},$$

$$k(p_n, D[p_n], t), k(p_{n+1}, D[p_{n+1}], t), k(p_n, p_{n+1}, t)\Big\}$$

$$= \max\left\{k(p_n, p_{n-1}, t)k(p_{n+1}, p_n, t), \frac{[k(p_{n+1}, p_n, t)]^2 + k(p_n, p_n, t)k(p_{n+1}, p_{n-1}, t)}{k(p_n, q, t)},\right.$$

$$\left. k(p_n, p_{n-1}, t), k(p_{n+1}, p_n, t), k(p_n, p_{n+1}, t)\right\}$$

$$\Rightarrow k(p_{n-1}, p_n, t) \leq \max[k(p_n, p_{n-1}, t), k(p_n, p_{n+1}, t)]$$

$$\Rightarrow k(p_{n-1}, p_n, t) \leq k(p_n, p_{n+1}, t)$$

A Cauchy sequence in G is $\{p_n\}$. As a result of G being complete, $\{p_n\}$ converges to p for some $p \in G$. Therefore, using D's continuity, we may write $D(p_n) = p_{n-1}$ as $\eta \to \infty$. Hence $D(p) = p$. This completes the proof.

Theorem 2.3.6

Assume that $(G, k, *_I)$ is an entire IVFMS and that (G, D) is a self-mapping of G. The mapping D meets the following requirement:

$$k(Dp, Dq, t) \geq \min$$

$$\left\{\frac{k(p, D[p], t)k(q, D[q], t) + k[(p, q, t)]^2 + k(p, D[p], t)k(p, q, t)}{1 + [k(p, D[p], t) + k(p, q, t) + k(q, D[q], t)]},\right.$$

$$\left.\frac{[1 + k(q, D[q], t)]k(p, D[p], t)}{1 + k(p, q, t)}, k(p, D[p], t), k(q, D[q], t), k(p, q, t)\right\}$$

$\forall p,\ q \in G$ with $p \neq q$ and D is onto. Then D admits a fixed point in G.

Proof: Let $p_0 \in G$, since D is onto, there is an element p_1 satisfying $p_1 \in D^{-1}p_0$. Similarly, if $p_{l-1} = p_l$ for-some l, then p_l is a fixed point of D, we can choose, $p_n \in D^{-1}p_{n-1}$, where $n = 2,3,4....$

Without losing generality, we can assume that $p_{n-1} \neq p_n$ for every n,

Therefore,

$$k(p_{n-1}, p_n, t) = k(D[p_n], D[p_{n+1}], t)$$

$$\geq \min\left\{\begin{array}{l}\dfrac{\begin{array}{l}k(p_n, D[p_n], t)\,k(p_{n+1}, D[p_{n+1}], t) + k[(p_n, p_{n+1}, t)]^2\\ +k(p_n, D[p_n], t)\,k(p_n, p_{n+1}, t)\end{array}}{1+\left[k(p_n, D[p_n], t) + k(p_n, p_{n+1}, t) + k(p_{n+1}, D[p_{n+1}], t)\right]},\\[2ex] \dfrac{\left[1+k(p_{n+1}, D[p_{n+1}], t)\right]k(p_n, D[p_n], t)}{1+k(p_n, p_{n+1}, t)},\\[2ex] k(p_n, D[p_{n+1}], t), k(p_{n+1}, D[p_{n+1}], t), k(p_n, p_{n+1}, t)\end{array}\right\}$$

$$\Rightarrow k(p_{n-1}, p_n, t) \geq k(p_n, p_{n+1}, t).$$

Thus, p_n is Cauchy sequence in G. Since G is complete, p_n converges to p, for some $p \in G.D(p_n) = p_{n-1}$, as $n \to \infty$. Hence, $D(p) = p$.

2.4 CONCLUSION

The concept of an interval-valued τ-norm was initially proposed, with the convergence of a sequence of interval numbers playing a crucial role in defining the continuity of such a τ-norm. This paved the way for the development of the continuous interval-valued τ-norm, which formed the cornerstone of the IVFMS (Interval-Valued Fuzzy Metric Space) concept. It's important to highlight that this definition necessitates modification by the condition (C6), ensuring its robustness and applicability.

Notably, distinctions arise between the traditional notion of fuzzy metric space and the novel IVFMS concept. These disparities underscore the unique characteristics and implications associated with IVFMS.

In the pursuit of establishing a fixed-point model for expansion mappings within the IVFMS framework, several theorems have been formulated and rigorously demonstrated. These developments signify a notable advancement in fixed-point model research, opening up new avenues for exploration and theoretical inquiry in this domain. All in all, the emergence of IVFMS marks a promising and innovative direction in the realm of fixed-point model research, with far-reaching implications for various theoretical and practical applications.

2.5 APPLICATIONS

The medical realm emerges as an exceptionally reliable and captivating domain where the utility of interval-valued fuzzy metric space finds profound significance. Within this context, the frequency distribution matrix assumes a pivotal role. This matrix offers a comprehensive depiction of the prevalence of symptoms among patients afflicted with a specific disease, relative to the total number of individuals diagnosed with that ailment. Extensively employed in discrimination analysis, it serves as a powerful tool for assessing and ranking symptoms based on their discriminatory efficacy in identifying distinct diseases.

However, it is imperative to recognize that while the frequency distribution matrix provides valuable insights, relying on it alone may fall short in delivering precise diagnoses, particularly in complex medical scenarios. Such complexities often arise when a single disease can be influenced by a multitude of factors, thereby manifesting in varied and sometimes unpredictable ways. In these nuanced situations, additional diagnostic methodologies and tools become indispensable to complement the insights garnered from the frequency distribution matrix, ensuring a more comprehensive and accurate understanding of the underlying medical conditions.

REFERENCES

[1] Agrawal, A. & Chouhan, P. (2005). Some fixed point theorems for expansion mappings. *Jnanabha*, 35, 197–199.

[2] Bhardwaj, R., Chouhan, S., Sanath Kumar, H. & Pandey, S. K. (2022). F contractive-type mappings in soft metric space. In Singh, M. (ed.), *Recent Trends in Design, Materials and Manufacturing*. Springer, pp. 3–14.

[3] Chakrabarty, K., Biswas, R. & Nanda, S. (1998). On fuzzy metric spaces. *Fuzzy Sets and Systems*, 99(1), 111–114.

[4] George, A. & Veeramani, P. (1994). On some results in fuzzy metric spaces. *Fuzzy Sets and Systems*, 64(3), 395–399.

[5] Jain, R. K. (1994). Some common fixed point theorems for expansion mappings. *Acta Ciencia Indica Mathematics*, 20, 217–220.

[6] Kramosil, I. & Mich´alek, J. (1975). Fuzzy metrics and statistical metric spaces. *Kybernetika*, 11(5), 336–344.

[7] Kaleva, O. & Seikkala, S. (1984). On fuzzy metric spaces. *Fuzzy Sets and Systems*, 12(3), 215–229.

[8] Li, C. (2009). Distances between interval-valued fuzzy sets. In *NAFIPS 2009-2009 Annual Meeting of the North American Fuzzy Information Processing Society*, pp. 1–3.

[9] Osman, M. A. (1983). Fuzzy metric space and fixed fuzzy set theorem. *Bulletin of the Malaysian Mathematical Sciences Society*, 6(1), 1–4.

[10] Popa, V. (1986). Some fixed-point theorems of expansion mappings. *Demonstratio Mathematica*, 19(3), 699–702.

[11] Sewani, G., Singh, A., Singh, R. & Bhardwaj, R.K. (2021). Constructing of some new results in intuitionistic fuzzy metric space by applying the abstraction of fuzzy iterated contraction. *Journal of Contemporary Issues in Business and Government*, 27(2), 4129.
[12] Shen, Y., Li, H. & Wang, F. (2012). On interval-valued fuzzy metric spaces. *International Journal of Fuzzy Systems*, 14(1), 35–44.
[13] Singh, R., Singh, A. & Goyal, A. (2015). Fixed point results in fuzzy Menger space with common property (EA). *Journal of Engineering Research and Applications*, 5(8, Part 1), 62–67.
[14] Singh, R., Singh, A. & Goyal, A. (2016). Fixed point results for p-1 compatible in fuzzy Menger space. *Advances in Fixed Point Theory*, 6(4), 520–527.
[15] Singh, U., Singh, N., Singh, R. & Bhardwaj, R. (2022). Common invariant point theorem for multi-valued generalized fuzzy mapping in b-metric space. In Singh, M. (ed.), *Recent Trends in Design, Materials and Manufacturing*. Springer, pp. 23–35.
[16] Wang, S. Z. (1984). Some fixed point theorems on expansion mappings. *Math. Japonica*, 29, 631–636.

3 Stability and Modified New Results for the Iterative Approximations

Abdolrahman Razani and Özen Özer

3.1 INTRODUCTION

The study of the convergence of an iterative sequence is one of the attractive subject in analysis, and many researchers are working on it (see [3,4,9–12,15–17]). In this chapter (based on [20] and [23]), we consider some generalized Lipschitizian mappings and study some iterative process for them. Furthermore, some numerical examples are also given.

At this step, we recall some definitions and Lemmas (also see [7–14, 20,21]).

Definition 3.1

Assume $Y^\star$ is the dual of the real Banach space Y. One can define the duality mapping $\mathcal{G} : Y \to Y^\star$ by

$$Gy := \left\{ g \in Y^\star : \langle y, g \rangle = \|y\|^2, \|g\| = \|y\| \right\}.$$

Lemma 3.1

[22] Assume $\{\nu_n \geq 0\}, \{\iota_n \geq 0\}$ and $\{\varphi_n \geq 0\}$ are satisfying $\sum_{n=1}^{\infty} \varphi_n < \infty$, $\delta_{n+1} \leq \iota_n \nu_n + \varphi_n$, and if $\iota_n \geq 1, \sum_{n=1}^{\infty} (\iota_n - 1) < \infty$ for all $n \geq 1$, then $\lim_{n\to\infty} \nu_n$ exists.

Lemma 3.2

[24] Let sequence $\{b_n \geq 0\}_{n=1}^{\infty}$ be such that

$$b_{n+1} \leq (1 - \rho_n) b_n + \epsilon_n, \text{ for all } n \geq 0$$

DOI: 10.1201/9781003638414-3

where $\{\epsilon_n\} \in \mathbb{R}$ and $\{\rho_n\} \in (0,1)$ are two sequences such that

I. $\lim_{n\to\infty} \rho_n = 0$ and $\Sigma_{n=1}^{\infty}\rho_n = \infty$.

II. $\sum_{n=1}^{\infty} |\epsilon_n| < \infty$ or $\limsup_{n\to\infty} \dfrac{\epsilon_n}{\rho_n} \le 0$.

Then $\lim_{n\to\infty} b_n = 0$.

Lemma 3.3

[18] Suppose Y is a uniformly convex Banach space, $\{\varphi_n\} \in (0,1)$ and $\{u_n\},\{v_n\} \in Y$ such that $\limsup_{n\to\infty}\|u_n\| \le \alpha, \limsup_{n\to\infty}\|v_n\| \le \alpha$ and $\limsup_{n\to\infty}\|\varphi_n u_n + (1-\varphi_n)v_n\| = \alpha$, for some $\alpha \ge 0$, then

$$\lim_{n\to\infty}\|u_n - v_n\| = 0.$$

Lemma 3.4

[1] Suppose $\{x_n\}$ is a bounded sequence on the uniformly convex Banach space Y. If $w_w(\{z_n\}) = \{z\}$, then $z_n \rightharpoonup z$, where $w_w(\{z_n\}) = \{z \in X : \exists z_{nj} \rightharpoonup z\}$ denotes the weak w-limit set of $\{x_n\}$.

A definition of Jung et al. [5] is recalled.

Definition 3.2

Let Y is Banach space and $D \ne \phi \subset Y$. A mapping $\mathcal{G} : D \to D$ is called generalized Lipschitzian mapping if for each $x, y \in D$ and $n \ge 1$

$$\begin{aligned}\|G^n x - G^n y\| \le a_n\|x-y\| + b_n\left(\|x - G^n x\| + \|G^n y - y\|\right)\\ + c_n\left(\|x - G^n y\| + \|G^n x - y\|\right)\end{aligned} \tag{3.1}$$

where the constants $a_n, b_n, c_n \ge 0$ are such that

$$\exists_{n0} \text{ such that } b_n + c_n < 1,$$

for all $n > n_0$.

Definition 3.3

Assume $D \subset X$, where X is Banach space. An uniformly generalized Lipschitzian mapping $\mathcal{F} : D \to D$ satisfying

$$\begin{aligned} \left\|\mathcal{F}^n x - \mathcal{F}^n y\right\| \le \alpha\|x-y\| + \beta\left(\left\|x - \mathcal{F}^n x\right\| + \left\|\mathcal{F}^n y - y\right\|\right) \\ + \gamma\left(\left\|x - \mathcal{F}^n y\right\| + \left\|\mathcal{F}^n x - y\right\|\right), \text{ for all } x, y \in D \end{aligned} \tag{3.2}$$

where constants $\alpha, \beta, \gamma \ge 0$ and $\beta + \gamma < 1$.

Lemma 3.5

Assume D is a closed, bounded, and convex subset of a uniformly convex Banach space X, and $\mathcal{F}$ is a uniformly generalized Lipschitzian self-map of D. Then, for all $x \in [x_0, x_1]$,

$$\forall_{\rho>0} \exists_{\xi(\rho)>0} \|\mathcal{F}x - x\| < \rho$$

whenever for $x_0, x_1 \in D, \|\mathcal{F}x_0 - x_0\| \le \xi(\rho)$ and $\|Fx_1 - x_1\| \le \xi(\rho)$, where $[x_0, x_1] = \{\xi x_1 + (1-\xi)x_0 : 0 \le \xi \le 1\}$.

Proof. Fix $\rho > 0$ and $x \in [x_0, x_1]$. Then, $x = \xi x_1 + (1-\xi)x_0$ for some ξ with $0 \le \xi \le 1$. It is obvious that if $\|x_1 - x_0\| < \rho_1$, then, $\|x - x_0\| < \rho_1$, where $\rho_1 = \dfrac{1-(b+c)}{1+a+2c}\dfrac{\rho}{2}$. Therefore, if $\xi(\rho) < \dfrac{1-(b+c)}{1+b+c}\dfrac{\rho}{2}$,

$$\begin{aligned} \|\mathcal{F}x - x\| &\le \|\mathcal{F}x - \mathcal{F}x_0\| + \|\mathcal{F}x_0 - x_0\| + \|x_0 - x\| \\ &\le a\|x_0 - x\| + b\left(\|\mathcal{F}x - x\| + \|\mathcal{F}x_0 - x_0\|\right) + c\left(\|x_0 - \mathcal{F}x\| + \|\mathcal{F}x_0 - x\|\right) \\ &\quad + \|\mathcal{F}x_0 - x_0\| + \|x_0 - x\| \\ &\le a\|x_0 - x\| + b\left(\| \mathcal{F}x - x \| + \|\mathcal{F}x_0 - x_0\|\right) \\ &\quad + c\left(\|x - \mathcal{F}x\| + \|\mathcal{F}x_0 - x_0\| + 2\|x_0 - x\|\right) \\ &\quad + \|\mathcal{F}x_0 - x_0\| + \|x_0 - x\| \end{aligned}$$

and

$$\|Fx - x\| \le \frac{1+a+2c}{1-(b+c)}\|x_0 - x\| + \frac{b+c+1}{1-(b+c)}\|x_0 - Fx_0\| < \rho. \tag{3.3}$$

Hence, we may assume $\|x_0 - x_1\| \geq \rho_1$. Let d_0 denote the diameter of D. For $0 \leq \xi < \frac{\rho_1}{d_0}$, $\|x_0 - x\| = \xi\|x_1 - x_0\| < \rho_1$. If $\xi(\rho) < \frac{1-(b+c)}{1+b+c}\frac{\rho}{2}$, we get $\|\mathcal{F}x - x\| < \rho$, from (1.3).

Let $\xi \geq \frac{\rho_1}{d_0}$. If $(1-\xi) < \frac{\rho_1}{d_0}$, so $\|x_1 - x\| = (1-\lambda)\|x_1 - x_0\| < \rho_1$, shows $\|Fx - x\| < \rho$.

One may assume that $\frac{\rho_1}{d_0} \leq \xi \leq 1 - \frac{\rho_1}{d_0}$ and $y = \mathcal{F}x$, thus $\|y - x_0\| = \|Fx - x_0\| \leq \|Fx - Fx_0\| + \|Fx_0 - x_0\|$. By (1.2) and (1.3),

$$\|y - x_0\| \leq \frac{a+b+c}{1-(b+c)}\|x - x_0\| + \frac{1+b+c}{1-(b+c)}\|Fx_0 - x_0\|$$

$$\leq r\xi\|x_1 - x_0\| + h\xi(\rho),$$

where $r = \frac{a+b+c}{1-(b+c)}$ and $h = \frac{1+b+c}{1-(b+c)}$, and also

$$\|y - x_1\| \leq \frac{a+b+c}{1-(b+c)}\|x - x_1\| + \frac{1+b+c}{1-(b+c)}\|\mathcal{F}x_1 - x_1|$$

$$\leq r(1-\xi)\|x_1 - x_0\| + h\xi(\rho)$$

Set

$$z_0 = \frac{y - x_0}{\xi\|x_1 - x_0\|} \tag{3.4}$$

and

$$z_1 = \frac{x_1 - y}{(1-\xi)\|x_1 - x_0\|} \tag{3.5}$$

then

$$\|z_0\| \leq \frac{r\xi\|x_1 - x_0\| + h\xi(\rho)}{\xi\|x_1 - x_0\|} \leq r + \frac{h\xi(\rho)d_0}{\rho_1^2}$$

and $\|z_1\| \leq r + \frac{h\xi(\rho)d_0}{\rho_1^2}$. On the other hand, for ξ with $\frac{\rho_1}{d_0} \leq \xi \leq 1 - \frac{\rho_1}{d_0}, \|\xi z_0 + (1-\xi)z_1\| = \frac{\|x_1 - x_0\|}{\|x_1 - x_0\|} = 1$. Therefore, using uniform convexity of E, we can choose $\xi(\rho)$ so small that $\|z_1 - z_0\| < \frac{\rho}{d_0}$. This implies from $x = \xi x_1 + (1-\xi)x_0$, (3.4) and (3.5) that

$$\|y - x\| = \|(1-\xi)(y - x_0) - \xi(x_1 - y)\| = \xi(1-\xi)\|x_1 - x_0\|\|z_1 - z_0\| < \rho$$

Since $y = \mathcal{F}x, \|\mathcal{F}x - x\| < \rho$.

Lemma 3.6

[21] Assume $D \neq \phi \subset X$ is closed, convex, and bounded, $\mathcal{F} : D \to D$ is a uniformly generalized Lipschitzian mapping, where X is uniformly convex Banach space, $\{x_j\} \in D$ is such that $x_j \rightharpoonup x_0$ and $x_j - \mathcal{F}x_j \to 0$, then $\mathcal{F}(x_0) = x_0$.

Proof. Notice that $\|(I - \mathcal{F})x_j\| = \varepsilon_j \to 0$, for each j. Let $\varepsilon_j \leq \xi(\varepsilon_{j-1}) < \varepsilon_{j-1}$, where $\xi(\varepsilon)$ for any $\varepsilon > 0$ is given by Lemma 1.8. If $x \in \overline{Co}\{x_j : j \geq k\}$ then $\|x - \mathcal{F}x\| \leq \varepsilon_{k-1}$. Thus $\overline{Co}\{x_j : j \geq k\}$ is weakly compact, $x_0 \in \overline{Co}\{x_j : j \geq k\}, k = 1,2,3,\cdots$ and thus $\|x_0 - \mathcal{F}x_0\| \leq \varepsilon_j$, $j = 1,2,3,\cdots$. Thus $\|\mathcal{F}x_0 - x_0\| = 0$.

Assume $D \neq \phi \subset X$ is closed, bounded, and convex, $\{\mathcal{F}_n\} : D \to D$ is a sequence of uniformly generalized Lipschitizian, such that the common fixed points set $\Gamma \neq \phi$ of $\{\mathcal{F}_n\}$, where X is a Banach space. Assume $k_n = \dfrac{a_n + b_n + c_n}{1-(b_n + c_n)}$ depend on $\mathcal{F}_n$ and $k_n \geq 1$ for all $n \geq 1$. One can define the sequence $\{x_n\}$ by $x_{n+1} = \mathcal{F}_n x_n$, for $n \geq 1$ and a given $x_1 \in D$.

Lemma 3.7

Let $\mathcal{F} : D \to D$ be a uniformly generalized Lipschitzian mapping, where X is a normed linear space and $D \neq \phi \subset X$ is convex and closed. Suppose $\{x_n\}$ is defined by $x_{n+1} = \mathcal{F}_n x_n$. One has $\lim_{n\to\infty}\|x_n - \mathcal{F}x_n\| = 0$ provided $\lim_{n\to\infty}\|x_n - \mathcal{F}^n x_n\| = 0$ and $\lim_{n\to\infty}\|x_n - x_{n+1}\| = 0$.

Proof. Let $r_n = \|\mathcal{F}^n x_n - x_n\|$. Then

$$\begin{aligned}\|x_{n+1}-\mathrm{F}x_{n+1}\| &\le \|x_{n+1}-\mathrm{F}^{n+1}x_{n+1}\|+\|\mathrm{F}x_{n+1}-\mathrm{F}^{n+1}x_{n+1}\| \\ &\le r_{n+1}+a\|x_{n+1}-\mathrm{F}^{n}x_{n+1}\| \\ &+b\left(\|x_{n+1}-\mathrm{F}x_{n+1}\|+\|\mathrm{F}^{n}x_{n+1}-\mathrm{F}^{n+1}x_{n+1}\|\right) \\ &+c\left(\|x_{n+1}-\mathrm{F}^{n+1}x_{n+1}\|+\|\mathrm{F}^{n}x_{n+1}-\mathrm{F}x_{n+1}\|\right) \\ &\le r_{n+1}+a\|x_{n+1}-\mathrm{F}^{n}x_{n+1}\| \\ &+b\left(\|\mathrm{F}^{n}x_{n+1}-x_{n+1}\|\right. \\ &\left.+\|x_{n+1}-\mathrm{F}x_{n+1}\|+\|\mathrm{F}^{n+1}x_{n+1}-x_{n+1}\|\right) \\ &+c\left(\|x_{n+1}-\mathrm{F}^{n+1}x_{n+1}\|\right. \\ &\left.+\|\mathcal{F}^{n}x_{n+1}-x_{n+1}\|+\|x_{n+1}-\mathcal{F}x_{n+1}\|\right),\end{aligned}$$

and

$$\|x_{n+1}-\mathcal{F}x_{n+1}\| \le \frac{1+b+c}{1-(b+c)}r_{n+1}+\frac{a+b+c}{1-(b+c)}\|x_{n+1}-\mathcal{F}^{n}x_{n+1}\|. \tag{3.6}$$

Thus $\|x_{n+1}-\mathcal{F}^{n}x_{n+1}\|$,

$$\|\mathcal{F}^{n}x_{n+1}-x_{n+1}\| \le \frac{1+b+c}{1-(b+c)}r_{n}+\frac{1+a+2c}{1-(b+c)}\|x_{n+1}-x_{n}\|. \tag{3.7}$$

From (3.6) and (3.7),

$$\begin{aligned}\|x_{n+1}-\mathcal{F}x_{n+1}\| &\le \frac{1+b+c}{1-(b+c)}r_{n+1}+\frac{a+b+c}{1-(b+c)}\|\mathcal{F}^{n}x_{n+1}-x_{n+1}\| \\ &\le \frac{1+b+c}{1-(b+c)}r_{n+1}+\frac{a+b+c}{1-(b+c)}\left(\frac{1+b+c}{1-(b+c)}r_{n}+\frac{1+a+2c}{1-(b+c)}\|x_{n+1}-x_{n}\|\right).\end{aligned}$$

3.2 CONVERGENCE OF THE ITERATIVE APPROXIMATION

The result presented here, generalizes the results in [2,20] and [23].

Let $D \neq \phi \subset X$, where X is a real Banach space and $T_i : D \to D$, for $i = 0, \cdots, m$, be generalized Lipschitizian mappings. Consider the iterative sequence $\{x_n\}$ which is

$$
\begin{cases}
x_1 \in D \\
u_n = \dfrac{1}{n+1}\sum_{j=0}^{n} T_0^j x_n \\
x_{n,1} = \left(1-\beta_{n,0}\right)x_n + \beta_{n,0}u_n \\
x_{n,2} = \left(1-\beta_{n,1}\right)x_n + \beta_{n,1}T_1^n x_{n,1} \\
\vdots \\
x_{n,m} = \left(1-\beta_{n,m-1}\right)x_n + \beta_{n,m-1}T_{m-1}^n x_{n,m-1} \\
x_{n+1} = \left(1-\beta_{n,m}\right)x_n + \beta_{n,m}T_m^n x_{n,m}
\end{cases}
\tag{3.8}
$$

where $\left\{\beta_{n,i}\right\}_{i=0}^{m} \in (0,1)$. We prove that if $T_i : D \to D$ for $i = 0,\cdots,m$, are generalized Lipschitzian mappings, then the sequence $\{x_n\}$ given by (3.8) in a uniformly convex Banach space, converges to $q \in \bigcap_{i=0}^{m} F(T_i)$.

Theorem 3.1

Let X be a uniformly convex Banach space and $T_i : D \to D$ for $i = 0,\cdots,m$ be generalized Lipschitizian mappings satisfying Opial's condition, where $D \neq \phi \subset X$ is closed and convex. Assume

$$
\begin{aligned}
\left\|T_i^n x - T_i^n y\right\| \le\ & a_n^i \left\|x-y\right\| + b_n^i\left(\left\|x - T_i^n x\right\| + \left\|T_i^n y - y\right\|\right) \\
& + c_n^i\left(\left\|x - T_i^n y\right\| + \left\|T_i^n x - y\right\|\right),
\end{aligned}
$$

where $x, y \in D \neq \phi, n \ge 1, k_n^i = \dfrac{a_n^i + b_n^i + c_n^i}{1-\left(b_n^i + c_n^i\right)}$ and $\left\{k_n^i\right\} \subset [1,\infty)$ for $i = 0,\cdots,$ m satisfy $k_n^i \to 1$ as $n \to \infty$.

In addition

$$
\sum_{n=1}^{\infty}\left(\eta_n k_n^1 k_n^2 \cdots k_n^m - 1\right) < \infty
$$

where $\eta_n = \max\left\{k_j^0, 1 \le j \le n\right\}$. Assume $\{x_n\}$ is given by (3.2.8). If $F = \bigcap_{i=0}^{m} F(T_i) \neq \phi$ and $0 < \liminf_{n\to\infty}\beta_{n,i} \le \limsup_{n\to\infty}\beta_{n,i} < 1$ for $i = 0,\cdots,m$, then

1. $\{x_n\}$ is weakly convergent to a common fixed point of T_i for $i = 0,\cdots,m$.
2. $\lim_{n\to\infty}\|x_n - T_i x_n\| = 0$ and $\lim_{n\to\infty}\|x_n - u_n\| = 0$, for $i = 1,\cdots,m$.
3. For all $q \in F$, $\lim_{n\to\infty}\|x_n - q\|$ exists.

Proof. Fixed $q \in F$ and $i = 0,\cdots,m$,

$$\begin{aligned}\|T_i^n x_n - q\| &\le a_n^i\|x_n - q\| + b_n^i\|T_i^n x_n - x_n\| + c_n^i\left(\|T_i^n x_n - q\| + \|x_n - q\|\right)\\ &\le a_n^i\|x_n - q\| + b_n^i\left(\|T_i^n x_n - q\| + \|x_n - q\|\right)\\ &+ c_n^i\left(\|T_i^n x_n - q\| + \|x_n - q\|\right),\end{aligned}$$

and so

$$\|T_i^n x_n - q\| \le \frac{a_n^i + b_n^i + c_n^i}{1-\left(b_n^i + c_n^i\right)}\|x_n - q\| = k_n^i\|x_n - q\|. \tag{3.9}$$

In order to find a bound for $\|x_{n+1} - q\|$, we compute $\|u_n - q\|$ and $\|x_{n,m} - q\|, i = 1,\cdots,m$, as follows:

i. Suppose $1 \le j \le n$ then we have

$$\|u_n - q\| \le \frac{1}{n+1}\left(\|x_n - q\| + \sum_{j=1}^{n} k_j^0\|x_n - q\|\right) \le \eta_n\|x_n - q\|, \tag{3.10}$$

by inequality (3.9).

ii. By (3.10)

$$\begin{aligned}\|x_{n,1} - q\| &\le (1-\beta_{n,0})\|x_n - q\| + \beta_{n,0}\eta_n\|x_n - q\|\\ &\le \left(1+\beta_{n,0}(\eta_n - 1)\right)\|x_n - q\|\\ &\le \eta_n\|x_n - q\|.\end{aligned} \tag{3.11}$$

iii. Using (3.11) we obtain

$$\begin{aligned}\|x_{n,i} - q\| &\le (1-\beta_{n,i-1})\|x_n - q\| + \beta_{n,i-1}k_n^{i-1}\|x_{n,i-1} - q\|\\ &\le k_n^{i-1}\cdots\eta_n\|x_n - q\|.\end{aligned} \tag{3.12}$$

Considering (i),(ii) and (iii), then

$$\begin{aligned}\|x_{n+1}-q\| &\le (1-\beta_{n,m})\|x_n-q\|+\beta_{n,m}k_n^m\|x_{n,m}-q\| \\ &\le \left(1+\beta_{n,m}\left(k_n^m\cdots k_n^1\eta_n-1\right)\right)\|x_n-q\|.\end{aligned} \tag{3.13}$$

Since $\sum_{n=1}^{\infty}\left(\eta_n k_n^1 k_n^2 k_n^3\cdots k_n^m-1\right)<\infty$ by a Lemma 3.1.2 and inequality (3.13) we get $\lim_{n\to\infty}\|x_n-q\|=r$ exists. Furthermore

$$\lim_{n\to\infty}\|x_{n+1}-q\|=\lim_{n\to\infty}\left\|(1-\beta_{n,m})(x_n-q)+\beta_{n,m}\left(T_m^n x_{n,m}-q\right)\right\|=r,$$

is satisfied. Beside form (3.12) we have result as follows:

$$\begin{aligned}\lim\sup_{n\to\infty}\|T_m^n x_{n,m}-q\| &\le \lim\sup_{n\to\infty}\left(k_n^m\|x_{n,m}-q\|\right) \\ &\le \lim\sup_{n\to\infty}\left(k_n^m\cdots k_n^1\eta_n\|x_n-q\|\right) \\ &\le \lim_{n\to\infty}\|x_n-q\|=r.\end{aligned}$$

Thus $\lim_{n\to\infty}T_m^n\|x_{n,m}-x_n\|=0$. Also

$$\frac{\|x_{n+1}-q\|-\|x_n-q\|+\beta_{n,m}\|x_n-q\|}{\beta_{n,m}k_n^m}\le\|x_{n,m}-q\|, \tag{3.14}$$

and finally, $\lim_{n\to\infty}\|x_n-q\|\le\lim\inf_{n\to\infty}\|x_{n,m}-q\|$. By (3.12),

$$\lim\sup_{n\to\infty}\|x_{n,m}-q\|\le\lim\sup_{n\to\infty}\left(k_n^{m-1}\cdots\eta_n\|x_n-q\|\right)=\lim_{n\to\infty}\|x_n-q\|. \tag{3.15}$$

(3.14) and (3.15) shows $\lim_{n\to\infty}\|x_n-q\|=\lim_{n\to\infty}\|x_{n,m}-q\|=r$. Using the same technique, we have $\lim_{n\to\infty}\|x_{n,i}-q\|=r, \lim_{n\to\infty}\|u_n-x_n\|=0$ and $\lim_{n\to\infty}T_i^n\|x_{n,i}-x_n\|=0$ for $i=1,\cdots m-1$.

Now, we prove $\|x_n-T_i^n x_n\|\to 0$. From equality (3.8), we obtain followings:

$$\|x_{n,i}-x_n\|=\left\|(1-\beta_{n,i-1})x_n+\beta_{n,i-1}T_{i-1}^n x_{n,i-1}-x_n\right\|=\beta_{n,i-1}\left\|T_{i-1}^n x_{n,i-1}-x_n\right\|\to 0$$

as $n\to\infty$. Notice that $\|x_n-T_i^n x_n\|\le\|x_n-T_i^n x_{n,i}\|+\|T_i^n x_{n,i}-T_i^n x_n\|$, where

$$\begin{aligned}\left\|T_i^n x_{n,i}-T_i^n x_n\right\| \le\ & a_n^i\left\|x_{n,i}-x_n\right\|+b_n^i\left(\left\|x_{n,i}-T_i^n x_{n,i}\right\|+\left\|T_i^n x_n-x_n\right\|\right)\\ &+c_n^i\left(\left\|T_i^n x_{n,i}-x_n\right\|+\left\|T_i^n x_n-x_{n,i}\right\|\right)\\ \le\ & a_n^i\left\|x_n-x_{n,i}\right\|+b_n^i\left(\left\|x_{n,i}-x_n\right\|+\left\|T_i^n x_{n,i}-x_n\right\|+\left\|T_i^n x_n-x_n\right\|\right)\\ &+c_n^i\left(\left\|T_i^n x_{n,i}-x_n\right\|+\left\|x_n-x_{n,i}\right\|+\left\|x_n-T_i^n x_n\right\|\right)\\ \le\ & \left(a_n^i+b_n^i+c_n^i\right)\left\|x_n-x_{n,i}\right\|\\ &+\left(b_n^i+c_n^i\right)\left(\left\|T_i^n x_{n,i}-x_n\right\|+\left\|T_i^n x_n-x_n\right\|\right).\end{aligned} \tag{3.16}$$

Since following inequity is satisfied,

$$\begin{aligned}\left\|x_n-T_i^n x_n\right\| \le & \left\|x_n-T_i^n x_{n,i}\right\|\\ &+\left(a_n^i+b_n^i+c_n^i\right)\left\|x_n-x_{n,i}\right\|\\ &+\left(b_n^i+c_n^i\right)\left(\left\|T_i^n x_{n,i}-x_n\right\|+\left\|T_i^n x_n-x_n\right\|\right).\end{aligned}$$

We get

$$\left\|x_n-T_i^n x_n\right\| \le k_n^i\left\|x_{n,i}-x_n\right\|+\left(1+k_n^i\right)\left\|x_n-T_i^n x_{n,i}\right\|. \tag{3.17}$$

By the inequality, we have (3.17), $\left\|x_n-T_i^n x_n\right\| \to 0$, and also, we obtain by Lemma 3.1.10 $\left\|x_n-T_i x\right\|_n \to 0$ as $n \to \infty$ for $i=0,\cdots,m$. Since

$$\lim_{n\to\infty}\left\|x_n-q\right\|$$

exists and the boundedness of $\{x_n\}$, there exists a subsequence $x_{n_k} \rightharpoonup x$. By Lemma 1.9$x \in F$.

Notice $w_w\left(\{x_n\}\right)$ just has an element.

Suppose there exists a subsequence $x_{n_j} \hookrightarrow w \neq x \in F$. Thus $\lim_{n\to\infty}\left\|x_n-q\right\|$ shows

$$\lim_{n\to\infty}\left\|x_n-x\right\| \text{ and } \lim_{n\to\infty}\left\|x_n-w\right\|.$$

This means

$$\lim_{n\to\infty}\left\|x_n-x\right\|=\lim_{k\to\infty}\left\|x_{n_k}-x\right\|<\lim_{k\to\infty}\left\|x_{n_k}-w\right\|=\lim_{n\to\infty}\left\|x_n-w\right\|$$

and

$$\lim_{n\to\infty}\|x_n - w\| = \lim_{j\to\infty}\|x_{n_j} - w\| < \lim_{j\to\infty}\|x_{n_j} - x\| = \lim_{n\to\infty}\|x_n - x\|.$$

This is a contradiction, and $w_w(\{x_n\})$ has only an element and by Lemma 3.1.5, $\{x_n\}$ is weakly convergent to x.

Theorem 3.2

Assume $\{x_n\}_{n\geq 0}$, is given by (2.8) and for all $i = 0, \cdots, m-1$,

$$0 < \liminf_{n\to\infty} \beta_{n,i} \leq \limsup_{n\to\infty} \beta_{n,i} < 1,$$

$\lim_{n\to\infty} \beta_{n,m} = 0$ and $\sum_{n=1}^{\infty} \beta_{n,m} = \infty$, where hypotheses Theorem 3.2.1 are satisfied. Then

$$x_n \overset{\text{Strongly}}{\rightarrow} q \in F.$$

Proof. Notice that

$$\begin{aligned}\|x_{n+1} - q\| &\leq (1-\beta_{n,m})\|x_n - q\| + \beta_{n,m}k_n^m\|x_{n,m} - q\| \\ &\leq \left(1 + \beta_{n,m}\left(k_n^m \cdots k_n^1\eta_n - 1\right)\right)\|x_n - q\|,\end{aligned}$$

By similar argument of Theorem 3.2.1. Since $\lim_{n\to\infty} \beta_{n,m} = 0$ (by Lemma 1.3), $\lim_{n\to\infty}\|x_n - q\| = 0$. Also $\lim_{n\to\infty}\|x_n - T_i x_n\| = 0$, for $i = 0, \cdots, m$ and $\lim_{n\to\infty}\|x_{n+1} - x_n\| = 0$.

Now present an example.

Example 3.1

Let $M = \mathcal{R}, D = [0, \infty)$. Let $\{x_n\}$ be the sequence given by (3.8), where $T_0 x = \frac{x}{100}$, $T_1 x = \frac{x}{1000}$ and $T_2 x = \frac{x}{10000}$. In addition, we assume that $\beta_{n,2} = \frac{n}{3n+1}, \beta_{n,1} = \frac{n}{4n+1}$ and $\beta_{n,0} = \frac{n}{5n+1}$. Thus

$$\begin{cases} u_n = \dfrac{1}{n+1}\sum_{j=0}^{n}\dfrac{x_n}{100^j}, \\ z_n = \dfrac{4n+1}{5n+1}x_n + \left(\dfrac{n}{5n+1}\right)u_n, \\ y_n = \dfrac{3n+1}{4n+1}x_n + \left(\dfrac{n}{4n+1}\right)\left(\dfrac{1}{1000^n}\right)z_n, \\ x_{n+1} = \dfrac{2n+1}{3n+1}x_n + \left(\dfrac{n}{3n+1}\right)\left(\dfrac{1}{10000^n}\right)y_n, \end{cases}$$

and $\bigcap_{i=0}^{2} F(T_i) = \{0\}$. If $x_1 = 1$, then we obtain the following table:

n	1	5	10	15	20	25	30
x_n	1	0.25962	0.03842	0.00541	0.00074	0.0001	0.00001

3.3 ITERATIVE APPROXIMATION OF GENERALIZED LIPSCHITIZIAN SEMIGROUPS

Assume $D \neq \phi$ is closed subset of a real Hilbert space $\mathcal{H}$ and $I : D \to D$ denotes the identity mapping.

Definition 3.4

Assume $D \neq \phi$ is a closed and convex subset of a real Hilbert space $\mathcal{H}$. A one-parameter family $\zeta = \{T(t) : 0 \leq t < \infty\}$ of self-mappings on D is called a nonexpansive semigroup on D if

I. For all $u \in D, T(0)u = u$.
II. For all $u \in C$ and $s, t \geq 0, T(s+t)u = T(s)T(t)u$.
III. For each $u \in D$, the mapping $t \to T(t)u$ is continuous on $[0, \infty)$.
IV. $\|T(t)u - T(t)v\| \leq \|u - v\|$, for all $u, v \in D$.

Set

$$\text{Fix } (\zeta) := \{u \in D : T(s)u = u, \text{ for all } s > 0\}.$$

Notice that Fix(ζ) is nonempty if D is bounded.

Plubtieng et al. [7] defined $\{x_n\}$, for $n \geq 0$, by

$$x_{n+1} = \alpha_n f(x_n) + \iota_n x_n + (1 - \iota_n - \alpha_n)\frac{1}{t_n}\int_0^{t_n} T(s)x_n\, ds,$$

where $x_0 \in D$ is arbitrary and $\{\alpha_n\},\{\iota_n\} \in (0,1)$ with $\alpha_n + \iota_n < 1$ and $\{t_n\} > 0$ is a divergent sequence where f is a contraction on $\mathcal{H}$. They proved $\{x_n\}$ is strongly convergent to a common fixed point ζ.

Assume the iteration process $\{x_n\}$ (see [8]), such that $x_0 \in D$ is arbitrary, A is a strongly positive bounded linear operator with coefficient λ and $0 < h < \frac{\lambda}{r}$, and

$$x_{n+1} = \alpha_n hf(x_n) + (1-\alpha_n A)\frac{1}{t_n}\int_0^{t_n} T(s)x_n\,ds, \tag{3.18}$$

It is proved $\{x_n\}$ is strongly convergent to a common fixed point ζ.

We recall the following lemmas.

Lemma 3.8

[19] Let $\zeta = \{T(t) : 0 \le t < \infty\}$ be a nonexpansive semigroup on $D \neq \phi$, a closed, bounded, and convex subset of $\mathcal{H}$. Then,

$$\limsup_{s\to\infty}\limsup_{t\to\infty}\sup_{x\in D}\left\| \frac{1}{t}\int_0^t T(u)x\,du - T(s)\left(\frac{1}{t}\int_0^t T(u)x\,du\right)\right\| = 0.$$

In this section, the convergent of the modified version of the iterative process is presented. This process are defined for each $x_1 \in D$

$$\begin{cases} x_1 \in D \\ u_n = \frac{1}{n+1}\sum_{j=0}^{n} T_0^j x_n, \\ x_{n,1} = \beta_{n,0}x_n + (1-\beta_{n,0})\frac{1}{t_n}\int_0^{t_n} K(s)u_n\,ds \\ x_{n,2} = \beta_{n,1}x_n + (1-\beta_{n,1})T_1^n x_{n,1}, \\ \vdots \\ x_{n,m} = \beta_{n,m-1}x_n + (1-\beta_{n,m-1})T_{m-1}^n x_{n,m-1}, \\ x_{n+1} = \alpha_n h\Phi(x_n) + \beta_{n,m}x_n + ((1-\beta_{n,m})I - \alpha_n\psi)T_m^n x_{n,m}, \end{cases} \tag{3.19}$$

where $\{\alpha_n\},\{\beta_{n,i}\}\in(0,1), i=0,\cdots,m$ are satisfying

1. $\lim_{n\to\infty}\alpha_n=0$ and $\sum_{n=1}^{\infty}\alpha_n=\infty$.
2. $0<\liminf_{n\to\infty}\beta_{n,i}\le\limsup_{n\to\infty}\beta_{n,i}<1, i=0,\cdots,m$.
3. Let $T_i:D\to D$ be a generalized Lipschitzian mapping such that

$$\|T_i^n x-T_i^n y\|\le a_n^i\|x-y\|+b_n^i\left(\|x-T_i^n x\|+\|T_i^n y-y\|\right)$$
$$+c_n^i\left(\|x-T_i^n y\|+\|T_i^n x-y\|\right)$$

for all $x,y\in D, i=0,\cdots,m$ and $n\ge1$, where $k_n^i=\dfrac{a_n^i+b_n^i+c_n^i}{1-\left(b_n^i+c_n^i\right)}$ and $\{k_n^i\}\subset[1,\infty)$ satisfy $k_n^i\to1$ as $n\to\infty$.

4. $\Phi:D\to D$ is a contraction with coefficient r.
5. ψ is a strongly positive bounded linear operator with coefficient λ on $\mathcal{H}$.

Lemma 3.9

[6] Let a strongly positive bounded linear operator ψ with coefficient λ on $\mathcal{H}$ and $0<p<\|\psi\|^{-1}$. Thus, $\|I-p\psi\|\le1-p\lambda$.

Theorem 3.3

Suppose

1. $D\ne\phi\subset\mathcal{H}$ is a closed, bounded, and convex set, where $\mathcal{H}$ is a real Hilbert space.
2. $T_i:D\to D$, for $i=0,\cdots,m$ are generalized Lipschitzian mappings, such that $\eta_n=\max\{k_j^0, 1\le j\le n\}$, and $\lim_{n\to\infty}\eta_n=1$. Also $\sum_{n=1}^{\infty}\left(\eta_n k_n^1 k_n^2\cdots k_n^m-1\right)<\infty$.
3. Consider the contraction $\Phi:D\to D$ with constant r and the strongly positive bounded linear operator ψ with coefficient λ on $\mathcal{H}$ with coefficient λ, while $0<h<\dfrac{\lambda}{r}$.
4. Assume $\zeta=\{K(t):1\le t<\infty\}$ is a nonexpansive semigroup on D.
5. Let $\{t_n>0\}$ be a real divergent sequence.
6. $\{x_n\}$ is given by (3.19).

If $F=\bigcap_{i=0}^{m}F(T_i)\cap F(\zeta)\ne\phi$, then

I. $\lim_{n\to\infty}\|x_n - x_{n+1}\| = 0.$

II. $\{x_n\} \to z \in F$ strongly, where z is a unique solution of

$$\langle(\psi - h\Phi)z, z - x\rangle \le 0, \quad \text{for all } x \in F. \tag{3.20}$$

Equivalently $P_F(I - \psi + h\Phi)z = z$.

Proof. Notice that $\{x_n\}$ is bounded. To prove this, assume $\|(1-\beta_{n,m})I - \alpha_n\psi\| \le 1 - \beta_{n,m} - \alpha_n\lambda$. By the same argument as in Theorem 2.1,

$$\begin{aligned}
\|x_{n+1} - q\| &= \left\|\alpha_n(h\Phi(x_n) - \psi q) + \beta_{n,m}(x_n - q) + ((1-\beta_{n,m})I - \alpha_n\psi)(T_m^n x_{n,m} - q)\right\| \\
&\le \alpha_n\|h\Phi(x_n) - \psi q\| + \beta_{n,m}\|x_n - q\| \\
&\quad + ((1-\beta_{n,m}) - \alpha_n\lambda)\|T_m^n x_{n,m} - q\| \\
&\le \alpha_n\|h\Phi(x_n) - h\Phi(q)\| + \alpha_n\|h\Phi(q) - \psi q\| + \beta_{n,m}\|x_n - q\| \\
&\quad + ((1-\beta_{n,m}) - \alpha_n\lambda)k_{n,m}\{\beta_{n,m-1}\|x_n - q\| \\
&\quad + (1-\beta_{n,m-1})\|T_{m-1}^n x_{n,m-1} - q\|\} \qquad (3.21) \\
&\ \vdots \\
&\le \alpha_n hr\|x_n - q\| + \beta_{n,m}\|x_n - q\| + \alpha_n\|h\Phi(q) - \psi(q)\| \\
&\quad + ((1-\beta_{n,m}) - \alpha_n\lambda)k_n^m \ldots k_n^1\eta_n\|x_n - q\| \\
&\le (1 - \alpha_n(\lambda - hr))k_n^m \ldots k_n^1\eta_n\|x_n - q\| + (\lambda - hr)\alpha_n\frac{\|h\Phi(q) - \psi q\|}{\lambda - hr},
\end{aligned}$$

for $q \in F$. An induction implies

$$\|x_{n+1} - q\| \le \max\left\{\|x_1 - q\|, (\lambda - hr)\frac{\|h\Phi(q) - \psi q\|}{\lambda - hr}\right\}.$$

Therefore, $\{x_n\}$ is bounded. In addition, $\{\Phi(x_n)\}, \{u_n\}$ and $\left\{\frac{1}{t_n}\int_0^{t_n} K(s)u_n ds\right\}$ are bounded.

Claim: $\lim_{n\to\infty}\|x_{n+1}-x_n\|=0$. From (3.21), we obtain

$$\begin{aligned}\|x_{n+1}-q\| &\leq \left(1-\alpha_n(\lambda-hr)\right)k_n^m\dots k_n^1\eta_n\|x_n-q\|+(\lambda-hr)\alpha_n\frac{\|h\Phi(q)-\psi q\|}{\lambda-hr}\\ &\leq k_n^m\dots k_n^1\eta_n\|x_n-q\|+(\lambda-hr)\alpha_n\frac{\|h\Phi(q)-\psi q\|}{\lambda-hr}.\end{aligned} \tag{3.22}$$

Since $\Sigma_{n=1}^{\infty}\left(k_n^m\dots k_n^1\eta_n-1\right)<\infty$ (also by Lemma 3.1.2 and inequality (3.22)) $\lim_{n\to\infty}\|x_n-q\|=m$ exists. If we consider $x_{n+1}=\beta_{n,m}x_n+\left(1-\beta_{n,m}\right)w_n$, where

$$w_n=\frac{\alpha_nh\Phi(x_n)+\left(\left(1-\beta_{n,m}\right)I-\alpha_n\psi\right)T_m^nx_{n,m}}{1-\beta_{n,m}},$$

and using (3.19), we have

$$\lim_{n\to\infty}\|x_{n+1}-q\|=\lim_{n\to\infty}\left\|\beta_{n,m}(x_n-q)+\left(1-\beta_{n,m}\right)(w_n-q)\right\|=m$$

and also,

$$\begin{aligned}\|w_n-q\| &\leq \frac{1}{1-\beta_{n,m}}\left\{\alpha_n\|h\Phi(x_n)-\psi q\|+\left(\left(1-\beta_{n,m}\right)-\alpha_n\lambda\right)\left\|T_m^nx_{n,m}-q\right\|\right\}\\ &\leq \frac{1}{1-\beta_{n,m}}\left\{\alpha_n\|h\Phi(x_n)-\psi q\|+\left\|T_m^nx_{n,m}-q\right\|\right\}\\ &\leq \frac{1}{1-\beta_{n,m}}\left\{\alpha_n\|h\Phi(x_n)-\psi q\|+k_n^m\dots k_n^1\eta_n\|x_n-q\|\right\}.\end{aligned} \tag{3.23}$$

Subsequently, we have $n\to\infty$, $\lim_{n\to\infty}\|w_n-q\|\leq m$, and so $\lim_{n\to\infty}\|w_n-x_n\|=0$. Since $\|x_{n+1}-x_n\|=\left(1-\beta_{n,m}\right)\|w_n-x_n\|$ then

$$\lim_{n\to\infty}\|x_{n+1}-x_n\|=0.$$

Let $x_{n,1}=\beta_{n,0}x_n+\left(1-\beta_{n,0}\right)\Omega_n$ where $\Omega_n=\frac{1}{t_n}\int_0^{t_n}K(s)u_nds$. By the same argument, $\lim_{n\to\infty}\|w_n-q\|=m$ and $\lim_{n\to\infty}\|x_{n,i}-q\|=m$ are hold. We show $\lim_{n\to\infty}\|\Omega_n-x_n\|=0$. By equality (3.19),

$$\left\|\Omega_n - q\right\| = \left\|\frac{1}{t_n}\int_0^{t_n} K(s)\,u_n\,ds - q\right\| \leq \eta_n \left\|x_n - q\right\|$$

shows $\lim_{n\to\infty}\left\|\Omega_n - q\right\| \leq m$. By Lemma 1.4,

$$\lim_{n\to\infty}\left\|\Omega_n - x_n\right\| = 0 \tag{3.24}$$

Notice that $P_F\left(I - \psi + h\Phi\right)$ is a contraction of D into itself. Thus, the existence $z \in D$ of as unique element such that

$$z = P_F\left(I - \psi + h\Phi\right)(z)$$

Now,

$$\limsup_{n\to\infty}\left\langle\left(\psi - h\Phi\right)z, z - T_m^n x_{n,m}\right\rangle \leq 0.$$

Let $z = P_F x_1$, and set

$$R = \left\{x \in \mathcal{H}, \left\|x - z\right\| \leq \left\|x_1 - z\right\| + \frac{\left\|h\Phi(z) - \psi z\right\|}{\lambda - hr}\right\}.$$

$R \neq \phi \subset \mathcal{H}$ is a bounded, convex and closed, and for each $s \in [0,\infty)$ and $\{x_n\} \subset R$, it is $K(s)$-invariant. Lemma 3.3.2 implies

$$\lim_{n\to\infty}\left\|\frac{1}{t_n}\int_0^{t_n} K(s)\,x_n\,ds - T(h)\left(\frac{1}{t_n}\int_0^{t_n} K(s)\,x_n\,ds\right)\right\| = 0.$$

There is a subsequence $\{x_k\}$ (since $\{x_n\}$ is bounded) such that

$$\limsup_{nc\infty}\left\langle\left(\psi - h\Phi\right)z, z - x_n\right\rangle = \lim_{kK(s)x_n ds\infty}\left\langle\left(\psi - h\Phi\right)z, z - x_k\right\rangle. \tag{3.25}$$

There exists a subsequence $x_{k_j} \hookrightarrow w \in \mathcal{H}$ (since $\{x_n\}$ is bounded). Assume that $x_k \rightharpoonup w$. Additionally, we know that $\left\|x_k - \Omega_k\right\| \to 0$, and

$$\left\|\Omega_k - w\right\| \leq \left\|x_k - \Omega_k\right\| + \left\|x_k - w\right\|,$$

shows $\Omega_k \rightharpoonup w$.

For $w \in F$, we should compute $w \in F(\zeta)$ and $w \in F(T_i), i = 0, \cdots, m$ as follows:

i. $w \in F(\zeta)$. Assume that $w \neq K(h)w$ for some $h \in [0, \infty)$. Thus Opial's condition implies

$$\begin{aligned}\liminf_{k\to\infty}\|\Omega_k - w\| &< \liminf_{k\to\infty}\|\Omega_k - K(h)w\| \\ &\leq \liminf_{k\to\infty}\left(\|\Omega_k - K(h)\Omega_k\| + \|K(h)\Omega_k - K(h)w\|\right) \\ &\leq \liminf_{k\to\infty}\|\Omega_k - w\|\end{aligned}$$

which is a contradiction. So, it is seen that $w = K(h)w$ for all $h \geq 0$ and $w \in F(\zeta)$.

ii. $w \in F(T_m)$. We have

$$\left\|T_m^k x_{k,m} - w\right\| \leq k_k^m \|x_{k,m} - w\| \leq k_k^m \ldots k_k^1 \eta_k \|x_k - w\|, \tag{3.26}$$

and also, by taking limit of (3.26) we have $T_m^k x_{k,m} \rightharpoonup w$. Since

$$\left\|x_k - T_m^k x_{k,m}\right\| \leq \|x_k - w\| + \left\|w - T_m^k x_{k,m}\right\|,$$

$\left\|x_k - T_m^k x_{k,m}\right\| \to 0$. So $\|x_k - x_{k+1}\| \to 0$ and Lemma 1.10 shows $\|x_k - T_m x_k\| \to 0$. Hence by Lemma 3.1.9, $w \in F(T_m)$ is hold.

iii. $w \in F(T_i), i = 1, \ldots m-1$. By the same argument (ii), $w \in F(T_i), i = 1, \ldots m-1$. Similarly, $\|u_k - w\| \leq \eta_k \|x_k - w\|$, implies that $u_k \rightharpoonup w$ and also $\|x_k - u_k\| \leq \|x_k - w\| + \|w - u_k\|$, $\|x_k - u_k\| \to 0$ and by Lemma 1.9, $w \in F(T_0)$

Considering (3.20) and $z = P_F(I - \psi + h\Phi)(z)$, show

$$\limsup_{n\to\infty}\left\langle(\psi - h\Phi)z, z - x_n\right\rangle = \left\langle(\psi - h\Phi)z, z - w\right\rangle \leq 0.$$

Additionally, since $\left\|x_n - T_m^n x_{n,m}\right\| \to 0$

$$\limsup_{n\to\infty}\left\langle(\psi - h\Phi)z, z - T_m^n x_{n,m}\right\rangle \leq \limsup_{n\to\infty}\left\langle(\psi - h\Phi)z, z - x_{n+1}\right\rangle \leq 0$$

Now, we show $x_n \to z$. By equality (3.19)

$$\|x_{n+1}-z\|^2 = \left\|\alpha_n h\Phi(x_n)+\beta_{n,m}x_n+\left((1-\beta_{n,m})I-\alpha_n\psi\right)T_m^n x_{n,m}-z\right\|^2$$

$$= \left\|\beta_{n,m}(x_n-z)+\left((1-\beta_{n,m})I-\alpha_n\psi\right)\left(T_m^n x_{n,m}-z\right)\right\|^2$$

$$+\alpha_n^2\|h\Phi(x_n)-\psi(z)\|^2$$

$$+2\alpha_n\beta_{n,m}\langle x_n-z,h\Phi(x_n)-\psi(z)\rangle$$

$$+2\alpha_n\left\langle\left((1-\beta_{n,m})I-\alpha_n\psi\right)\left(T_m^n x_{n,m}-z\right),h\Phi(x_n)-\psi(z)\right\rangle$$

$$\le\left\{\beta_{n,m}\|x_n-z\|+\left((1-\beta_{n,m})-\alpha_n\lambda\right)\left\|T_m^n x_{n,m}-z\right\|\right\}^2$$

$$+\alpha_n^2\|h\Phi(x_n)-\psi(z)\|^2$$

$$+2\alpha_n\beta_{n,m}\langle x_n-z,h\Phi(x_n)-h\Phi(z)\rangle$$

$$+2\alpha_n\beta_{n,m}\langle x_n-z,h\Phi(z)-\psi(z)\rangle$$

$$+2\alpha_n(1-\beta_{n,m})\left\langle T_m^n x_{n,m}-z,h\Phi(x_n)-h\Phi(z)\right\rangle$$

$$+2\alpha_n(1-\beta_{n,m})\left\langle T_m^n x_{n,m}-z,h\Phi(z)-\psi(z)\right\rangle$$

$$-2\alpha_n^2\left\langle\psi\left(T_m^n x_{n,m}-z\right),h\Phi(x_n)-\psi(z)\right\rangle$$

$$\le\left[\left\{\beta_{n,m}+\left((1-\beta_{n,m})-\alpha_n\lambda\right)k_n^m\ldots k_n^1\eta_n)\right\}^2+2\alpha_n\beta_{n,m}hr\right.$$

$$\left.+2\alpha_n(1-\beta_{n,m})k_n^m\ldots k_n^1\eta_n hr\right]\|x_n-z\|^2$$

$$+\alpha_n^2\|h\Phi(x_n)-\psi(z)\|^2+2\alpha_n\beta_{n,m}\langle x_n-z,h\Phi(z)-\psi(z)\rangle$$

$$+2\alpha_n(1-\beta_{n,m})\left\langle T_m^n x_{n,m}-z,h\Phi(z)-\psi(z)\right\rangle$$

$$+2\alpha_n^2\left\|\psi\left(T_m^n x_{n,m}-z\right)\right\|\|h\Phi(x_n)-\psi(z)\|$$

$$\le\left[1-2(\lambda-hr)\alpha_n\right]\left(k_n^m\ldots k_n^1\eta_n\right)^2\|x_n-z\|^2+\alpha_n\left\{\alpha_n\left(\lambda^2\left(k_n^m\ldots k_n^1\eta_n\right)^2\|x_n-z\|^2\right.\right.$$

$$\left.+\|h\Phi(x_n)-\psi(z)\|^2+2\lambda\left\|T_m^n x_{n,m}-z\right\|\|h\Phi(x_n)-\psi(z)\|\right)$$

$$+2\beta_{n,m}\langle x_n-z,h\Phi(z)-\psi(z)\rangle$$

$$\left.+2(1-\beta_{n,m})\left\langle T_m^n x_{n,m}-z,h\Phi(z)-\psi(z)\right\rangle\right\}.$$

Since $\{x_n\}$ is bounded, we get $\{\Phi(x_n)\},\{u_n\}$ and $\left\{\frac{1}{t_n}\int_0^{t_n} K(s)u_n ds\right\}$ are bounded. By $\lim_{n\to\infty} k_n^i = 1, i = 0,\cdots m$ and $\lim_{n\to\infty}\eta_n = 1$, set a constant $L > 0$ such that

$$L \ge \lambda^2 \left(k_n^m \ldots k_n^1\eta_n\right)^2 \|x_n - z\|^2 + \|h\Phi(x_n) - \psi(z)\|^2 + 2\lambda\|T_n^m x_{n,m} - z\|\|h\Phi(x_n) - \psi(z)\|$$

for every $n \ge 0$. Thus,

$$\|x_{n+1} - z\|^2 \le \left[1 - 2(\lambda - hr)\alpha_n\right]\left(k_n^m \ldots k_n^1\eta_n\right)^2 \|x_n - z\|^2 + \alpha_n\Theta_n, \tag{3.27}$$

Where

$$\Theta_n = 2\beta_{n,m}\langle x_n - z, h\Phi(z) - \psi(z)\rangle + 2(1 - \beta_{n,m})\langle T_m^n x_{n,m} - z, h\Phi(z) - \psi(z)\rangle + \alpha_n L.$$

We get $\limsup_{n\to\infty}\Theta_n \le 0$. Applying Lemma 1.3, $x_n \to z$.

Finally, based on [12,13], we present some examples.

Example 3.2

Let $\mathcal{H} = \mathbb{R}$ and $C = [0,\infty)$, and

1. $T_0 x = \frac{x}{100}$ and $T_1 x = \frac{x}{1000}$ are hold.
2. $\phi(x) = \frac{x}{700}$ (where $r = \frac{1}{700}$ is the contraction constat of ϕ.
3. $\psi(x) = \frac{x}{1000}$ (where $\lambda = \frac{1}{1000}$ is the coefficient related to ψ on $\mathcal{H}$ and $0 < h = \frac{1}{1000} < \frac{\lambda}{r}$).
4. $\alpha_n = \frac{3}{14n}, \beta_{n,1} = \frac{n}{14n+1}$ and $\beta_{n,0} = \frac{n}{13n+1}$ are hold.
5. $K(s) = e^{-s}$ and $t_n = n$ are satisfied.

   Assume $\{x_n\}$ is the sequence given by (3.19). Thus, we consider followings:

$$\begin{cases} u_n = \dfrac{1}{n+1}\displaystyle\sum_{j=0}^{n}\dfrac{x_n}{100^j}, \\ y_n = \dfrac{n}{13n+1}x_n + \dfrac{12n+1}{13n+1}\left(\dfrac{1-e^{-n}}{n}\right)u_n, \\ x_{n+1} = \dfrac{3}{9800000n}x_n + \dfrac{n}{14n+1}x_n + \dfrac{1}{1000^n}\left(\dfrac{13n+1}{14n+1} - \dfrac{3}{14000n}\right)y_n, \end{cases}$$

and $F=\{0\}$. If $x_1=1$, we can prepare following table,

n	1	5	10	20	25	30
x_n	1	2.262×10^{-5}	3.989×10^{-11}	1.31×10^{-22}	2.397×10^{-28}	4.398×10^{-34}

Using the same condition, we can explain the main result in [8] as follows:

Example 3.3

Let $\mathcal{H}=\mathbb{R}, C=[0,\infty)$ and $f(x)=\frac{x}{700}$ such that f is a contraction with constant $r=\frac{1}{700}$. Besides, consider that $13\alpha_n=\frac{3}{14n}$ and $A(x)=\frac{x}{1000}$, where A is a strongly positive bounded linear operator with coefficient $\lambda=\frac{1}{1000}$ and $0<h=\frac{1}{1000}<\frac{\lambda}{r}$ on $\mathcal{H}$.

Suppose that $T(s)=e^{-s}$ and $t_n=n$.

Additionally, $\{x_n\}$ is the sequence defined by (3.18). Therefore,

$$x_{n+1}=\frac{3}{9800000n}x_n+\left(1-\frac{3}{14000n}\right)\left(\frac{1-e^{-n}}{n}\right)x_n$$

and $\text{Fix}(T)=\{0\}$. If $x_1=1$, we have following table

n	1	5	10	20	25	30
x_n	1	0.0212342	0.0000013	4.143×10^{-18}	8.123×10^{-25}	5.7×10^{-32}

The above table compares the rate of convergence of the approximation method with the main result in reference [8].

3.4 STABILITY RESULTS FOR ITERATION SEQUENCES

Let (X,d) be a complete metric space, $x_{n+1}=T(x_n)$ is an iteration procedure, where $T:X\to X$ is a map. Assume $T(x^*)=x^*$ and $x_n\to x^*\in X$.

Definition 3.5

Assume $\{y_n\}\in X$ and $\varepsilon_n=d(y_{n+1},T(y_n))$. The iteration $x_{n+1}=T(x_n)$ is called T-stable, if

$$\lim_{n\to\infty}\varepsilon_n=0\Rightarrow\lim_{n\to\infty}y_n=x^*$$

The iteration $x_{n+1}=T(x_n)$ is called bounded T-stable, if $\{y_n\}$ is a bounded sequence and

$$\lim_{n\to\infty} \varepsilon_n = 0 \Rightarrow \lim_{n\to\infty} y_n = x^*$$

Now, consider the modified iteration $\{x_n\}$ defined by (3.8) and (3.19).

Theorem 3.4

The sequence $\{x_n\}_{n\geq 0}$ defined by (3.8) is (T_i)-stable for $i = 0,\cdots,m$, under the hypotheses Theorem 3.2.2.

Proof. Assume that $\{x_n\}_{n\geq 0}$ generated by (3.8) is defined by $x_{n+1} = x_n$. Let $\{y_n\}$ be an arbitrary sequence in X with $\epsilon_n = \|y_{n+1} - \mathfrak{F}y_n\|$ such that $\lim_{n\to\infty}\epsilon_n = 0$. Since $\{x_n\}$ is bounded, then $\{y_n\}$ is bounded. By Theorem 3.2.2, the sequence $\{x_n\}_{n\geq 0} \to q \in F = \bigcap_{i=0}^{m} F(T_i)$ strongly in D. We will prove that $\lim_{n\to\infty} y_n = q$. Thus,

$$\begin{aligned}\|T_i^{n+1}y_{n+1} - q\| &\leq a_{n+1}^i\|y_{n+1} - q\| + b_{n+1}^i\|T_i^{n+1}y_{n+1} - y_{n+1}\|\\ &\quad + c_{n+1}^i\left(\|T_i^{n+1}y_{n+1} - q\| + \|y_{n+1} - q\|\right)\\ &\leq a_{n+1}^i\|y_{n+1} - q\| + b_{n+1}^i\left(\|T_i^{n+1}y_{n+1} - q\|\right.\\ &\quad \left.+ \|y_{n+1} - q\|\right) + c_{n+1}^i\left(\|T_i^{n+1}y_{n+1} - q\| + \|y_{n+1} - q\|\right),\end{aligned}$$

for $i = 0,\cdots,m$ and

$$\|T_i^{n+1}y_{n+1} - q\| \leq \frac{a_{n+1}^i + b_{n+1}^i + c_{n+1}^i}{1 - \left(b_{n+1}^i + c_{n+1}^i\right)}\|y_{n+1} - q\| = k_{n+1}^i\|y_{n+1} - q\| \tag{3.28}$$

for $i = 0,\cdots,m$. By inequality (3.13), we have

$$\begin{aligned}\|\mathfrak{F}y_{n+1} - q\| &\leq (1 - \beta_{n+1,m})\|y_{n+1} - q\| + \beta_{n+1,m}\|T_m^{n+1}y_{n+1,m} - q\|\\ &\leq (1 - \beta_{n+1,m})\|y_{n+1} - q\| + \beta_{n+1,m}k_{n+1}^m\cdots k_{n+1}^1\eta_{n+1}\|y_{n+1} - q\| \qquad (3.29)\\ &\leq \mu_{n+1}\|y_{n+1} - q\|\end{aligned}$$

where $\mu_n = 1 + \beta_{n,m}\left(k_n^m\cdots k_n^1\eta_n - 1\right)$. From inequalities (3.28) and (3.29),

$$\begin{aligned}\|y_{n+2} - q\| &\leq \|y_{n+2} - \mathfrak{F}y_{n+1}\| + \|\mathfrak{F}y_{n+1} - q\|\\ &\leq \epsilon_{n+1} + \mu_{n+1}\|y_{n+1} - q\|\end{aligned}$$

is hold. Additionally,

$$\begin{aligned}\|y_{n+3}-q\| &\leq \|y_{n+3}-\mathfrak{F}y_{n+2}\|+\|\mathfrak{F}y_{n+2}-q\| \\ &\leq \epsilon_{n+2}+\mu_{n+2}\|y_{n+2}-q\| \\ &\leq \epsilon_{n+2}+\mu_{n+2}\epsilon_{n+1}+\mu_{n+2}\mu_{n+1}\|y_{n+1}-q\|\end{aligned}$$

Then, we show that the statement is true for $j+1$ as follows:

$$\begin{aligned}\|y_{n+j+1}-q\| &\leq \|y_{n+j+1}-\mathfrak{F}y_{n+j}\|+\|\mathfrak{F}y_{n+j}-q\| \\ &\leq \epsilon_{n+j}+\|\mathfrak{F}y_{n+j}-q\| \\ &\epsilon_{n+j}+\mu_{n+j}\|y_{n+j}-q\| \\ &\leq \epsilon_{n+j}+\mu_{n+j}\epsilon_{n+j-1}+\mu_{n+j}\mu_{n+j-1}\|y_{n+j-1}-q\| \\ &\leq \cdots \\ &\leq \epsilon_{n+j}+\mu_{n+j}\epsilon_{n+j-1}+\mu_{n+j}\mu_{n+j-1}\epsilon_{n+j-1} \\ &\quad +\cdots+\mu_{n+j}\mu_{n+j-1}\cdots\mu_{n+1}\epsilon_{n} \\ &\quad +\mu_{n+j}\mu_{n+j-1}\cdots\mu_{n}\|y_{n}-q\|.\end{aligned}$$

By Lemma 3.1.3 and since $\lim_{n\to\infty}\epsilon_n=0$, then $\lim_{n\to\infty} y_n=q$.

Theorem 3.5

The sequence $\{x_n\}_{n\geq 0}$ defined by (3.19) is $T_i(\zeta)$-stable for $i=0,\cdots,m$, where the hypotheses of Theorem 3.3.4.

Proof. Suppose that the sequence $\{x_n\}_{n\geq 0}$ generated by (3.19) is defined by $x_{n+1}=\mathfrak{F}x_n$. Let $\{y_n\}$ be an arbitrary sequence in $\mathcal{H}$ with $\epsilon_n=y_{n+1}-\mathfrak{F}y_n$ such that $\lim_{n\to\infty}\epsilon_n=0$. Since $\{x_n\}$ is bounded, then $\{y_n\}$ is bounded. By Theorem 3.3.4, the sequence $\{x_n\}_{n\geq 0}\to z\in F=\bigcap_{i=0}^{m}F(T_i)\cap F(\zeta)$ strongly in D. By the same argument as in Theorem 3.3.4,

$$\begin{aligned}\|y_{n+2}-z\|^2 &= \|y_{n+2}-\mathfrak{F}y_{n+1}+\mathfrak{F}y_{n+1}-z\|^2 \\ &= \epsilon_{n+1}^2+\langle\epsilon_{n+1},\mathfrak{F}y_{n+1}-z\rangle \\ &\quad +\langle\mathfrak{F}y_{n+1}-z,\epsilon_{n+1}\rangle+\|\mathfrak{F}y_{n+1}-z\|^2\end{aligned}$$

From inequality (3.27),

$$\left\|\mathfrak{F}y_{n+1}-z\right\|^2 \leq \mu_{n+1}\left\|y_{n+1}-z\right\|^2+\alpha_{n+1}\Theta_{n+1},$$

where

$$\mu_{n+1}=\left[1-2(\lambda-hr)\alpha_{n+1}\right]\left(k_{n+1}^m\ldots k_{n+1}^1\eta_{n+1}\right)^2,$$

and

$$\Theta_{n+1}=2\beta_{n+1,m}\left\langle y_{n+1}-z,h\Phi(z)-\psi(z)\right\rangle+2\left(1-\beta_{n+1,m}\right)\left\langle T_m^{n+1}y_{n+1,m}-z,h\Phi(z)-\psi(z)\right\rangle$$
$$+\alpha_{n+1}L.$$

Continuing this process, we obtain

$$\begin{aligned}
\left\|y_{n+j+1}-z\right\|^2 &= \left\|y_{n+j+1}-\mathfrak{F}y_{n+j}+\mathfrak{F}y_{n+j}-z\right\|^2\\
&= \epsilon_{n+j}^2+\left\langle \epsilon_{n+j},\mathfrak{F}y_{n+j}-z\right\rangle\\
&+\left\langle \mathfrak{F}y_{n+j}-z,\epsilon_{n+j}\right\rangle+\left\|\mathfrak{F}y_{n+j}-z\right\|^2\\
&\leq\cdots\\
&\leq \epsilon_{n+j}^2+\mu_{n+j}\epsilon_{n+j-1}^2\mu_{n+j}\mu_{n+j-1}\epsilon_{n+j-2}^2+\cdots\\
&\quad+\mu_{n+j}\mu_{n+j-1}\cdots\mu_n\epsilon_{n+}^2\\
&\quad+\mu_{n+j}\mu_{n+j-1}\cdots\mu_n\left\|y_n-z\right\|^2\\
&\quad+\alpha_{n+j}\Theta_{n+j}+\mu_{n+j}\alpha_{n+j-1}\Theta_{n+j-1}+\cdots+\mu_{n+j}\mu_{n+j-1}\cdots\alpha_n\Theta_n,
\end{aligned}$$

for $j+1$. We get $\limsup_{n\to\infty}\Theta_n\leq 0$. Applying Lemma 1.3, $\lim_{n\to\infty}y_n=z$.

3.5 CONCLUSION

In conclusion, this paper presents novel insights into the stability and modified results for iterative approximations, particularly focusing on generalized Lipschitzian mappings in Hilbert spaces. Through theoretical analysis and numerical examples, we demonstrate the efficacy of modified iterative processes in approximating common fixed points within iterative sequences.

The stability of these iterations has been rigorously proven, providing a theoretical foundation for their practical applications. The findings of this study contribute to the advancement of iterative approximation techniques and offer valuable insights into the dynamics of fixed-point approximation in mathematical contexts.

Future research could explore further extensions and applications of the proposed methods in various problem domains.

ACKNOWLEDGEMENT

We extend our appreciation to editors and referees of the book for their valuable contributions and feedback during the course of this study.

CONFLICT OF INTEREST

The authors declare no conflict of interest regarding the publication of this paper.

REFERENCES

[1] Agarwal, R. P., O'Regan, D. and Sahu, D. R. (2007). Iterative construction of fixed points of nearly asymptotically nonexpansive mappings. *J. Nonlinear Convex Anal.*, 8: 61–79.

[2] Cho, Y. J., Kang, J. I. and Zhou, H. (2005). Approximating common fixed points of asymptotically nonexpansive mappings. *Bull. Korean Math.* Soc., 42: 661–670.

[3] Homaeipour, S. and Razani, A. (2014). Convergence of an iterative method for relatively non- expansive multi-valued mappings and equilibrium problems in Banach spaces. *Optim. Lett.*, 8: 211–225.

[4] Homaeipour, S. and Razani, A. (2012). Convergence theorems for equilibrium problems and relatively nonexpansive multi-valued mappings in Banach spaces. *Numer. Funct. Anal. Optim.*, 33: 1386–1398.

[5] Jung, J. S. and Thakur, B. S. (2001). Fixed point theorems for generalized Lipschitzian semi- groups. *Int. J. Math. Math. Sci.*, 28: 41–50.

[6] Marino, G. and Xu, H. K. (2007). Weak and strong convergence theorems for strict pseudo- contractions in Hilbert spaces. *J. Math. Anal. Appl.*, 329: 336–346.

[7] Plubtieng, S. and Punpaeng, R. (2008). Fixed point solutions of variational inequalities for nonexpansive semigroups in Hilbert space. *Math. Comput. Modell.*, 48: 279–286.

[8] Plubtieng, S. and Wangkeeree, R. (2008). A general viscosity approximation method of fixed point solutions of variational inequalities for nonexpansive semigroups in Hilbert space. *Bull. Korean. Math. Soc.*, 45: 717–728.

[9] Razani, A. (2010). *Results in Fixed Point Theory.* Andisheh Zarin Publisher, Qazvin.

[10] Razani, A. and Moradi, R. (2006). *Fixed Point Theory in Modular Space.* Saieh Ghostar Publisher, Qazvin.

[11] Razani, A. and Moradi, R. (2015). On the iterates of asymptotically k-strict pseudo-contractive mappings in Hilbert spaces with convergence analysis. *Fixed Point Theory Appl.*, 1–17. doi: 10.1186/s13663-015-0270-0

[12] Razani, A. and Moradi, R. (2016). Nonlinear iterative algorithms for quasi contraction mapping in modular space. *Georgian Math. J.*, 23: 105–119.

[13] Razani, A. and Salahifard, H. (2010). Invariant approximation for CAT (0) spaces. *Nonlinear Anal.*, 72: 2421–2425.

[14] Razani, A. and Salahifard, H. (2011). Approximating fixed points of generalized nonex-pansive mappings. *Bull. Iranian Math. Soc.*, 37: 235–246.

[15] Razani, A. and Yazdi, M. (2012). A new iterative method for generalized equilibrium and fixed point problems of nonexpansive mappings. *Bull. Malays. Math. Sci. Soc.,* 35: 1049–1061.

[16] Razani, A. and Yazdi, M. (2012). A new iterative method for nonexpansive mappings in Hilbert Spaces. *J. Nonlinear Funct. Anal.: Theory Appl.*, 3: 85–92.

[17] Razani, A. and Yazdi, M. (2014). A new iterative method for a family of nonexpansive mappings. *Math. Rep.*, 16: 7–23.

[18] Schu, J. (1991). Iterative construction of fixed points of asymptotically nonexpansive mappings. *J. Math. Anal. Appl.,* 158: 407–413.

[19] Shimizu, T. and Takahashi, W. (1997). Strong convergence to common fixed points of nonexpansive mappings. *J. Math. Anal. Appl.,* 211: 71–83.

[20] Shioji, T. and Takahashi, W. (1999). A strong convergence theorem for asymptotically nonexpansive mappings in Banach spaces. *Arch. Math.*, 72: 354–359.

[21] Takahashi, W. (2000). *Nonlinear Functional Analysis: Fixed Point Theory and Its Applications.* Yokohama Publishers, Yokohama.

[22] Tan, K. K. and Xu, H. K. (1992). The nonlinear ergodic theorem for asymptotically nonexpansive mappings in Banach spaces. *Proc. Amer. Math. Soc.,* 114: 399–404.

[23] Tan, K. K. and Xu, H. K. (1994). Fixed point iterative processes for asymptotically nonexpansive mappings. *Proc. Amer. Math. Soc.*, 122: 733–739.

[24] Xu, H. K. (2004). Viscosity approximation methods for nonexpansive mappings. *J. Math. Anal. Appl.,* 298: 279–291.

4 QSPR Modelling with Topological Indices

Sanghita Dutta, Jibitesh Dutta, Gunajyoti Saharia, and Diamond Kharkongor

4.1 INTRODUCTION

In various applications, chemists encounter the daunting task of performing numerous experiments to find the most potent compounds. For example, they need to test around ten thousand compounds to place ten substituents in four positions of an asymmetric benzene ring for the desired compound [1]. Further increasing the complexity, each compound's physical and biological activity must then be evaluated for any meaningful application. The traditional experimental studies to evaluate each compound are not only time-consuming and costly but also often impractical. Quantitative Structure–Property Relationship (QSPR) modelling with topological indices can alleviate these challenges with an intelligent approach.

Quantitative Structure–Activity/Property Relationship (QSAR/QSPR) [2] studies are powerful mathematical methods for predicting the relationship between a molecule's structure and biological activity or physicochemical properties. There is a subtle difference between QSPR and QSAR methods. While the QSAR method predicts the relationship between molecular structure and biological activity, QSPR analysis indicates the relationship between a chemical compound and its physical properties. These methods are used by researchers to predict the physical and biological activity of various beneficial chemical compounds [3]. It is worth mentioning that QSAR methods are helpful in filtering active pharmaceutical molecules to expedite drug discovery [4].

Hansch [5] is recognized as the father of QSAR/QSPR, and it first gained traction in pharmaceutical chemistry in the middle of the 20th century. It is worth noting that predictive modelling has much older origins, and the first studies were performed in 1841 to investigate how biological activities were correlated with physicochemical characteristics like molecular weight and water solubility [6]. The field of QSPR has evolved significantly since then, and one can read a thorough review of the most recent developments in the theory of molecular descriptors and their applications in an excellent review by Devillers and Balaban [7]. Furthermore, more advanced models have been developed to predict more precisely the properties of chemical compounds with improvements in computational methods, and their application has started penetrating various fields [8]. The effectiveness of QSAR/QSPR models depends on various factors, such as the choice of molecules in the training set, the choice of molecular descriptors, and the methods used in model development.

DOI: 10.1201/9781003638414-4

A molecular descriptor is a numerical representation of a chemical structure computed using mathematical and cheminformatic tools. As described by Todeschini and Consonni [9], "molecular descriptors are the final result of a logical and mathematical procedure which transforms chemical information encoded within a symbolic representation of a molecule into a useful number or the result of some standardized experiment". There are various types of molecular descriptors based on their properties. Some common types of molecular descriptors are topological geometrical, electrostatic, and quantum chemical descriptors.

Topological indices provide valuable insights into the structural characteristics of molecules. These indices, derived from the graph representation of molecular structures, provide a numerical description of a molecule's topology. Topological indices were first introduced by Weiner [10] in 1947 to predict the boiling points of some isomers of alkanes. In a seminal paper, Dearden first emphasized the application of topological indices in QSPR analysis in [11]. Subsequently, Syed et al. [12] studied various degree-based topological indices and constructed a QSPR model for multiple antiviral drugs to treat Covid-19 patients. In 2022, Adnan and co-researchers [13] developed a QSPR model for select topological indices and explored the physical characteristics of antituberculotic drugs. Additionally, in 2022, Syed, in collaboration with other authors [14], conducted QSPR analysis on degree-based topological indices, demonstrating their strong correlation with the physical properties of drugs used in breast cancer treatment. For a more comprehensive understanding of the topic, we suggest referring to citations [15–17].

Chemical compounds are first converted into graphs for the calculation of topological indices. While modelling molecules as graphs, atoms and bonds represent vertices and edges, respectively, forming what is known as molecular graphs. The vertex's degree is the number of edges incident to a vertex in a graph. This notion of degree is analogous to valence in chemistry, and both concepts explain connectivity within structures. One can refer to reference [18] for more details about graph theory. Chemical graph theory is a captivating branch of graph theory that provides the mathematical framework to convert any molecular structure into a graph. It first appeared when the chemical graphs were created by Cayley [19] and Sylvester [20], and nowadays it has become crucial in understanding multiple aspects of compound behaviour.

Degree-based indices are commonly employed in chemical graph theory among the massive list of over 3,000 topological indices in the literature. They are used in predicting physicochemical properties like boiling points, surface area, reactivity, etc. To simplify the calculation process of degree-based indices, Deutsch and Klavžar introduced the notion of M-polynomials, which are algebraic polynomials broadly employed to calculate various degree-based topological indices [21]. The Degree Connectivity Index (DCI) [22], Zagreb indices [23], Wiener Index (WI) [24], Randić Index (RA) [25], and Balaban Index (J) [26] are some prominent degree-based indices that capture essential particulars concerning the connectivity and topology of molecules.

Nanostructures constructed from silicon dioxide, such as nanotubes, nanotori, and nanosheets, have multifarious applications in nanotechnology. It is worth noting that materials display distinct qualities at the nanoscale. Therefore, it is essential to understand and predict their behaviours theoretically. QSPR analysis fosters this by correlating the molecular structure of silicon dioxide nanostructures with their physical properties. In this context, to further improve our knowledge of these nanostructures, Arockiaraj et al. derived some degree and distance-based topological indices of nanosheets, nanotubes and nanotori of SiO_2 [27].

In this chapter, our objective is to show how QSPR modelling with topological indices can be used to predict some physical properties of nanostructures. We will illustrate the power of QSPR modelling with two examples. The chapter is organized as follows: Section 4.2 is devoted to calculating topological indices using the *M*-polynomial, while Section 4.3 is dedicated to the QSPR analysis of two models related to silicon dioxide nanostructures. Finally, Section 4.4 is devoted to discussion.

4.2 COMPUTATION OF TOPOLOGICAL INDICES WITH *M*-POLYNOMIAL

In this section, we derive the expressions of the *M*-polynomial and subsequently compute various topological indices of nanostructures like SiO_2 nanosheets, nanotubes and nanotori. While multiple methods are available for computing topological indices, the use of *M*-polynomial simplifies the process considerably. We can write the general expression for the degree-based topological index of a graph G as:

$$\mathrm{TI}(G) = \sum_{uv \in E(G)} f(d_u, d_v),$$

where d_u and d_v denotes the degrees of the vertices u and v respectively [28]. Further, the general expression for a *M*-polynomial of a graph G is defined as

$$M(G; x, y) = \sum_{i \leq j} m_{ij} x^i y^j,$$

where, m_{ij} denotes the number of edges $e = uv$ in G with $\{d_u, d_v\} = \{i, j\}$ and x, y as arbitrary variables [29].

In what follows, Table 4.1 lists some of the well-known degree-based topological indices, their definitions in terms of degree of vertices, and the corresponding formulas for deriving the topological indices from the *M*-polynomial.

Here,

$$D_x = x\frac{\partial(f(x,y))}{\partial x}, \quad D_y = y\frac{\partial(f(x,y))}{\partial y}, \quad S_x = \int_0^x \frac{f(t,y)}{t}\partial y, \; D_y = y\frac{\partial(f(x,y))}{\partial y}$$

$$S_x = \int_0^x \frac{f(t,y)}{t}\partial y$$

$$S_y = \int_0^y \frac{f(t,x)}{t}\partial x, \quad J(f(x,y)) = f(x,x). \qquad J(f(x,y)) = f(x,x)$$

TABLE 4.1
Degree-Based Topological Indices: Definitions and Derivations

Sl. No.	Topological Indices	Formula	Derivations from $M(G; x,y)$
1	First Zagreb index $M_1(G)$	$\sum_{uv\in E(G)} d_u + d_v$	$(D_x + D_y)(M(G;x,y))\|_{x=1=y}$
2	General Randić index $R_\beta(G)$, $\beta \in N$	$\sum_{uv\in E(G)} \frac{1}{(d_u d_v)^\beta}$	$(D_x^\beta D_y^\beta)(M(G;x,y))\|_{x=1=y}$
3	Forgotten index $F(G)$	$\sum_{uv\in E(G)} d_u^2 + d_v^2$	$(D_x^2 + D_y^2)(M(G;x,y))\|_{x=1=y}$
4	Harmonic index $H(G)$	$\sum_{uv\in E(G)} \frac{2}{d_u + d_v}$	$2S_xJ(M(G;x,y))\|_{x=1=y}$
5	Second modified Zagreb index $M_2^*(G)$	$\sum_{uv\in E(G)} \frac{1}{d_u d_v}$	$S_xS_y(M(G;x,y))\|_{x=1=y}$
6	Symmetric division index SDD(G)	$\sum_{uv\in E(G)} \frac{d_u^2 + d_v^2}{d_u d_v}$	$(D_xS_y + D_yS_x)(M(G;x,y))\|_{x=1=y}$

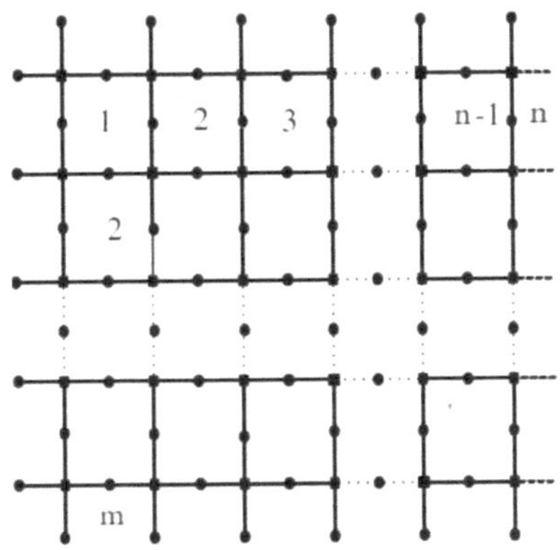

FIGURE 4.1 $SiO_2[m, n]$ nanosheet.

Figures 4.1–4.4 represent the molecular graphs of the SiO_2 nanosheet, nanotorus, nanotube and C8 layered structure, respectively. The following theorems help us compute the M-polynomial and corresponding degree-based topological indices of various SiO_2 nanostructures, including nanosheets, nanotubes and nanotori.

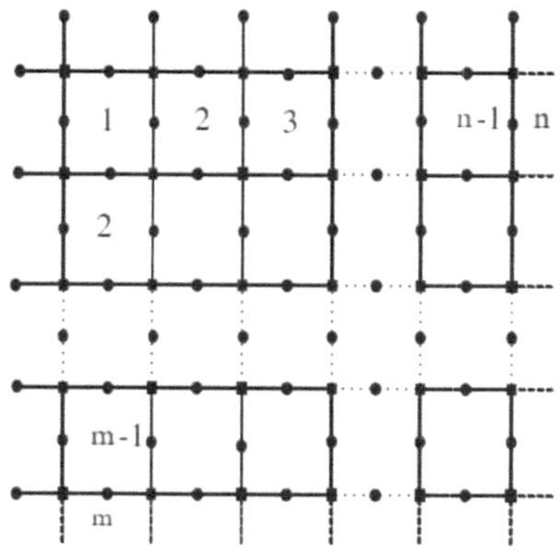

FIGURE 4.2 SiO_2[m, *n*] nanotorus.

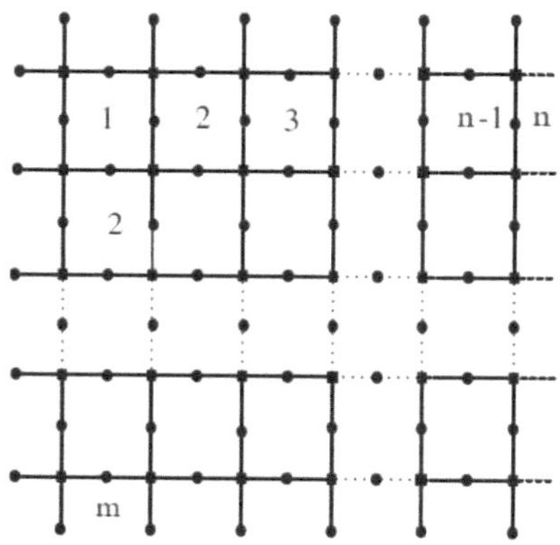

FIGURE 4.3 SiO_2[*m*, *n*] nanotube.

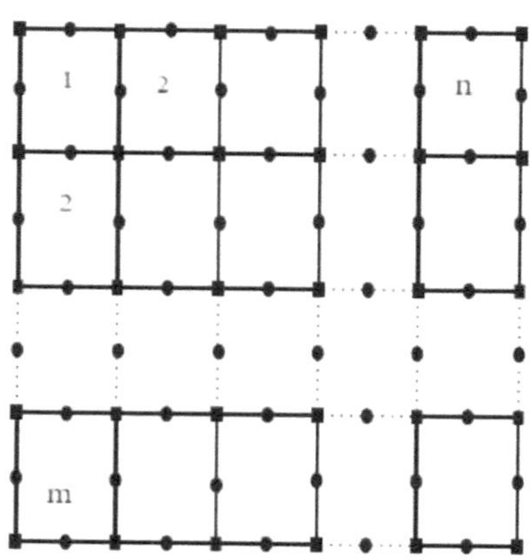

FIGURE 4.4 SiO_2[*m*, *n*] C_8 layered structure.

Theorem 4.2.1

The M-polynomial of the graph G representing a $SiO_2[m, n]$ nanosheet, where m and n are positive integers, is given by [30]:

$$M(G;x,y) = (2m+2n+4)xy^4 + (2m+2n+4mn)x^2y^4.$$

Corollary 4.2.1

Some of the degree-based TIs of the graph G representing $SiO_2[m, n]$ nanosheet are [30]:

1. $M_1(G) = 22(m+n) + 24mn + 20.$
2. $R_\beta(G) = \frac{2}{4^\beta} + \frac{2}{8^\beta}(m+n) + 2^{(3\beta+1)}\ mn + 2^{(2\beta+1)}.$
3. $F(G) = 292m^2 + 1160mn + 560m + 292n^2 + 560n + 688m^2n + 608mn^2 + 320m^2n^2 + 27.$
4. $H(G) = \frac{22(m+n)}{15} + \frac{4mn}{3} + \frac{8}{5}.$
5. $M_2^*(G) = \frac{2m+2n+4}{4} + \frac{2m+2n-4mn}{8}.$
6. $\text{SDD}(G) = \frac{m+n-2}{2} + 13m + 13n - 10mn + 16.$

Theorem 4.2.2

Let m and n be positive integers. If G is the graph of $SiO_2[m, n]$ nanotube, then the M-polynomial of G is given by:

$$M(G;x,y) = 2nxy^4 + (4mn+2n)x^2y^4.$$

Proof: The graph of the SiO_2 nanotube is given in Figure 4.3. We partition the edge set of this graph into edges of the type $E(d_u,d_v)$ as given in the following Table 4.2.

Hence, the M-polynomial of the $SiO_2[m, n]$ nanotubes is given by,

TABLE 4.2
Edge Partition of the Edge Set of *G* Representing SiO_2 [*m, n*] Nanotube

Edge Partition (d_u, d_v)	Number of Edges
(1, 4)	$2n$
(2, 4)	$4mn+2n$

$$M(\mathrm{SiO_2};x,\ y)=\sum m_{i,j}\,x^i y^j$$

$$=m_{1,4}xy^4+m_{2,4}x^2y^4$$

$$=2nxy^4+(4mn\ +\ 2n)\ x\ \ y^4\ .$$

$=m_{1,4}xy^4+m_{2,4}x^2y^4=2\mathrm{nxy}^4+\ (4\mathrm{mn}\ +\ 2\mathrm{n})\ \mathrm{x\ y}^4$. Corollary 4.2.2

Some of the degree-based TIs of the graph *G* representing SiO_2[*m, n*] nanotube are:

1. $M_1(G) = 22n+24mn.$
2. $R_\beta(G)=\left[16^\beta\beta(2mn+n)\right]^\beta(mn+n)^{(\beta^2-\beta)}.$
3. $F(G)=320m^2n^2+292n^2+584mn^2.$
4. $H(G)=\dfrac{4n}{5}+\dfrac{2n+4mn}{3}.\dfrac{2n+4mn}{3}.$
5. $M_2^*(G)=\dfrac{n}{2}+\dfrac{2mn+n}{4}.$
6. $\mathrm{SDD}(G)=\ \dfrac{n}{2}+m+12n+10mn.$

Proof: Let $M(G;x,\ y)=2nxy^4+(4mn+2n)x^2y^2.$

Then,

$$D_x\big(M(G;x,\ y)\big)=2nxy^4+(8mn+4n)x^2y^4\ .$$

$$D_y\big(M(G;x,\ y)\big)=8nxy^4+(16mn+8n)x^2y^4.$$

$$J(M(G;x,y)=2nx^5+(4mn+2n)x^6.$$

$$S_x\left(J\left(M\left(G;x,y\right)\right)\right)=2n\ \frac{x^5}{5}+\left(4mn+2n\right)\ \frac{x^6}{6}.$$

By using Table 4.1 we get,

$$M_1(G)=\left(D_x\ +\ D_y\right)\left(M\left(G;x,y\right)\right)\Big|_{\{x=1=y\}}.$$

$$=22n+24mn.$$

$$R_\beta(G)=D_x^\beta D_y^\beta\left(M\left(G;x,y\right)\right)\Big|_{\{x=1=y\}}$$

$$=\left[16^\beta\beta(2mn+n)\right]^\beta\left(mn+n\right)^{\left(\beta^2-\beta\right)}.$$

$$F(G)=\left(D_x^2\ +\ D_y^2\right)\left(M\left(G;x,y\right)\right)\Big|_{\{x=1=y\}}$$

$$=320m^2n^2+292n^2+584mn^2\ .$$

$$H(G)=2S_xJ\left(M\left(G;x,y\right)\right)\Big|_{\{x=1\}}$$

$$=\frac{4n}{5}+\frac{2n+4nm}{3}.$$

$$M_2^*(G)=S_xS_y\left(M\left(G;x,y\right)\right)\Big|_{\{x=1=y\}}$$

$$=\frac{n}{2}+m+12n+10mn.$$

$$SDD(G)=(D_xS_y+D_yS_x)\left(M\left(G;x,y\right)\right)\Big|_{\{x=1=y\}}$$

$$=\frac{n}{2}+m+12n+10mn.$$

Theorem 4.2.3

Let m and n be positive integers. If G is the graph of $SiO_2[m, n]$ C_8 layered structure, then the M-polynomial of G is given by:

$$M\left(G;x,y\right)=8x^2y^2+\left[6(m-1)+6(n-1)\right]x^2y^3+\left[4(m-1)(n-1)\right]x^2y^4.$$

Proof: The graph of the $SiO_2[m, n]$ C_8 layered structure is given in the Figure 4.4. We partition the edge set of this graph into edges of the type $E(d_u,d_v)$ as given in the following Table 4.3.

Hence, the M-polynomial of the $SiO_2[m, n]$ C_8 layered structure is given by,

TABLE 4.3
Edge Partition of The Edge Set of G Representing SiO_2 [m, n] C_8 Layered Structure

Edge Partition (d_u, d_v)	Number of Edges
(2, 2)	8
(2, 3)	$6(m-y)+6(n-1)$
(2, 4)	$4(m-1)(n-1)$

$$M(SiO_2; x, y) = \sum m_{i,j} x^i y^j$$

$$= m_{2,2}x^2y^2 + m_{2,3}x^2y^3 + m_{2,4}x^2y^4$$

$$= 8x^2y^2 + [6(m-1)+6(n-1)]x^2y^3 + [4(m-1)(n-1)]x^2y^4.$$

Corollary 4.2.3

Some of the degree-based TIs of the graph G of $SiO_2[m, n]$ C_8 layered structure are as follows:

1. $M_1(G) = 18mn + 12m + 12n - 10.$
2. $R_\beta(G) = 2^{\beta^2+\beta}\beta^\beta(9m+9n-1)^\beta(m+n+8mn-2)^{\beta^2+\beta}$.
3. $F(G) = 104m^2 + 104n^2 + 104m^2n + 364mn^2 + 56mn - 136m - 136n + 52.$
4. $H(G) = \frac{12}{5}[(m-1)+(n-1)] + \frac{4}{3}(m-1)(n-1) + 4.$
5. $M_2^*(G) = \frac{m+n+1}{2}.$
6. $SDD(G) = 3(m+n) + 10mn + 13.$

Proof: Let $M(G; x, y) = 8x^2y^2 + [6(m-1)+6(n-1)]x^2y^3.$
$+[4(m-1)(n-1)]x^2y^4$

Then,

$$D_x(M(G; x, y)) = 16x^2y^2 + [12(m-1)+12(n-1)]x^2y^3 + 8(m-1)(n-1)x^2y^3.$$

$$D_y\left(M\left(G;x,y\right)\right)=16x^2y^2+\left[18(m-1)+18(n-1)\right]x^2y^3+16(m-1)(n-1)x^2y^3.$$

$$J\left(M;x,y\right)=8x^4+\left[6(m-1)+6(n-1)\right]x^5+\left[4(m-1)(n-1)\right]x^6.$$

$$S_x\left(J\left(M\left(G;x,y\right)\right)\right)=\frac{8}{5}x^4+\left[(m-1)+(n-1)\right]x^5+\frac{4(m-1)(n-1)}{7}x^6.$$

Hence,

1. $M_1(G)=18mn+12m+12n-10.$

2. $R_\beta(G)=2^{\beta^2+\beta}\beta^\beta(9m+9n-1)^\beta(m+n+8mn-2)^{\beta^2+\beta}$.

3. $F(G)=104m^2+104n^2+104m^2n+364mn^2+56mn-136m-136n+52.$

4. $H(G)=\frac{12}{5}\left[(m-1)+(n-1)\right]+\frac{4}{3}(m-1)(n-1)+4.$

5. $M_2^*(G)=\frac{m+n+1}{2}.$

6. $\mathrm{SDD}(G)=3(m+n)+10mn+13.$

Theorem 4.2.4

Let m and n be positive integers. If G is the graph of $SiO_2[m, n]$ nanotorus, then the M-polynomial of G is given by

$$M\left(G;x,y\right)=4mnx^2y^4.$$

Proof: The graph of the SiO_2 nanotube is given in Figure 4.2. In the graph of the SiO_2 nanotube, there is only one type of edge based on the degrees of the end vertices, which is of type (2,4). There are $4mn$ edges in the SiO_2 nanotube. Hence, the M-polynomial of the SiO_2 nanotorus is given by,

$$\begin{aligned} M\left(\mathrm{SiO_2};x,y\right) &= \sum m_{i,j}\,x^iy^j,\\ &= m_{2,4}x^2y^4,\\ &= 4mnx^2y^4. \end{aligned}$$

Corollary 4.2.4

Some of the degree-based TIs of the graph G of $SiO_2[m, n]$ nanotorus are:

1. $M_1(G) = 24mn.$
2. $R_\beta(G) = \left[24^{\beta+1}\beta(mn)^\beta\right]^\beta.$
3. $F(G) = 320m^2n^2.$
4. $H(G) = \frac{4mn}{3}.$
5. $M_2^*(G) = \frac{mn}{2}.$
6. $\mathrm{SDD}(G) = 10mn.$

Proof: Let $M(SiO_2; x, y) = 4mnx^2y^4$.

Then,

$$D_x\left(M(G;x,\ y)\right) = 8mnx^2y^4$$

$$D_y\left(M(G;x,\ y)\right) = 16mnx^2y^4$$

$$J(M(G;x,y) = 4mnx^6$$

$$S_x\left(J\left(M(G;x,y)\right)\right) = 4mn\ \frac{x^6}{7}$$

By using Table 4.1, we have

1. $M_1(G) = 24mn.$
2. $R_\beta(G) = \left[24^{\beta+1}\beta(mn)^\beta\right]^\beta.$
3. $F(G) = 320m^2n^2.$
4. $H(G) = \frac{4mn}{3}.$
5. $M_2^*(G) = \frac{mn}{2}.$
6. $\mathrm{SDD}(G) = 10mn.$

4.2.1 Surface Plot of *M*-Polynomials

In what follows, surface plots of the M-polynomials for $SiO_2[1, 1]$ nanosheets, nanotori and nanotubes, using maple, are depicted to visualize their structural characteristics (Figures 4.5–4.7).

4.3 QSPR ANALYSIS

This section examines various regression techniques, such as linear and multiple linear regression, to analyse our dataset on nanoparticle properties and associated indices. We aim to identify 'the most suitable regression approach for our specific dataset, shedding light on nanoparticle behaviour and predictive accuracy.

The selected indices, namely the Zagreb index, modified Zagreb index, Randić index, forgotten index, harmonic index and symmetric division deg index, serve as crucial descriptors for characteriszing the structural properties of nanoparticles. These indices encapsulate essential features of nanoparticle structures that influence their physical properties and capture different aspects of molecular topology, strengthening the analysis by providing useful information about nanostructures. They are widely studied and applied in QSPR analyses across various chemical compounds. Furthermore, these indices have been extensively used in the literature to

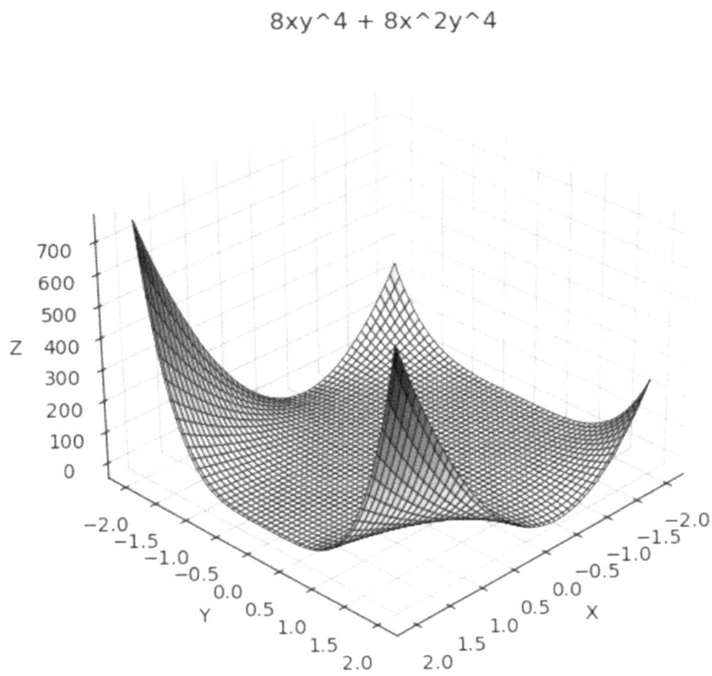

FIGURE 4.5 Surface plot of SiO_2 [1, 1] Nanosheet using $M(G; x, y)$ derived in Theorem 4.2.1.

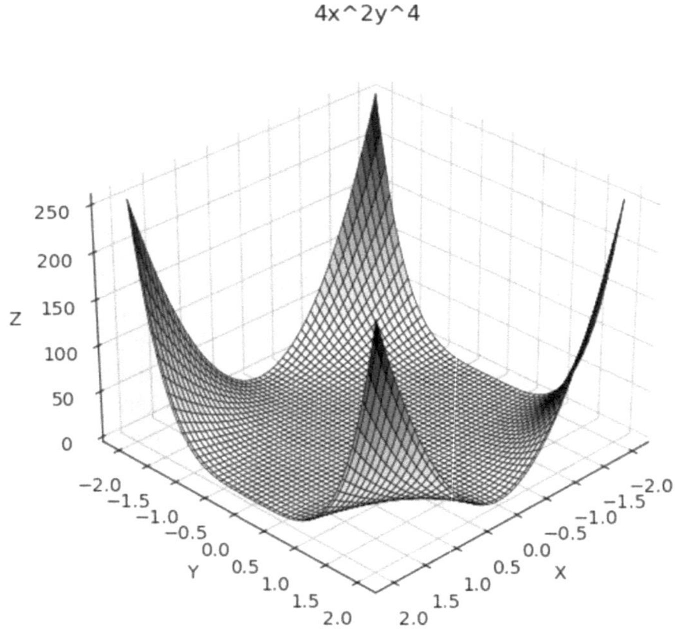

FIGURE 4.6 Surface plot of SiO_2 [1, 1] Nanotorus using $M(G; x, y)$ derived in Theorem 4.2.4.

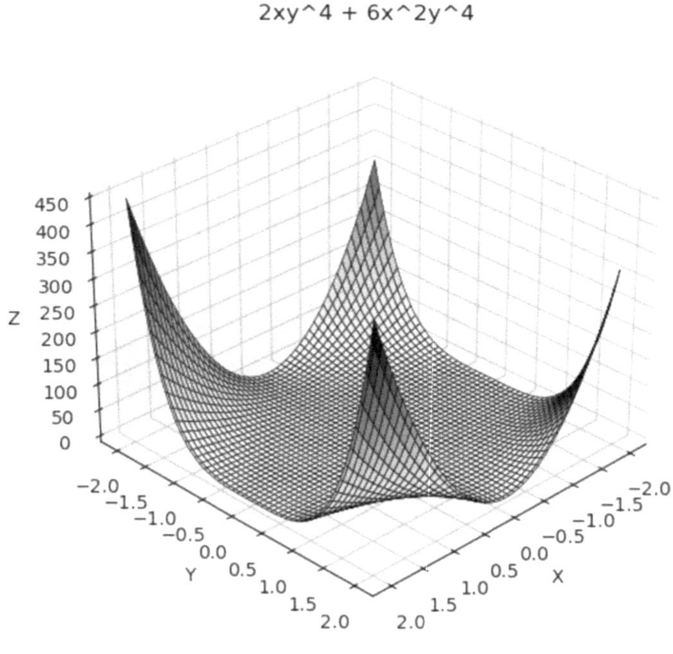

FIGURE 4.7 Surface plot of SiO_2 [1, 1] Nanotube using $M(G; x, y)$ derived in Theorem 4.2.2.

correlate topological features with physical properties in multiple domains, making it easier to derive significant insights from the QSPR model.

The primary steps of QSPR modelling using topological indices include the following:

(i) selection of suitable descriptors that encode structural information, gathering a compound dataset with known properties, (ii) computing descriptors for each compound, (iii) pre-processing the data, (iv) creating statistical or machine learning models, (v) assessing model performance, (vi) interpreting the outcomes of the model, (vii) validating and adjusting the models as necessary, and ultimately (viii) using the validated models to predict properties and design compounds in relevant domains.

Quantitative Structure–Property Relationship modelling includes a range of methods for understanding the link between molecular structure and properties. While simple linear regression (SLR) remains a fundamental approach for establishing linear connections between descriptors and target properties, various advanced versions of SLR are also used for more sophisticated modelling. SLR is ideal for exploring primary connections and involves predicting a dependent variable using a single independent variable. On the other hand, multiple linear regression (MLR) offers a more in-depth insight into variable relationships and improves predictive precision by considering multiple predictors simultaneously. More precisely, MLR overcomes the limitations of SLR and increases accuracy. Furthermore, advanced methods like deep learning, kernel methods and quantum machine learning are used to understand the complex connections between how molecules are structured and their properties.

In what follows, we choose two specific examples and illustrate the method of QSPR using simple and multiple regression techniques with the assistance of R to understand the process better.

4.3.1 Example: Simple Linear Regression (SLR) Model for SiO_2 Nanosheet

SLR is a basic statistical technique used to model the relationship between two variables: one independent variable (x) and one dependent variable (y). Here, we assume that there is a linear relationship between y and x. Thus, the equation of an SLR model is given as:

$$y = c_0 + c_1 x + \epsilon,$$

where c_0, c_1 are constants, while ϵ denotes the error term.

We employ linear regression modelling to analyse the data related to SiO_2 nanosheets. By applying statistical techniques, our aim is to uncover relationships and patterns within the dataset, enabling a comprehensive understanding of nanosheet characteristics and behaviours. We refer the reader to [31] for undefined terminologies in statistics.

From Corollary 4.2.1, by varying the values of m and n, we have obtained the following values of M_1 for the SiO_2 nanosheet (Table 4.4).

Analysis of existing literature revealed that no systematic studies have been conducted on the density (and porosity) of colloidal SiO_2 particles synthesized under identical

TABLE 4.4
Presentation of M_1 for Different Values of $[m, n]$

$[m,n]$	[1,1]	[1,2]	[1,3]	[2,2]	[2,3]	[2,4]	[3,3]
M_1	88	134	180	204	274	344	368

TABLE 4.5
Generated Values of Density within the Given Range with the Help of SPSS

$M_1(x_i)$	88	134	180	204	274	344	368
Density $(y_i)(\text{g/cm}^3)$	1.8411	1.8354	1.8265	1.8198	1.8115	1.8021	1.7832

conditions across a wide size range. Consequently, specific data for the density of these nanosheets were not available. To fill this gap, we referred to density ranges mentioned in reference [32]. Using statistical analysis software SPSS, we created seven hypothetical data points within this range, as shown in Table 4.5. For more details about SPSS software, one can refer to [33]. These simulated data points help facilitate our QSPR analysis. This approach allows us to estimate the density of silicon dioxide nanosheets using statistical modelling techniques, despite the lack of direct experimental data. We also note that the density of nanoparticles decreases as the size increases [34].

Correlation Analysis: Correlation between two variables x and y, can be expressed mathematically using Pearson's correlation coefficient, denoted by r. The formula for calculating r is:

$$r = \frac{\sum_{i=1}^{n}(x_i - \overline{x})(y_i - \overline{y})}{\sqrt{\left(\sum_{i=1}^{n}(x_i - \overline{x})^2\right)\left(\sum_{i=1}^{n}(y_i - \overline{y})^2\right)}},$$

where $\overline{x}$ and $\overline{y}$ are the mean of x_i's and y_i's respectively.

This equation quantifies the strength and direction of the linear relationship between the variables. A correlation coefficient closer to 1 indicates a strong positive correlation, while a value closer to −1 indicates a strong negative correlation. A value near 0 suggests little to no linear relationship between the variables

The correlation matrix of the above dataset is given by Table 4.6:

For the given dataset, the correlation coefficient (r_{xy}) is calculated as −0.974423, indicating a strong negative correlation between M_1 and the density of the SiO_2 nanosheet, where x and y denotes M_1 and the density of the SiO_2 nanosheet, respectively. Note that a negative correlation means that as one variable increases, the other

TABLE 4.6
Correlation Matrix of the Dataset from Table 4.5

	M_1	M_1	Density
		1.0000000	
	-0.974423		
Density	-0.974423	1.0000000	

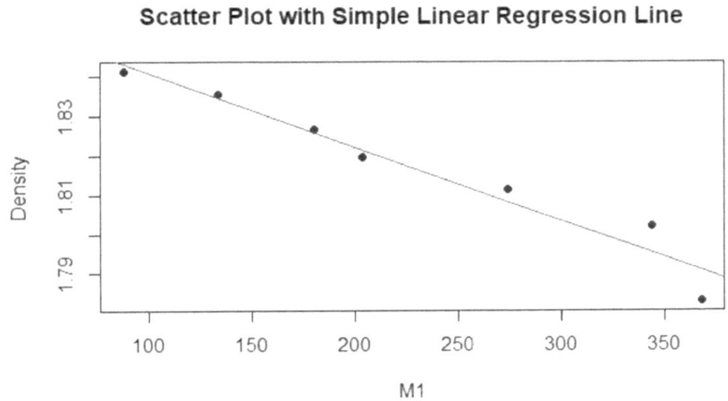

FIGURE 4.8 Scatterplot of M_1 vs density of SiO_2 with linear regression line. This scatterplot shows the relationship between SiO_2 density and M_1, with a guiding red line representing the best-fit line derived from linear regression analysis.

tends to decrease, while a positive correlation suggests that as one variable increases, the other also increases.

Model Development:
Based on the dataset, the developed linear regression model using the software R is given by:

$$\text{Density} = -0.0001856M_1 + 1.859.$$

For further insights into the R software, the reader may refer to reference [35]. It should be noted that the density remains positive for sufficiently large values of M_1. In Figure 4.8, we present the plot of M_1 versus SiO_2 density alongside the aforementioned linear regression line. Here, we employ a topological index to represent chemical compounds. Through linear regression, we predict the physical property of the compound. The essence lies in how the topological index correlates with the property of interest (in our case, density). The graph typically displays a straight line that optimally fits the data points, indicating the relationship between the index and the property. This line aids us in predicting the properties of other compounds based on their topological indices.

Model Evaluation:

The coefficients and other parameters of the least square model are presented in Table 4.7. It gives the details of the coefficients and statistical significance of the regression model. Here, we can see that both the coefficients are relatively high compared to their respective standard errors, resulting in large t-values. Hence, we can conclude that the null hypothesis is rejected.

Model Accuracy Assessment: After rejecting the null hypothesis, a natural follow-up question arises: To what extent can the model accurately represent or fit the data? To answer this question, we need some additional statistical parameters.

The quality of a linear regression fit is typically assessed using two related metrics: the residual standard error (RSE) and the coefficient of determination (R^2). The RSE is calculated as follows: $\text{RSE} = \sqrt{\frac{\text{RSS}}{n-2}}$, where $\text{RSS} = \sum_{i=1}^{n} (y_i - \hat{y}_i)^2$ is the usual reduced sum square. Here, the predicted value of y $y(\hat{y}_i)$ for each observation i is calculated using the equation $\hat{y}_i = \hat{c}_0 + \hat{c}_1 x_i$, where $\hat{c}_0$ and $\hat{c}_1$ are the estimated intercept and slope coefficients, respectively.

Table 4.8 presents the values of RSE, R^2 and the F-value for the linear regression model. RSE serves as an approximation of the standard deviation of the residual term ϵ. In simpler terms, it represents the typical deviation of the response variable from the actual regression line. A small RSE value indicates that the model fits the data very well. Furthermore, the high R^2 value (approximately 95%) suggests that the linear regression model explains a significant portion of the variability in density based on TI (M_1). The elevated F-value indicates the overall significance of the regression model (See Appendix A for the relevant statistical details).

In summary, the developed SLR model effectively captures the relationship between TI (M_1) and the density of SiO_2 nanosheet, providing valuable insights into nanoparticle behaviour and predictive accuracy.

TABLE 4.7
Coefficients and Various Parameters of the Least Square Model

	Coefficient	Standard error (SE)	t-value	$P_r(>\lvert t\rvert)$
Intercept	1.859	0.004736	392.556	2.04×10^{-12}
M_1	−0.0001856	0.00001914	−9.696	0.000198

TABLE 4.8
Statistical Parameters for SLR of SiO_2 Nanosheet

Residual Standard Error (RSE)	R-Squared Value (R^2)	F − value
0.004936	0.9495	78.01

4.3.2 Example of Multiple Linear Regression (MLR) Model for SiO_2 Nanosheet

Multiple linear regression is a statistical technique used to model the relationship between two or more independent variables and a dependent variable. It assumes that the relationship between the independent variables and the dependent variable is linear. The general form of multiple linear regression can be expressed mathematically as:

$$y = c_0 + c_1 x_1 + \cdots + c_n x_n + \epsilon,$$

where y is the dependent variable, x_i's are the independent variables and n is the number of independent variables, c_i's are the coefficients and ϵ is the error term.

Here, we have demonstrated the MLR technique by employing various topological indices as predictors, with the density of SiO_2 nanosheet serving as the response variable. As before, from Corollary 4.2.1, we obtained selected topological indices in Table 4.9 for different values of m and n.

Similar to Example 4.3.1, due to the absence of actual data, we are resorting to generating hypothetical data as shown in Table 4.10 using SPSS for Multiple Linear Regression (MLR), leveraging the known range of density values [32].

Correlation Analysis: Table 4.11 presents the correlation between various topological indices and the density of SiO_2.

TABLE 4.9
Selected Topological Indices for Different Values of *m* and *n*

$[m, n]$	[1, 1]	[1, 2]	[1, 3]
M_1	88	138	180
R_1	25	41.25	57.5
F	4507	10575	19083
H	3.8	3.25	3.07
M_2^*	2	2.25	2.5
SDD	32	35.5	39

TABLE 4.10
SPSS Generated Values of Density for SiO_2 Nanosheet

$[m,n]$	[1,1]	[1,2]	[1,3]
Density $(y_i)(\text{g/cm}^3)$	1.8411	1.8354	1.8265

TABLE 4.11
The Correlation Matrix of the Several Topological Indices and the Density of SiO_2 Nanosheet

	M_1	R_1	F	H	M_2^*	SDD	Density
M_1	1.0000000	0.9987421	0.9892865	−0.9726262	0.9987421	0.9987421	−0.9845457
R_1	0.9987421	1.0000000	0.9953621	−0.9597512	1.0000000	1.0000000	−0.9920884
F	0.9892865	0.9953621	1.0000000	−0.9282822	0.9953621	0.9953621	−0.9995642
H	−0.9726262	−0.9597512	−0.9282822	1.0000000	−0.9597512	−0.9597512	0.9168995
M_2^*	0.9987421	1.0000000	0.9953621	−0.9597512	1.0000000	1.0000000	−0.9920884
SDD	0.9987421	1.0000000	0.9953621	−0.9597512	1.0000000	1.0000000	−0.9920884
Density	−0.9845457	−0.9920884	−0.9995642	0.9168995	−0.9920884	−0.9920884	1.0000000

Model Development:
Based on the dataset, the developed multiple linear regression model using the software R is given by:

$$\text{Density} = 1.845438 + 0.0004M_1 - 0.001582R_1 + 0.00069F + 0.016254H$$

$$+0.004521M_2^* - 0.005246SDD.$$

Model Evaluation: Similar to SLR model, the following table(Table 4.12) shows the coefficients and the relevant statistical parameters, which are obtained by using R software

In the above table, the coefficients are quite large compared to the standard error values, signifying that the null hypothesis is rejected (Table 4.12).

Model Accuracy Assessment:
In the MLR scenario, we can also assess the model using some statistical parameters, such as RSE, R^2, F-value etc. from Table 4.13.

Here the RSE value is 0.003236, which is a small value, signifying that the MLR model better fits the data. Additionally, the greater R^2 and FF-value (i.e. 0.955286 and 81.625, respectively) guarantee that the model developed is accurate to a certain extent. The smaller RSE value and the greater R^2 and FF-statistics in MLR model, compared to SLR model, signify that the MLR model can be used for better prediction.

TABLE 4.12
The Coefficients and the Statistical Parameters for MLR Model

	Coefficient	Standard Error (SE)	t – value	$P_r(>\lvert t\rvert)$
Intercept	1.845438	0.004352	390.125	2.01×10^{-12}
M_1	0.0004	0.00001714	1.021	0.00125
R_1	−0.001582	0.00000856	−8.693	0.00065
F	0.00069	0.00001984	1.119	0.00209
H	0.016254	0.00254136	3.562	0.01295
M_2^*	0.004521	0.00052146	1.937	0.00852
SDD	−0.005246	0.00000354	−9.712	0.00042

TABLE 4.13
The Statistical Parameters of MLR Model for SiO_2 Nanosheet

Residual Standard Error (RSE)	R-Squared Value (R^2)	F–F-value
0.003236	0.955286	81.625

It is worth mentioning that the models presented in this study are based on hypothetical data generated for illustrative purposes. Furthermore, our purpose was to illustrate the process of QSPR analysis through two simple examples, while real-world models require thorough validation and consideration of various factors.

4.4 DISCUSSION

In this chapter, we provided an elementary introduction to QSPR analysis using topological indices. After highlighting the power of QSPR in predicting the physical properties, we computed the relevant topological indices using the *M*-polynomial. To better understand the analysis, we focused on two examples of silicon dioxide nanoparticle regression models, namely (i) SLR and (ii) MLR models of silicon dioxide. We found that the MLR model provides a more accurate prediction than the SLR model in terms of the statistical parameters. Furthermore, the statistical parameters obtained from both models confirm the effectiveness of our approach, though there is room for further improvements. We also note that similar analysis can be conducted to predict properties like boiling point, melting point, enthalpy and so on for different nanostructures.

We also acknowledge that model validation is essential to QSPR analysis. However, it is not covered in this chapter, as it is beyond the scope of the present discussion. Comprehensive QSPR modelling should focus on model validation techniques to improve the robustness of the models. To improve model dependability for practical applications, our future research may concentrate on thorough validation techniques, including k-fold cross-validation and assessment metrics like mean squared error and the coefficient of determination (R^2). With advanced research significantly advancing model validation techniques, QSPR and its role in understanding and predicting the physical properties of molecules have a bright future.

The evolution of explainable AI will enhance transparency and trustworthiness in our analyses, ensuring that regression models meet the highest standards of reliability [36]. Finally, we look forward to a bright future where further progress is made in QSPR analysis with machine learning and data analysis advancements.

APPENDIX A

Given that RSE is expressed in the units of y, determining what qualifies as a favourable RSE can be ambiguous. The R^2 value can remove this ambiguity by providing an alternative measure of data fitting. It is represented as

$$R^2 = \frac{\text{TSS} - \text{RSS}}{\text{TSS}},$$

where $\text{TSS} = \sum_{i=1}^{n} (y_i - \bar{y})^2$ is the total sum of squares.

TSS represents the overall variance present in the response variable y, depicting the inherent variability before any regression analysis is conducted. Conversely, RSS quantifies the remaining variability in the response after regression analysis. Therefore, the difference TSS − RSS illustrates the amount of variability in the response that the regression analysis accounts for or removes. R^2, on the other hand, signifies the proportion of variability in y using the predictor variable x. An R^2 value approaching 1 suggests that a significant portion of the response variability is explained by the regression model. Conversely, a value close to 0 indicates that the regression model explains little of the response variability, which could arise due to inaccuracies in the linear model. In Table 4.8, the R^2 value of 0.9495 indicates that approximately 95% of the variability can be attributed to the linear regression model based on M_1. In simple linear regression, the R^2 value and the squared correlations are identical (i.e. $R^2 = r^2 = Cor(x,y)$). The F-value is used to test the overall significance of the regression model. It compares the variability explained by the model to the variability not explained. A higher F-value suggests that the model is more significant overall. In Table 4.8, the F-value is significantly high, indicating that the model is effective.

REFERENCES

[1] Rouvray, D. H. & Bonchev, D. (1991). *Chemical Graph Theory: Introduction and Fundamentals*. Tunbridge Wells, Kent, England: Abacus Press.

[2] Roy, K., Kar, S. & Das, R. N. (2015). *Statistical Methods in QSAR/QSPR. A Primer on QSAR/QSPR Modeling*. Springer., Cham.

[3] Tetko, I. V., Surko, I., Pandey, A. K., Zhu, H., Tropsha, A., Papa, E., Oberg, T., Todeschini, R., Fourches, D. & Vamek, A. (2008). Critical assessment of QSAR models of environmental toxicity against tetrahymena pyriformis: Focusing on applicability domain and overfitting by variable selection. *Journal of Chemical Information Model* 48, no. 9: 1733–1746.

[4] Kwon, S., Bae, Ho., Jo, J. & Yoon, S. (2019). Comprehensive ensemble in QSAR prediction for drug discovery. *BMC Informatics* 20, no. 521: 1–12.

[5] Hansch, C. (1993). Quantitative structure-activity relationships and the unnamed science. *Accounts of Chemical Research* 26, no. 4: 147–153.

[6] Blake, J. (1884). On the connection between physiological action and chemical constitution. *The Journal of Physiology* 5, no. 1: 36–44.

[7] Devillers, J. & Balaban, A. T. (1999). *Topological Indices and Related Descriptors in QSAR and QSPR*. Gordon and Breach Science Publishers: Amsterdam.

[8] Hsu, H. H., Hsu, Y. C., Chang, L. J. & Yang, J. M. (2017). An integrated approach with new strategies for QSAR models and lead optimization. *BMC Genomics* 18, no. 2: 104.

[9] Todeschini, R. & Consonni, V. (2000). *Handbook of Molecular Descriptors*. Wiley-VCH: Weinheim.

[10] Wiener, H. (1947). Structural determination of paraffin boiling points. *Journal of the American Chemical Society* 69, no. 1: 17–20.

[11] Dearden, J. C. (2017). The use of topological indices in QSAR and QSPR modeling. In Roy, K. (ed.), *Advances in QSAR Modeling*, Springer Nature: Cham. pp. 57–88.

[12] Kirmani, S. A. K., Ali P. & Azam, F. (2021). Topological indices and QSPR/QSAR analysis of some antiviral drugs being investigated for the treatment of COVID-19 patients. *International Journal of Quantum Chemistry* 121, no. 9: e26594.

[13] Adnan, M., Bokhary, S. A. U. H., Abbas, G. & Iqbal, T. (2022). Degree-based topological indices and QSPR analysis of antituberculosis drugs. *Journal of Chemistry* 2022, 5748626.

[14] Bokhary, S. A. U. H., Adnan, M. K. S. & Cancan, M. (2022). On topological indices and QSPR analysis of drugs used for the treatment of breast cancer. *Polycyclic Aromatic Compounds* 42, no. 9: 6233–6253.

[15] Shanmukha, M. C., Basavarajappa, N. S., Shilpa, K. C. & Usha, A. (2020). Degree-based topological indices on anticancer drugs with QSPR analysis. *Heliyon* 6, no. 6: e04235.

[16] Zhang, X., Saif, M. J., Idrees, N., Kanwal, S., Parveen, S. & Saeed, F. (2023). QSPR analysis of drugs for treatment of schizophrenia using topological indices. *ACS Omega* 8, no. 44: 41417–41426.

[17] Balasubramaniyan, D. & Chidambaram, N. (2023). On some neighbourhood degree-based topological indices with QSPR analysis of asthma drugs. *The European Physical Journal Plus* 138, no. 9: 823.

[18] West, D. B. (2009). *Introduction to Graph Theory.* PHI Learning Private. Ltd: New Delhi.

[19] Cayley, A. (1857). XXVIII. On the theory of the analytical forms called trees. *The London, Edinburgh, and Dublin Philosophical Magazine and Journal of Science* 13, no. 85: 172–176.

[20] Sylvester, J. J. (1878). On an application of the new atomic theory to the graphical representation of the invariants and covariants of binary quantics, with three appendices. *American Journal of Mathematics* 1, no. 1: 64.

[21] Deutsch, E. & Klavžar, S. (2015). M-polynomial and degree-based topological indices. *Iranian Journal of Mathematical Chemistry* 6, no. 2: 93–102.

[22] Gutman, I. & Aleksandar, I. (2010). Degree-based topological indices. *MATCH Communications in Mathematical and in Computer Chemistry.* 63: 259–276.

[23] Gutman, I., Trinajstić, N. (1972). Graph theory and molecular orbitals. Total π-electron energy of alternant hydrocarbons. *Chemical Physics Letters* 17: 535–538.

[24] Wiener, H. (1947). Structural determination of paraffin boiling points. *Journal of the American Chemical Society* 69, no. 1: 17–20.

[25] Randić, M. (1975). On characterization of molecular branching. *Journal of the American Chemical Society* 97, no. 23: 6609–6615.

[26] Balaban, A. T. (2001). Topological indices based on topological distances in molecular graphs. *Pure and Applied Chemistry* 73, no. 7: 1247–1252.

[27] Arockiaraj, M., Klavžar, S., Mushtaqa, S. & Balasubramaniane, K. (2019). Distance-based topological indices of nanosheets, nanotubes and nanotori of SiO_2. *Journal of Mathematical Chemistry* 57: 343–369.

[28] Mondala, S., De, N. & Pal, A. (2020). Topological indices of some chemical structures applied for the treatment of COVID-19 patients. *Polycyclic Aromatic Compounds* 42, no. 4: 1220–1234.

[29] Deutsch, E. & Klavžar, S. (2015). M-polynomial and degree-based topological indices. *Iranian Journal of Mathematical Chemistry* 6, no. 2: 93–102.

[30] Thayamathy, P. J. N., Elango, P. & Koneswaran, M. (2018). M-polynomial and degree based topological indices for silicon oxide. *International Research Journal of Pure & Applied Chemistry* 16, no. 4: 1–9.

[31] James, G., Witten, D., Hastie, T. & Tibshirani, R. (2013). *An Introduction to Statistical Learning with Applications in R.* Springer Text in Statistics: New York.

[32] Masalov, V. M., Sukhinina, N. S., Kudrenko, E. A & Emelchenko, G. A. (2011). Mechanism of formation and nanostructure of Stöber silica particles. *Nanotechnology* 22, no. 27: 275718.

[33] Morgan, G. A., Leech, N. L., Gloeckner, G. W. & Barrett, K. C. (2004). *SPSS for Introductory Statistics: Use and Interpretation.* Psychology Press: New York.

[34] Yao, Y., Wei, Y. & Chen, S. (2015). Size effect of the surface energy density of nanoparticles. *Surface Science* 636, 19–24.

[35] Venables, W. N., Smith, D. M. & R Development Core Team (2009). *An Introduction to R.* Network Theory Limited: Clifton, United Kingdom. 2nd edition.

[36] Soares, T. A., Nunes-Alves, A., Mazzolari, A., Ruggiu, F., Wei, G. W. & Merz, K. (2022). The (re)-evolution of quantitative structure–activity relationship (QSAR) studies propelled by the surge of machine learning methods. *Journal of Chemical Information and Modeling* 62, no. 22: 5317–532.

5 First-Order Phase Transition and Heat Capacity Calculation of a Transuranic Compound at Room Temperature

Purvee Bhardwaj and Bhavna Pandey

5.1 INTRODUCTION

In the early years, many theories were developed to explain americium telluride, and these theories calculate the phase transition (PT) using different methods. Currently, we calculate the phase transition and other properties of americium telluride using The Modified three-body interaction Potential Model (MTBIM). The PT of AmTe occurs at a high pressure of 15 GPa. AmTe shows a NaCl structure at ambient temperature, but at high pressure, the structure transforms from B_1 (NaCl) to B_2 (CsCl). AmTe is a combination of two different types of elements, each with its own properties. The first element is americium (Am), a radioactive chemical element and a transuranic member of the actinide series, while tellurium (Te) is a chemical element in the chalcogen group of the periodic table. Americium is a silvery, shiny radioactive metal, whereas tellurium is a silver-white metalloid [1–5]. Americium is the third element of the transuranic series. It is produced by the bombarding uranium or plutonium with neutrons. Am is not a naturally occurring element, meaning it is artificial. In natural processes, Te is harmless, but during the decay process of tellurium chloride, it can discharge toxic fumes. At present, many reports are available on AmTe.

Calculations through the I_{ST} principle for structural PT and electronic properties in AmTe were performed using the TB- LMTO method by J. Pataiya and his group. PT and elastic properties [EP] of americium telluride were studied using MIPT method by Ayanyas and his group [1–5]. Calculations through I_{ST} principle of Pressure-Induced Structural and Electronic Properties in Americium Monochalcogenides were also done using the TB- LMTO method by Pataiya and his group. Many-body interactions in binary ionic solids are explained by R.K. Singh [6]. High-pressure phase (HP)-PT and EP of AmTu using Breathing Shell Models were studied by Jha and Sanyal [7]. The electronic and EP of the light actinide tellurides were studied by Wachter et al. [8]. HP- structural PT and elastic properties of yttrium pnictides were studied using the interionic potential theory by Bhardwaj and Singha [9]. Structural, electronic,

DOI: 10.1201/9781003638414-5

elastics, and HP- properties of barium chalcogenides were studied using the ab initio method by Bouhemadou et al. [10]. Rukamangad et al. considered the structural and EP of rare earth nitrides at high pressure, studied using the theory of two-body ion potential with transfer ion charge [11]. Varshaney et al. calculated Pressure-induced PT and mechanical phenomena of uranium monochalcogenides with the help of long-range Columb force and Hafemeister and Flygare type short-range forces [12]. Here, we calculated the electronic structure of actinides metals with the help of the full LMTO method [13]. The electronic charge densities and for PbSe and PbTe were derived from recent band-structure pseudopotential calculations [14]. We also calculate electronic- structure and optical- properties of lead chalcogenides using the full LMTO method, respectively [15].

The electrical structure, electrical properties, and dielectric strength of PbSe and PbTe were studied with the help of full potential linearized augmented plane wave (FPLAPW) method and scalar relativistic summation, respectively, which linearizes the potential to improve the plane wave potential [16,17]. In previous theories, electronic spectra were calculated using different approximation methods and potential methods, but in this theory, optical spectra were calculated considering the spin-orbit interaction [18]. Here, we present the Pressure and Temperature dependence of energy levels, calculated using pseudopotential calculations [19]. Cardona et al. studied the optical properties of IV–VI and V materials [20]. Anderson studied the liner approach of band theory using two methods: augmented plane wave and muffin-tin orbitals in linear combinations [21]. Anderson et al. also studied the first principle of tight binding theory [22]. Vogt et al. studied Magnetism of monochalcogenides of rare earth elements and actinides, as well as both rare earths and actinides [23].

Recalculations of the electronic structures of rare earth nitrides using the ab initio method were done by Aerts et al. [24]. Petit studied electronic structure of Am monochalcogenides and monopnictides with t the help of ab initio self-interaction method [25].We conducted a study on the lattice dynamics of uranium rock salt structural compounds [26]. There are many theories that explain the phase transition of AmTe Here we discuss the phase transition of americium telluride with the help of the MTIBM method.

5.2 METHOD OF CALCULATION

Model Configuration and Data: When pressure is applied to a body, it causes a change in its volume. It also creates a change in the electron shell distribution of a crystal. This deformation or change in distribution leads to the transfer of charge among the adjacent electrons.

These transitions interact with all neighboring electrons in the lattice, leading to many physical interactions [27–31]. The mechanism of ion overlap and charge transfer among the electrons can be explained by the following fact: when the shells of the ions are displaced with respect to each other, an extra charge develops in the lattice because of the overlapping of the ions. This extra charge is defined by two equations: the Gibbs free energy and three-body interaction. Here, three-body interaction parameter of the lattice, which depends on the distance between the adjacent ions, is denoted by $f(r)$. Notations are taken as

U = internal energy for system at 0 temperature
S = entropy. At temperature $T = 0$ K

Gibbs free energies [GFS] for NaCl (B_1) and CsCl (B_2) are

$$GB_1(r) = UB_1(r) + PVB_1 - TS_1 \quad (5.1)$$

$$GB_2(r') = UB_2(r') + PVB_2 - TS_2 \quad (5.2)$$

Here P is the pressure of the lattice, and VB_1 (=2.00 r_3) and VB_2 (=1.51 r_3) are the unit cell volumes of phases B_1 and B_2, respectively. S_1 and S_2 are the entropies of phases B_1 and B_2, respectively.

$$\begin{aligned} U\,B_1(r) = &-\alpha_M Z^2e^2/r - 12\,\alpha_M\,Ze^2\,f(r)/r - e^2(\alpha_1+\alpha_2)/r_4 \\ &+ 6b\beta_{ij}\,\exp\left[(r_i+r_j-r)/\rho\right] \\ &+ 6b\beta_{ij}\,\exp\left[(2r_i-1.414r)/\rho\right] \\ &+ 6b\beta_{ij}\,\exp\left[(2r_j-1.414r)/\rho\right] \end{aligned} \quad (5.3)$$

$$\begin{aligned} UB_2(r') = &-\alpha'_M Z^2e^2/r' - 16\,\alpha'_M Ze^2\,f(r)/r - e^2(\alpha_1+\alpha_2)/2r_4 \\ &+8b\beta_{ij}\,\exp\left[(r_i+r_j-r_s)/\rho\right] \\ &+3b\beta_{ij}\,\exp\left[(2r_i-1.414r)/\rho\right] \\ &+3b\beta_{ij}\,\exp\left[(2r_j-1.414r)/\rho\right] \end{aligned} \quad (5.4)$$

αM and α'M are the Medlung constant of the sodium chloride and cesium chloride structure, respectively. e is the charge of the electron, and r and r' are the interionic separations for NaCl and CsCl, respectively. β_{ij} is the Pauling constant, defined as $\beta_{ij} = 1 + (Z_i/n_i) + (Z_j/n_j)$, where Z_i (Z_j) and $n_i(n_j)$ are the numbers of valence electrons and the number of electrons of $i(j)$th ion. b is hardness, and ρ is the range parameter. The first term in equations (5.3) and (5.4) represents the long-range Coulomb energy. The second term is calculated for interactions in three bodies corresponding to the nearest-neighbor separations for B_1 to B_2, and the third term represents the van der Waals interactions. The fourth term represents the energy term corresponding to Born Mayer interactions, and the fifth and sixth term represents are the overlay nauseating terms denoted by Hafemeister and Flygare potential.

The entropy difference of the last term is described by the following relation:

$$S_1 - S_2 = \int_1^2 \left[{(C_1 - C_2)}/{T} \right]$$

Here, 1 corresponds to the B_1 phase, 2 corresponds to the B_2 phase, C_1 and C_2 represent the specific heat of the two phases.

Their values can be calculated using the Grunsien parameter ϒ and linear isothermal coefficients of temperature β as

$$C_1 = \{\beta ViB\} i/\Upsilon i$$

Here, the Grunsien parameter (ϒ) can be calculated using the following formula

$$\Upsilon = -r_0/6\left[U'''(r_0)/U''(r_0)\right] = -r_0/6\rho$$

In the GFE, three parameters – b, ρ, $fm(r)$ – are involved and are obtained with the help of lattice parameters. If U represents the first derivative of lattice energy at the state of equilibrium, then:

$$\left[dU/dr\right]r = r_0 = 0$$

$$\left[d2U/dr2\right]r = r_0 = 9k\ r_0 BT$$

where k is a constant that depends on crystal structure,

r_0 is the equilibrium nearest-neighbor distance,
a BT is the bulk modulus of the crystal.

It is clear that from the second-order derivative of lattice energy U, we also calculate the bulk modulus of the AmTe.

5.3 RESULTS AND DISCUSSIONS

The calculations for parameters of model – b, ρ and $f(r)$ – are done with the help of input parameters r(Å), ri (Å), and rj (Å) (see Table 5.1).

Now, we calculate the phase transition with the help of the model parameter. The phase transition of AmTe occurs at a high pressure of 15 GPa and the volume collapse at this phase transition is 10.2% (see Table 5.2).

We also conclude that other authors have calculated the volume collapse [VCP] and PT using different methods. In Table 5.2, it is shown that Pataiya et al. [2] calculated the volume collapse, and their value is 11.37%. In the previous year, Ayanyas et al. calculated the phase transition, which occurred at 13.9 GPa [1]. Additionally, other authors have calculated the phase transitions using both experimental and theoretical theories [1–5].

TABLE 5.1
Parameters Taken for Input [IP] and Parameters Taken for Model [MT] of AmTe

Crystal		IP			MP	
AmTe	r(Å)	ri (Å)	rj (Å)	b (10^{-12} ergs)	ρ	$f(r)$
	2.89	1.09	1.52	10.4233	0.232	−0.00132

TABLE 5.2
PT and Volume Change of AmTe

	PT			VCP		
Solid	**Present**	**Expt.**	**Others**	**Present**	**Expt.**	**Others**
AmTe	15	14.4	13.9 [1] 15 [2]	10.2%	10.2%	11.37% [2]

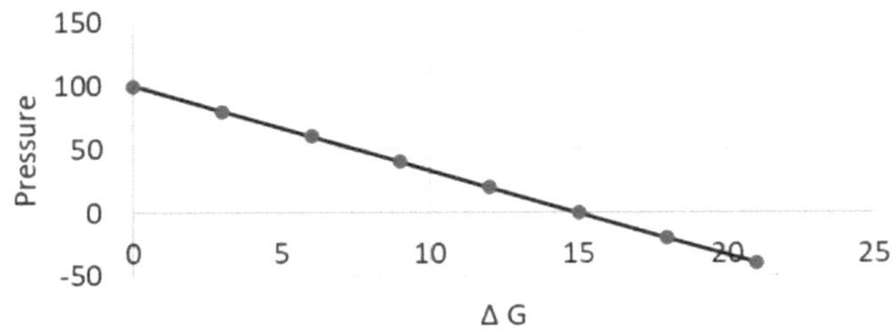

FIGURE 5.1 The graph between pressure and the difference of Gibb's free energies ΔG.

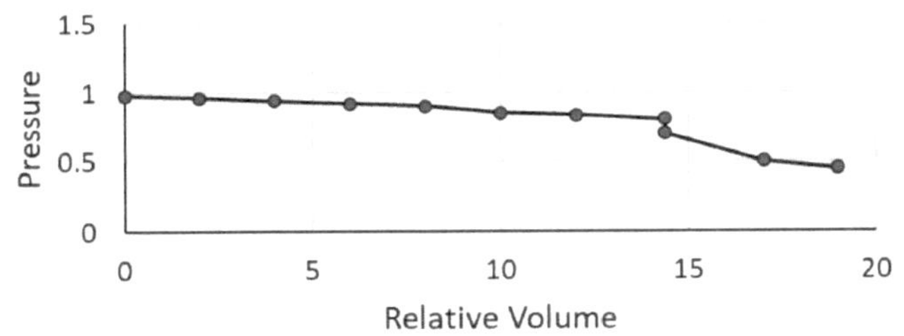

FIGURE 5.2 The graph between pressure (P) and Relative volume (V).

Here, we plot a graph of phase transition versus ΔG, which represents the difference in Gibbs free energies between the two phases The result of this graph is a straight line, indicating a direct relationship between the phase transition and the change in energy (Figure 5.1).

The next graph plots pressure against its relative change in volume. In the graph (Figure 5.2), we clearly observe that P gives a smooth curve with respect to V. However, at point where $V = 14.4$, the compound undergoes a phase transition from NaCl to CsCl. After this phase transition, the curve becomes smooth again (Figure 5.2).

The phase transition [PT] of AmTe occurs at 15 GPa, which is more consistent with experimental data compared to previous theoretical theories [1–5]. This data is also compared with other experimental and theoretical data. Additionally, we calculate the lattice constant of AmTe and compare it with the lattice constants of other transuranic compounds. The graph in Figure 5.3 shows that some compounds have lower lattice constants, while others exhibit higher lattice constant (Figure 5.3).

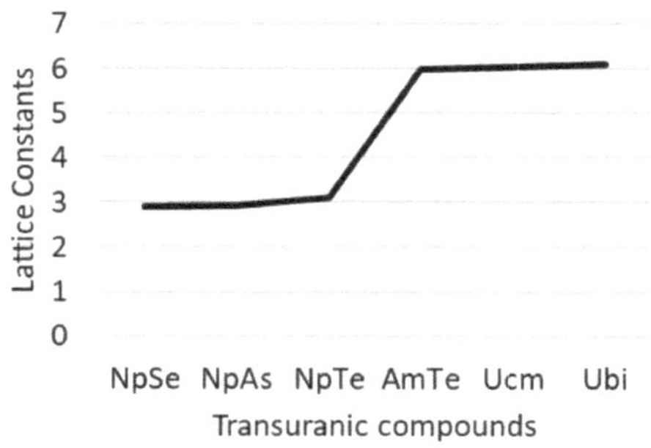

FIGURE 5.3 Comparison of the lattice constant of AmTe to other elements of the transuranic series.

Here, 1, 2, 3, 4, 5, 6, 7 represent the approximate values of lattice constants for various transuranic compounds, which include NpAs, NpSe, NpTe, AmTe, UCm, and UBi. It is important to note that here we have only compared the lattice constants of the B1 phases for all the transuranic compounds.

5.4 CONCLUSIONS

From the study, it can be concluded that the phase transition of AmTe occurs at high pressure of 15 GPa. AmTe shows NaCl structure at ambient temperature, but at high pressure, the structure transforms from B_1 (NaCl) to B_2 (CsCl). This model effectively explains the phase transition of americium telluride and its associated properties in a highly appropriate manner. The results of this theory align well with other experimental theories.

ACKNOWLEDGMENTS

The authors are thankful to the Department of Physical Science, Rabindranath Tagore University Bhopal Madhya Pradesh for the useful help for the research work.

REFERENCES

[1] Ayanyas, M., Rukmangad A., and Arya S.B., and Sanyal P.S. (2013). High pressure phase transition and elastic properties of americium telluride, *AIP Conference Proceedings*, 1536(1), 927–928. https://doi.org/10.1063/1.4810533

[2] Pataiya J., Ayanyas M., Makode C., Singh A., and Sanyal P.S. (2015). First principle calculations of structural phase transition and electronic properties in AmTe, *AIP Conference Proceedings*, 1665, 030001. https://doi.org/10.1063/1.4917576

[3] Pataiya J., Ayanyas M., Makode C., Singh A., and Sanyal P.S. (2014). Forstprinciple calculations of pressure induced structural and electronic properties in americium monochalcogenides, *Advance Material Research*, 1047, 35. https://doi.org/10.4028/www.scientific.net/AMR.1047.35

[4] Jha K.P. and Sanyal P.S. (1995). Lattice vibrational properties of uranium chalcogenides, *Physica B*, 216, 125–131. https://doi.org/10.1016/0921-4526(95)00438-6

[5] Pagare G., Jha K.P., and Sanyal P.S. (2005). High pressure behaviour of lanthanum mono pnictides, *Journal of Alloys and Compounds*, 398, 16–20. https://doi.org/10.1016/j.jallcom.2005.02.044
[6] Singh K.R. (1982). Many body interactions in binary ionic solids, *Physics Reports*, 85, 259–350. https://doi.org/10.1016/0370-1573(82)90020-5
[7] Jha P.K. and Sanyal P.S. (1995). Lattice vibrations in Yb pnictide compounds, *Physical Review B*, 52(22), 15898. https://doi.org/10.1103/PhysRevB.52.15898
[8] Wachter P., Filzmoser M., and Rebizant J. (2001). Electronic and elastic properties of the light actinide tellurides, *Physica B* 293, 199–223. https://doi.org/10.1016/S0921-4526(00)00575-5
[9] Bhardwaj P. and Singha S. (2015). Phase transition and bulk properties of some polonide compounds: A comparative study, *Phase Transitions*, 88(9), 859876, https://doi.org/10.1080/01411594.2015.1020312
[10] Bouhemadou A., Khenata R., Zegrar F., Sahnoun M., Baltache H., and Reshak A.H. (2006). Ab initio study of structural, electronic, elastic and high pressure properties of barium chalcogenides, *Computational Materials Science*, 38, 263–270. https://doi.org/10.1016/j.commatsci.2006.03.001
[11] Rukmangad A., Aynyas M., and Sanya P.S. (2009). Motion and lifetime of dust grains in a tokamak plasma, *Journal of Pure & Applied Physics*, 47, 114–118. https://www.sciencedirect.com/science/article/abs/pii/S0022311504009547
[12] Varshney D., Kaurav N., Kinge R., and Singh R. (2007). High pressure phase transition and elastic properties of cerium chalcogenides and pnictides, *Journal of Condensed Matter Physics*, 19, 236204–236218. https://doi.org/10.1088/0953-8984/19/23/236204
[13] Bhardwaj P. and Singh S. (2016). First principle calculation of structural, electronic and elastic properties of rare earth nitride, *Materials Science-Poland*, 34(4), 715–720. https://doi.org/10.1515/msp-2016-0123
[14] Tsang Y.W. and Cohen M.L. (1972). Electronic charge density in PbTe, *Solid State Communications*, 10(9), 871–874.
[15] Okoye C.M.I. (2004). Structural, electronic, and optical properties of beryllium monochalcogenides, *European Physical Journal B-Condensed Matter and Complex Systems*, 39, 5–17. https://doi.org/10.1140/epjb/e2004-00164-3
[16] Albanesi E.A., Okoye C.M.I., Rodríguez C.O., y Blanca E.P., and Petukhov A.G. (2000). Electronic structure, structural properties, and dielectric functions of IV-VI semiconductors: PbSe and PbTe, *Physical Review B*, 61, 16589.
[17] Lach-hab M., Papaconstantopoulos D.A., and Mehl M.J. (2002). Electronic structure calculations of lead chalcogenides PbS, PbSe, PbTe, *Journal ofPhysics and Chemistry of Solids*, 63, 833. https://doi.org/10.1016/S0022-3697(01)00237-2
[18] Godsall M. and Chilton N.F. (2022). Investigation of the electronic structure and optical spectra of uranium (IV), (V), and (VI) complexes using multiconfigurational methods, *The Journal of Physical Chemistry A*, 126(36), 6059–6066. https://doi.org/10.1021/acs.jpca.2c03314
[19] Bhardwaj P., Mishra A.P., and Bhardwaj R. (2021). Computational study of polycrystalline intermetallics, *Materials Today: Proceedings*, 47(13). https://doi.org/10.1016/j.matpr.2021.05.126
[20] Cardona M. and Greenaway L.D. (1964). Optical properties and band structure of group IV-VI and group V materials, *Physics Review*, 133, 1685. https://doi.org/10.1103/PhysRev.133.A1685
[21] Andersen K.O. (1975). Linear methods in band theory, *Physics Review B*, 12, 3060–3083 https://doi.org/10.1103/PhysRevB.12.3060
[22] Andersen K.O. and Jepsen O. (1984). Explicit, first-principles tight-binding theory, *Physical Review Letters*, 53. https://doi.org/10.1103/PhysRevLett.53.2571

[23] Vogt O. and Mattenberger K. (1995). The magnetism of localized or nearly localized 4f and 5f shells, *Journal of Alloys and Compounds*, 223, 226–236 https://doi.org/10.4236/ampc.2015.53013

[24] Aerts M.C., Strange P., Horne M., Temmerman M.W., Szotek Z., and Svane A. (2004). Half-metallic to insulating behavior of rare-earth nitrides, *Physics Review B*, 69, 045115. https://journals.aps.org/prb/issues/69/4

[25] Petit L., Svane A., Temmerman M.W., and Szotek Z. (2001). Scientific and technological issues related to rare earth oxides: An introduction, *Physics Review B*, 63, 165107–165113. https://doi.org/10.1126/science.204.4395.832

[26] Jackman A.J., Holden M.T., Buyers W.J.L., DuPlessis P.D.V., Vogt O., and Genossar J. (1986). Systematic study of the lattice dynamics of the uranium rocksalt-structure compounds, *Physics Review B*, 33, 7144–7153. https://doi.org/10.1103/PhysRevB.33.7144

[27] Bhardwaj P. and Singh S. (2014). Temperature and pressure behaviour of narrow-gap semiconductors including galena, *Current Applied Physics*, 14, 496–507. https://doi.org/10.1016/j.cap.2014.01.006

[28] Bhardwaj P. and Singh S. (2015). Temperature–pressure behavior and equation of state of praseodymium pnictides, *Journal of Material Science in Semiconductor Processing*, 31, 44–51. https://doi.org/10.1016/j.mssp.2014.11.013

[29] Bhardwaj P. and Singh S. (2003). Structural, mechanical properties and stability of yttrium chalcogenides under high pressure, *European Physical Journal B*, 86, 144. https://doi.org/10.1140/epjb/e2013-30893-9

[30] Bhardwaj P. and Singh S. (2012). Phase transition, structural, mechanical and thermal properties of erbium pnictides under pressure, *Journal of Phase Transition*, 85(9), 791–802. https://doi.org/10.1080/01411594.2012.661861

[31] Bhardwaj P. and Singh S. (2012). Structural phase stability and elastic properties of refractorycarbides, *International Journal of Refractory Metals and Hard Materials*, 35, 115–121. https://doi.org/10.1016/j.ijrmhm.2012.04.017

6 Introduction of Modal Analysis to Study the Hydroelastic Response of a Submerged Elastic Structure

Tapas Mal, Sourav Gupta, and Souvik Kundu

6.1 INTRODUCTION

In the last few decades, many researchers have attempted to investigate various wave-structure interaction problems involving floating or submerged flexible structures, as these are both important in real-world applications and mathematically challenging. In practical situations, these types of problems are highly relevant. These studies include examining the structural response of thin, elastic sea ice due to incident waves in the ice zone area, VLFS (very large floating structures), and floating breakwaters. A detailed study on ice-wave interaction and solution methods to address these problems is described in Squire [1,2]. Another important application of flexible floating breakwaters their use in designing various models of wave energy converters from the ocean (cf. [3,4]).

The hydrodynamic behaviors of flexible structures are crucial for designing various breakwaters in practical scenarios. Many researchers have investigated the wave-structure interaction by floating structures over the last few years, which has enriched the literature. Evans and Davies [5] investigated the scattering problem by a floating thin semi-elastic plate by employing the Wiener–Hopf method in finite-depth water. The same problem was solved by Fox and Squire [6] using the eigenfunction expansion method. A theoretical model was developed by Meylan and Squire [7] to investigate the behavior of ice- floes in long-crested sea waves. In the context of linear water wave theory, Cho and Kim [8] studied the oblique wave scattering with a flexible membrane. Cho and Kim [8] examined the scattering of oblique waves by a horizontal flexible membrane using linear hydroelastic theory. In this study, they used the eigenfunction expansion method to solve diffraction as well as radiation problems analytically. A multi-domain method (BEM) was developed in their study to validate the analytical solution. Sahoo et al. [9] analyzed the elastic nature of a thin floating structure (semi-infinite plate) in finite-depth water. In the two-dimensional domain, Gayen and Mandal [10] investigated the wave-structure interaction with

DOI: 10.1201/9781003638414-6

a floating elastic plate in the deep ocean. In their study, they reduced the governing boundary value problem (BVP) into Carleman-type singular integral equations. Later, a theoretical and numerical model was advanced by Meylan et al. [11] to study the wave interaction with a poroelastic floating disc. Koley et al. [12] examined the scattering of oblique waves by using the Fredholm integral equation technique for the floating thin porous plate. A theoretical model was developed by Zheng et al. [13] to study wave-structure interaction with an arbitrarily shaped porous elastic plate. A thorough literature review involving floating structures is available in Zheng et al. [13] and Squire [2,14].

On the other hand, studying the submerged structures is highly relevant, as they have crucial applications in coastal regions. There are many advantages to using submerged structures, such as (a) better water circulation, (b) ship navigation, (c) coastal views, and (d) cost-effectiveness. Yu and Chwang [15] investigated the scattering problem by a horizontally submerged rigid disc in finite-depth water employing the eigenfunction matching technique. A detailed survey on scattering problems with a horizontally submerged structure (rigid permeable and impermeable) was given by Yu [16]. Hu and Wang [17] studied a model involving a permeable vertical wall and a horizontally submerged plate. Porter [18] investigated wave interaction with a rigid circular disc horizontally submerged in infinite-depth water using the Fourier transform technique. Meylan [19] investigated wave diffraction by an elastic vertical plate, utilizing the same method developed in Meylan and Squire [7]. This work was further extended in the time domain by Peter and Meylan [20]. The interaction of a wave with a vertical plate was examined by Chakraborty and Mandal [21,22]. Islam et al. [23] developed a theoretical model based on an integral equation for solving the wave–structure interaction with a rigid disc in a two-layer fluid, where the disc is situated in the lower layer fluid horizontally. More recently, Das et al. [24] extended this work to a three-layer fluid using the same method described by Islam et al. [23].

In all the above studies, researchers developed different mathematical techniques and tools to solve water wave interaction problems with various submerged or floating structures over the last few years. The authors are unaware of any work that examines the hydroelastic response of a horizontal elastic structure submerged in water using the concept of modal analysis along with the hypersingular integral equation method.

This study includes the investigation of water wave scattering as well as radiation by a thin, horizontally submerged elastic disc using the hypersingular integral equation method. The hydroelastic response of the elastic disc due to incident waves has been analyzed by using the concept of modal analysis. Modal analysis is an efficient technique for studying the dynamic behavior of any structure. The primary motive for using modal analysis is to analyze modes of vibration, natural frequencies, and damping factors of the system. A system is modeled as a combination of damping, stiffness, and mass matrices through modal analysis theory. The equation of motion corresponding to the system is solved to obtain the eigenfrequencies and eigenmodes of the system. The natural frequencies represent the eigenvalues of the system, and the modes of vibration represent the eigenvectors of the system. When a system is excited at certain natural frequencies, the modes of vibration describe how the system will vibrate. The concept of modal analysis has wide applications in designing aircraft, buildings, automobiles, etc. This concept can also be used to identify faults or damage in existing structures or machinery. Mathematically, we need to find two things to analyze a system: these are (a) the natural frequencies (eigenvalues of the system) and (b) the mode of vibration (eigenvectors of the system). The natural frequency, or eigen-frequency, stands for the frequency at which a system oscillates in the absence of any driving or damping forces. The motions of a system vibrating at its natural frequency are referred to as its normal mode or mode of vibration, and the technique for determining these natural frequencies and modes of vibration is known as modal analysis (cf. [7]). The hypersingular integral equation approach is analytically and computationally an efficient method compared to other existing methods, such as the eigenfunction expansion or Wiener–Hopf methods, for handling problems involving wave-structure interactions. The eigenfunction expansion method provides the infinite series form of a solution, while a direct analytical expression incorporating integrals is yielded by the integral equation technique. Evaluating solutions from integral representations requires comparatively fewer iterations than using the eigenfunction expansion approach (cf. [24,25]). The Wiener–Hopf technique is a very powerful mathematical method for solving hydroelastic problems, although it has some major limitations. This method cannot be used for solving problems involving non-uniform, inclined, or curved plates/discs (or arbitrary geometries). Here, the hypersingular integral equation method provides a useful alternative to the Wiener–Hopf technique for solving problems mentioned above. Thus, the integral equation technique is often used for solving the problems associated with the fields of applied mathematical physics (cf. [26]). In this study, we derive a two-dimensional integral equation with a hypersingular kernel from the boundary value problem (BVP) by employing Green's second identity. We then convert it from a two-dimensional hypersingular integral equation to a one-dimensional Fredholm integral equation by employing the technique described in Martin and Farina [27], Guidera and Lardner [28]. We solve the one-dimensional integral equation using the Nyström approach with Gauss-Legendre quadrature nodes. The concept of modal analysis has been used to analyze the motions of the submerged elastic structure and its effect on wave deformation.

6.2 MATHEMATICAL FORMULATION

A horizontally submerged elastic disc with negligible thickness and radius a is considered (see Figure 6.1). We use a cylindrical coordinate system (r,θ,z), for structural motion, with the origin on the free surface and the z axis measured in the upward direction. By incorporating the assumption that the motion is time-harmonic with linear water wave theory, we can write the potential function Φ as

$$\Phi(x,y,z,t) = Re\left[e^{-i\omega t}\phi(x,y,z)\right], \tag{6.1}$$

where ϕ is the complex spatial velocity potential, and it satisfies the Laplace equation, ω is the angular frequency, t stands for time, i for the imaginary unit, and Re for the real part.

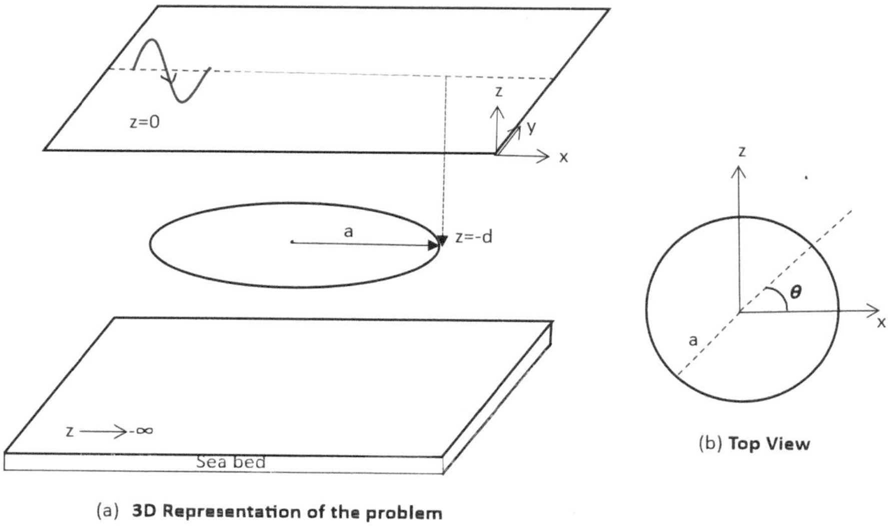

FIGURE 6.1 Sketch of the thin elastic disc submerged in deep water.

The time-independent potential ϕ can be decomposed as

$$\phi = \phi^S + \phi^R + \phi^I, \tag{6.2}$$

where ϕ^S and ϕ^R denote the scattered and the radiated parts, respectively, and ϕ^I is the known incident potential, which can be expressed as (cf. Farina and Martin [29])

$$\phi^I(x, y, z) = i\left(\frac{gA}{\omega}\right)e^{iK(x-iz)}, \quad \text{A being the incident wave amplitude} \tag{6.3}$$

and

$$\phi^I(r,\theta,z) = \sum_{n=0}^{\infty} \frac{gA}{\omega} e^{Kz} \varepsilon_n i^{n+1} J_n(Kr)\cos n\theta, \tag{6.4}$$

where g is the gravitational acceleration, $K = \frac{\omega^2}{g}$, ε_n represents the Neumann symbol, and J_n stands for the n-th order Bessel function.

The vertical displacement of the disc (ζ) satisfies the condition

$$\frac{\partial \zeta}{\partial t} = \frac{\partial \Phi}{\partial z} \quad \text{on} \quad z = -d. \tag{6.5}$$

The disc's displacement about $z = -d$ can be decomposed as

$$\zeta(r,\theta,t) = Re\left[e^{-i\omega t}\xi(r,\theta)\right], \tag{6.6}$$

where $\xi(r,\theta)$ represents the disc's displacement in $(\theta,r)-$ direction.

Along with Laplace equation, ϕ satisfies the following

$$\frac{\partial\phi}{\partial z}-K\phi=0 \quad \text{on} \quad z=0, \tag{6.7}$$

$$|\nabla\phi|\to 0 \quad \text{as} \quad z\to-\infty, \tag{6.8}$$

and

$$\lim_{r\to\infty} r^{1/2}\left(\frac{\partial}{\partial r}-ik_0\right)\phi^S=0, \tag{6.9}$$

where k_0 is a positive real number (the root of dispersion relation) and $r=\|(x,y)\|$.

We have considered a flexible circular disc with negligible thickness; therefore, it satisfies the equation of motion for thin-plate theory:

$$D\nabla^4\zeta+\tilde{\rho}h\frac{\partial^2\zeta}{\partial t^2}=p \quad \text{on} \quad z=-d, \tag{6.10}$$

where $\tilde{\rho}$ is the disc's density, h is its thickness, p is the pressure over the disc's surface, $D=\dfrac{Eh^3}{12(1-\nu^2)}$ represents the flexibility of the disc, with E standing for Young's modulus and ν being the Poisson's ratio .

The position or orientation of a submerged disc depends on the different edge conditions of the disc. If the edge of the disc is clamped, then the displacement and the rotation along the edge of the disc are zero. Hence,

$$\zeta=\frac{\partial\zeta}{\partial r}=0 \quad \text{at} \quad z=a. \tag{6.11}$$

The complex displacement function $\xi(r,\theta)$ of the disc, including natural modes, can be expressed as

$$\xi(r,\theta)=\sum_{n=0}^{\infty}\sum_{j=1}^{\infty}\varsigma_{n,j}w_{n,j}(r,\theta), \tag{6.12}$$

where $\varsigma_{n,j}$ denotes the magnitude of (n,j)th mode of motion of the disc, and $w_{n,j}(r,\theta)$ represents the natural functions, which can be obtained by using the edge conditions of the disc (cf. Itao and Crandall [30]).

Now, the radiated potential ϕ^R can be expressed by using the amplitude of the disc's motion ($\varsigma_{n,j}$) as follows

$$\phi^R=\sum_n\sum_j\varsigma_{n,j}\phi^R_{n,j}, \tag{6.13}$$

where $\phi^R_{n,j}$ is the radiated velocity potential corresponding (n,j)th mode of motion of the disc.

6.3 SOLUTION METHODOLOGY

We first derive a two-dimensional integral equation with a hypersingular kernel from the governing boundary value problem (BVP) by using Green's second identity to $\mathcal{G}$ (source potential) and ϕ. The $\mathcal{G}(P,Q)$'s in the fluid region, except at the (α,β,γ) (source point), is given by (cf. [27])

$$\mathcal{G}(P,Q) \equiv \frac{1}{R_1} + \int_0^\infty J_0(k\mathcal{R})e^{-k(z+\gamma)}\frac{k+K}{k-K}\mathrm{d}k, \tag{6.14}$$

where $\mathcal{R} = \sqrt{(x-\alpha)^2+(y-\beta)^2}$, $R_1 = \sqrt{\mathcal{R}^2+(z-\gamma)^2}$, and P, Q denotes the points on the fluid region, and the above integral has a singularity in it. Now, by employing Green's theorem to $\mathcal{G}(P,Q)$ and ϕ in the fluid region, we obtain (cf. [27])

$$\phi(P) = (4\pi)^{-1}\int\int_S [\phi(q)]\frac{\partial}{\partial\gamma}\mathcal{G}(P,q)\,\mathrm{d}S_q, \quad q \in S \tag{6.15}$$

where S represents the disc's surface, and $\left(\phi(q^+)-\phi(q^-)\right) = [\phi(q)]$ and q^+, q^- stands for the two opposite sides of the disc.

The two-dimensional integral equation with a hypersingular kernel can be obtained as follows

$$(4\pi)^{-1}\int\int[\phi(q)]\frac{\partial^2}{\partial z\partial\gamma}\mathcal{G}(p,q)dS_q = \dot{u}(p), \quad p \in S, \tag{6.16}$$

where $\dot{u}(p) = \dfrac{\partial\phi(p)}{\partial z}$ is the normal velocity of the structure. The above integral is taken over the surface of the disc (S) and in Hadamard finite part sense, whereas the expression for $\dfrac{\partial^2\mathcal{G}(p,q)}{\partial z\partial\gamma}$ is given by

$$\frac{\partial^2\mathcal{G}}{\partial z\partial\gamma}(p;q) = \frac{1}{\mathcal{R}^3} + \mathcal{H}(p;q), \quad \text{on } S, \tag{6.17}$$

and

$$\mathcal{H}(p;q) = \int_0^\infty e^{-2kd}\frac{k^2(k+K)}{k-K}J_0(k\mathcal{R})\mathrm{d}k$$

having a singularity in the integrand of the above integral.

To minimize the numerical computational cost of solving the equation (6.16), we convert it from a two-dimensional to a one-dimensional Fredholm integral equation by employing the method described in Martin and Farina [27]. Thus, the one-dimensional integral equation can be obtained in the form:

$$\mathcal{N}_n(\tau) = \aleph_n(\tau) + \int_0^1 \mathcal{T}_n(\tau;\tau')\mathcal{N}_n(\tau')\mathrm{d}\tau', \quad 0 \le \tau \le 1 \tag{6.18}$$

where

$$\aleph_n(\tau) = -\frac{4}{\pi\mathfrak{C}_n}\tau^{-n}\int_0^\tau \frac{r^{n+1}}{(\tau^2 - r^2)^{1/2}}\dot{u}_n(r)dr, \tag{6.19}$$

$$\mathcal{T}_n(\tau,\tau') = \frac{2}{\pi}\tau\tau'\int_0^\infty e^{-kd}k^2\frac{k+K}{k-K}j_n(k\tau)j_n(k\tau')dk, \tag{6.20}$$

$\mathfrak{C}_n$ is the normalizing factor, and j_n represents the spherical Bessel function of order n. The above integral (6.20) has a singularity in it.

Now, the Fredhlom integral equation (6.18) for the scattering problem takes the form:

$$\mathcal{N}_n^S(\tau) = \aleph_n^S(\tau) + \int_0^1 \mathcal{T}_n(\tau;\tau')\mathcal{N}_n^S(\tau')d\tau', \quad 0 \le \tau \le 1, \tag{6.21}$$

where the expression for $\aleph_n^S$ can be obtained by substituting the expression for $\dot{u}_n^S$ and using the Bessel function property

$$\tau^{-n}\int_0^\tau \frac{r^{n+1}}{(\tau^2 - r^2)^{1/2}}J_n(kr)dr = \tau j_n(k\tau), \tag{6.22}$$

into the equation (6.19).

In the case of the radiation problem, the integral equation (6.18) becomes:

$$\mathcal{N}_{n,j}^R(\tau) - \int_0^1 \mathcal{T}_n(\tau;\tau')\mathcal{N}_{n,j}^S(\tau')\mathrm{d}\tau' = \aleph_{n,j}^R, \quad \tau \in [0,1], \tag{6.23}$$

where $\aleph_{n,j}^R$ can be obtained from the boundary condition

$$\frac{\partial\phi_{n,j}^R}{\partial z} = -i\omega w_{n,j} \quad \text{on } S, \tag{6.24}$$

and the Bessel function property (6.22) in equation (6.19). The expression for $w_{n,j}$ is given by

$$w_{n,j}(r,\theta) = \left[A_{n,j}J_n(\upsilon_{n,j}r) + B_{n,j}I_n(\upsilon_{n,j}r)\right]\cos n\theta, \tag{6.25}$$

where $A_{n,j}$ stands for the amplitude parameter, $B_{n,j}$ for the mode shape parameter, and $\upsilon_{n,j}$ represents the eigenfrequencies of the disc motion, which can be obtained from the edge condition of the disc. The concept of modal analysis has been employed to express the disc motion in the form (6.25).

The integral equations (6.21) and (6.23) can be solved numerically by the Nyström approach along with the Gauss-quadrature rule. The kernel $\mathcal{T}_n(\tau;\tau')$ for both the integral equations (6.21) and (6.23) can be found in Martin and Farina ([27]) for axi-symmetric (heave i.e. $n = 0$) motion of the disc (Figure 6.2).

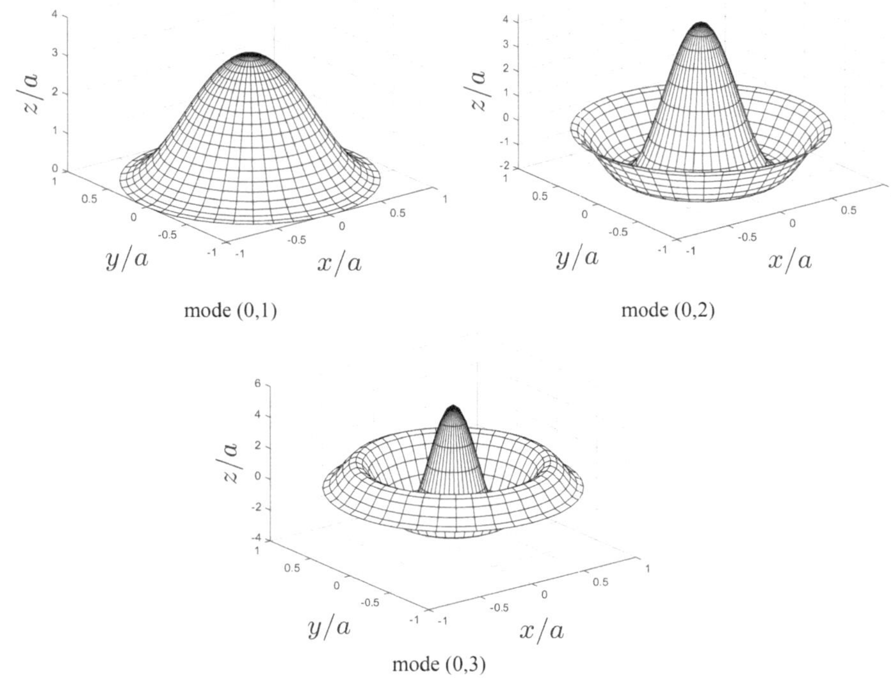

FIGURE 6.2 Different modes of axisymmetric motion of a Clamped-edge circular disc.

6.4 RESULTS AND DISCUSSIONS

The comparison between present and published results is depicted in Figure 6.3.

Results corresponding to the hydrodynamic force and the surface elevation are derived numerically, and related discussions are elaborated in this section. Here, we have considered the combined heave motion ($n = 0,\ j = 4$) of the disc to generate these numerical results.

The dimensionless vertical exciting force exerted on the disc can be evaluated as:

$$F^T = \left| \frac{1}{\rho g a^2 A} \left(F^D + \sum_{j=1}^{N} \varsigma_{0,j} F_{0,j}^R \right) \right|. \tag{6.26}$$

where

$$F^D = \int_S \left[\phi^S(q) \right] dS_q = \frac{8iAga^2}{\omega} e^{-Kd} \int_0^a s\mathcal{N}_0^S(s)\,ds, \quad \mathfrak{C}_0^S = \frac{4gA}{\pi\omega} ie^{-Kd} \tag{6.27}$$

and

$$F_{0,j}^R = \int_S \left[\phi_{0,j}^R(q) \right] dS_q = -8a^2 \int_0^a s\mathcal{N}_{0,j}^R(s)\,ds, \tag{6.28}$$

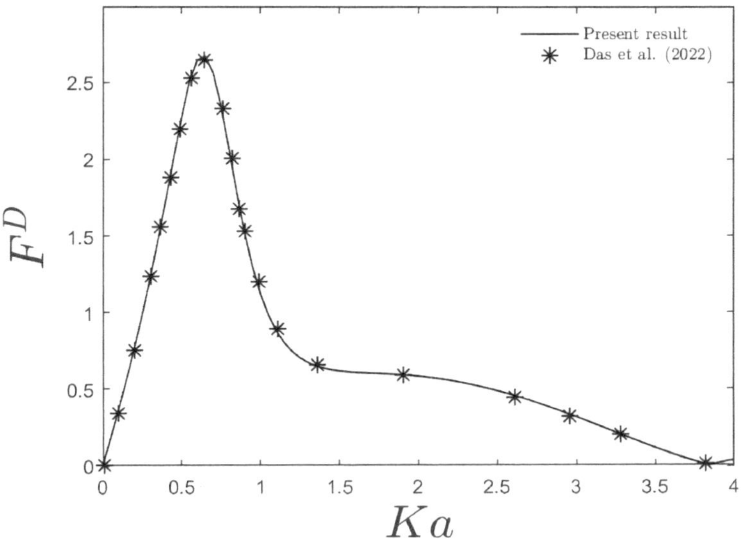

FIGURE 6.3 Comparison of the hydrodynamic force between the results of Das et al. [31] for ice − cover rigidity = ice − cover thickness = porous effect parameter = 0 and our results for $\varsigma_{0,j} = 0$ with the submergence depth $d/a = 0.3$.

by choosing $\mathfrak{C}_0^R = -\frac{4}{\pi}$ for heave (axis-symmetric) motion of the disc.

The free surface elevation (η_f) is obtained as follows:

$$\eta_f(x, y, z = 0, t) = Re\left[e^{-i\omega t}\eta_1(x, y, 0)\right] = Re\left[e^{-i\omega t}\frac{i\omega}{g}\phi(x, y, 0)\right], \qquad (6.29)$$

where η_1 is the time-independent water wave amplitude, and the expression for $\phi(x, y, 0)$ can be obtained in the fluid region from the equation (6.15) in terms of $[\phi]$.

6.4.1 Observation on Hydrodynamic Force

- **Effect of Submergence Depth on Hydrodynamic Force:** Figure 6.4 demonstrates the relationship between the hydrodynamic force and the disc's submergence depth. This figure shows that as the submergence depth decreases, the peak magnitude of the hydrodynamic force increases. The reason for this observation is that as the disc's submergence depth decreases, the intensity of the wave increases, i.e., the wave energy concentration increases, and as a result, the wave load on the structure increases.
- **Effect of Disc Flexibility on Hydrodynamic Force:** The deviation of the peak magnitude of the hydrodynamic force with rigidity is depicted in Figure 6.5. This figure reveals that as the flexibility of the disc decreases, or in other words, as the rigidity of the disc increases, the peak magnitude

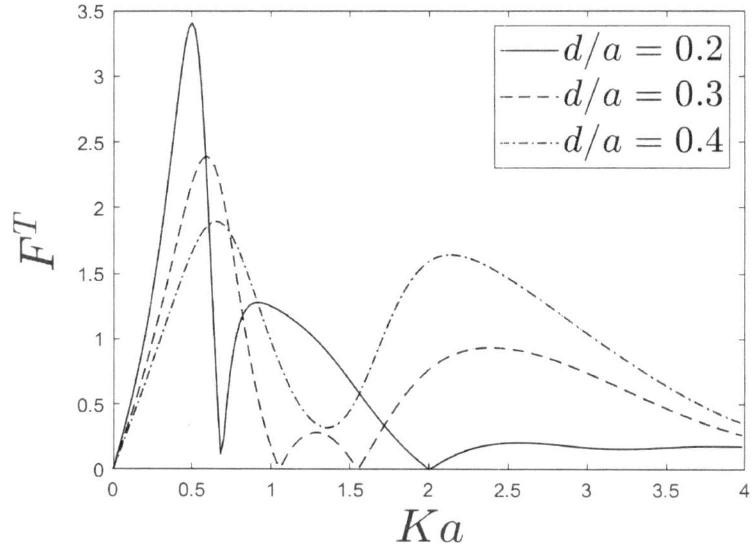

FIGURE 6.4 Deviation of the magnitude of hydrodynamic force with the changes of d/a for the rigidity $D/a^4 = 0.1$.

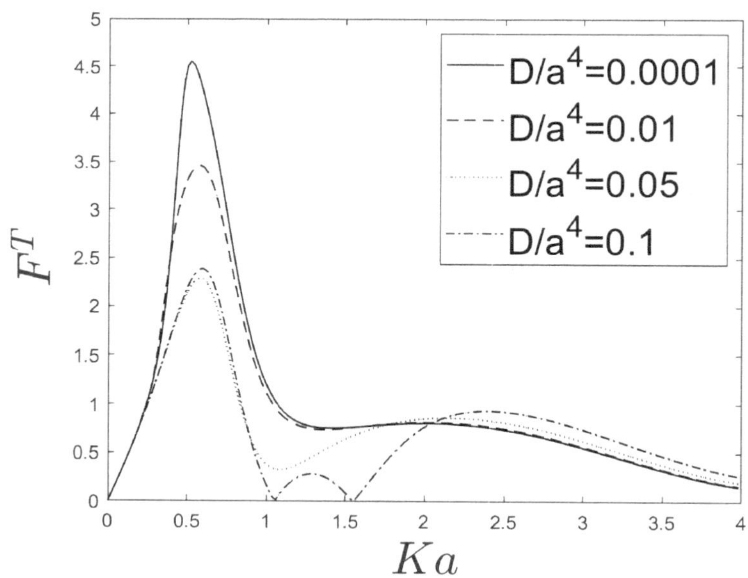

FIGURE 6.5 Deviation of the magnitude of hydrodynamic force with the changes of D/a^4 for the submergence depth $d/a = 0.3$.

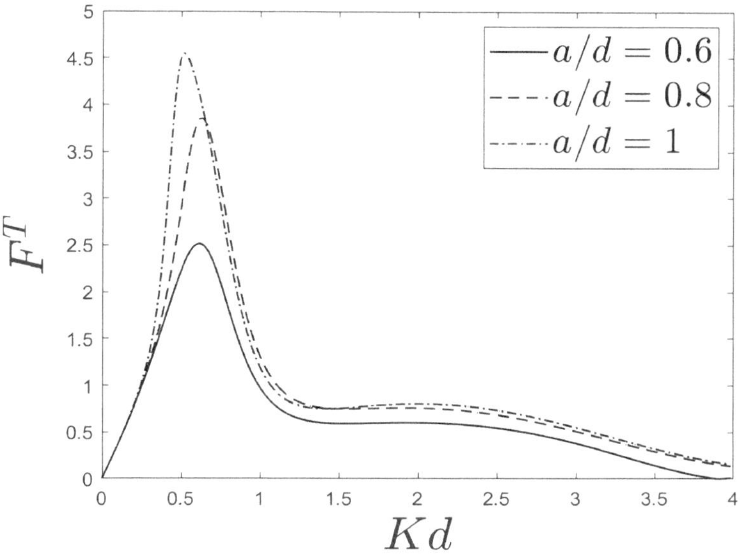

FIGURE 6.6 Variation of hydrodynamic force with the changes of disc's radius a with $D/d^4 = 0.0001$, $d/a = 0.3$.

of the vertical exciting force (hydrodynamic force) decreases. This is because as the disc becomes more flexible (less rigid), the impact of large motion-induced waves generated by the disc's motion becomes more significant. Furthermore, the radiation effect also contributes to this observation, as the radiation effect becomes prominent with increasing flexibility of the disc.

- **Effect of Disc Radius on Hydrodynamic Force:** The effect of disc radius on the hydrodynamic force is shown in Figure 6.6. It is observed that as the disc's radius increases, the peak magnitude of the hydrodynamic force increases. The possible reason is that with the increasing disc surface area, the wave load increases, and as a result, the vertical exciting force (hydrodynamic force) also increases on the disc.

6.4.2 Observation on Surface Elevation

The contour representation of surface elevation around the structure is presented in Figure 6.7a–c. These figures reveal that as the submergence depth increases, the peak magnitude of the surface elevation decreases. It is also observed from these figures that the multiple sharp peaks disappear as the submergence depth increases. This occurs because the wave energy concentration gradually increases with the decreasing value of the submergence depth. Additionally, as the disc rigidity is less $\left(D/a^4 = 0.0001\right)$, large motion-induced waves may also contribute to this observation.

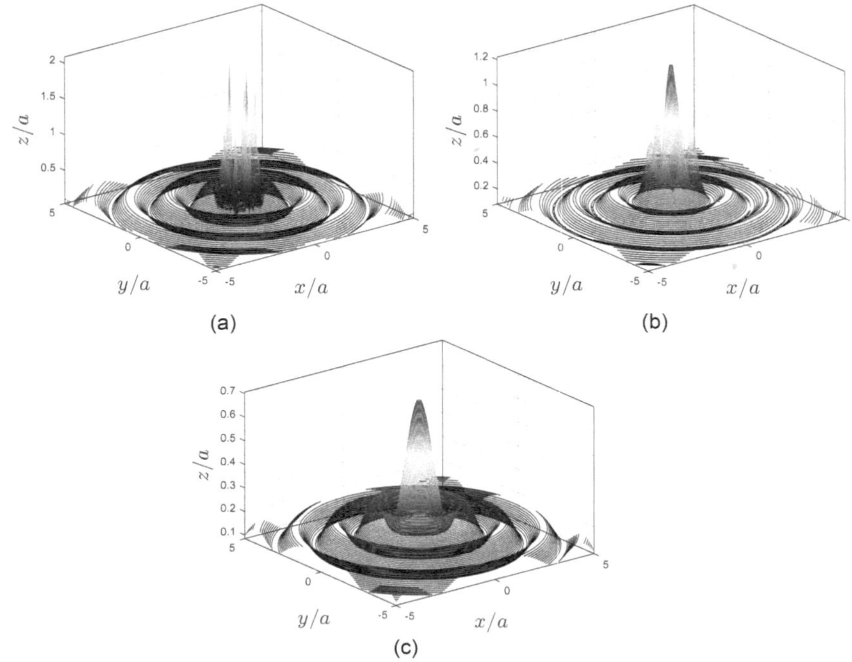

FIGURE 6.7 Flow distribution (η_1/A) around the disc for three submergence depth (d/a) (a) 0.1, (b) 0.2 and (c) 0.3 with $Ka = 2$, $D/a^4 = 10^{-4}$.

6.5 CONCLUSION

Here, we present a mathematical method to consider the effects of an elastic disc's motion on incoming and generated waves based on modal analysis and the hypersingular integral equation approach. The governing BVP is handled by converting it into a two-dimensional integral equation with a hypersingular kernel for potential differences on both sides of the disc. This equation is then reduced to a one-dimensional integral equation by expressing the unknown and known functions in Fourier series expansions. The Nyström approach with the Gauss-quadrature rule has been used to numerically solve the one-dimensional integral equation. Numerical results for the free surface deformation and the hydrodynamic force are presented. Some interesting results are observed regarding these physical quantities. From the above results, we find that the disc's properties and its submergence depth have a significant impact on the physical quantities. The magnitude of the vertical exciting force for the disc with high flexibility dominates the hydrodynamic force for the disc with lower flexibility. The vertical exciting force (hydrodynamic force) increases with the increasing disc radius. Interestingly, it is noticed that in the contour representation for the free surface pattern, the peak magnitude of the surface elevations directly above the disc decreases, and multiple peaks disappear as the submergence depth of the disc increases.

REFERENCES

[1] Squire, V.A., 2007. Of ocean waves and sea-ice revisited. *Cold Regions Science and Technology, 49*(2), pp. 110–133.
[2] Squire, V.A., 2020. Ocean wave interactions with sea ice: A reappraisal. *Annual Review of Fluid Mechanics, 52*, pp. 37–60.
[3] Michailides, C. and Angelides, D.C., 2012. Modeling of energy extraction and behavior of a Flexible Floating Breakwater. *Applied Ocean Research, 35*, pp. 77–94.
[4] Michailides, C. and Angelides, D.C., 2015. Optimization of a flexible floating structure for wave energy production and protection effectiveness. *Engineering Structures, 85*, pp. 249–263.
[5] Evans, D.V. and Davies, T.V., 1968. *Wave-ice interaction* (No. R1313). Castle Point Station, Davidson Laboratory, Stevens Institute of Technology.
[6] Fox, C. and Squire, V.A., 1994. On the oblique reflexion and transmission of ocean waves at shore fast sea ice. *Philosophical Transactions of the Royal Society of London. Series A: Physical and Engineering Sciences, 347*(1682), pp. 185–218.
[7] Meylan, M.H. and Squire, V.A., 1996. Response of a circular ice floe to ocean waves. *Journal of Geophysical Research: Oceans, 101*(C4), pp. 8869–8884.
[8] Cho, I.H. and Kim, M.H., 1998. Interactions of a horizontal flexible membrane with oblique incident waves. *Journal of Fluid Mechanics, 367*, pp. 139–161.
[9] Sahoo, T., Yip, T.L. and Chwang, A.T., 2001. Scattering of surface waves by a semi-infinite floating elastic plate. *Physics of Fluids, 13*(11), pp. 3215–3222.
[10] Gayen, R. and Mandal, B.N., 2009. Scattering of surface water waves by a floating elastic plate in two dimensions. *SIAM Journal on Applied Mathematics, 69*(6), pp. 1520–1541.
[11] Meylan, M.H., Bennetts, L.G. and Peter, M.A., 2017. Water-wave scattering and energy dissipation by a floating porous elastic plate in three dimensions. *Wave Motion, 70*, pp. 240–250.
[12] Koley, S., Mondal, R. and Sahoo, T., 2018. Fredholm integral equation technique for hydroelastic analysis of a floating flexible porous plate. *European Journal of Mechanics-B/Fluids, 67*, pp. 291–305.
[13] Zheng, S., Meylan, M.H., Fan, L., Greaves, D. and Iglesias, G., 2020. Wave scattering by a floating porous elastic plate of arbitrary shape: A semi-analytical study. *Journal of Fluids and Structures, 92*, p. 102827.
[14] Squire, V.A., 2008. Synergies between VLFS hydroelasticity and sea ice research. *International Journal of Offshore and Polar Engineering, 18*(04), pp. 241–253.
[15] Yu, X. and Chwang, A.T., 1993. Analysis of wave scattering by submerged circular disk. *Journal of Engineering Mechanics, 119*(9), pp. 1804–1817.
[16] Yu, X., 2002. Functional performance of a submerged and essentially horizontal plate for offshore wave control: A review. *Coastal Engineering Journal, 44*(02), pp. 127–147.
[17] Hu, H.H. and Wang, K.H., 2005. Damping effect on waves propagating past a submerged horizontal plate and a vertical porous wall. *Journal of Engineering Mechanics, 131*(4), pp. 427–437.
[18] Porter, R., 2015. Linearised water wave problems involving submerged horizontal plates. *Applied Ocean Research, 50*, pp. 91–109.
[19] Meylan, M.H., 1994, April. A flexible vertical sheet in waves. In *ISOPE International Ocean and Polar Engineering Conference* (pp. ISOPE-I). ISOPE.
[20] Peter, M.A. and Meylan, M.H., 2010. A general spectral approach to the time-domain evolution of linear water waves impacting on a vertical elastic plate. *SIAM Journal on Applied M*athematics, 70(7), pp. 2308–2328.
[21] Chakraborty, R. and Mandal, B.N., 2013. Water wave scattering by an elastic thin vertical plate submerged in finite depth water. *Journal of Marine Science and Application, 12*, pp. 393–399.

[22] Chakraborty, R. and Mandal, B.N., 2014. Scattering of water waves by a submerged thin vertical elastic plate. *Archive of Applied Mechanics*, *84*, pp. 207–217.
[23] Islam, N., Kundu, S. and Gayen, R., 2019. Scattering and radiation of water waves by a submerged rigid disc in a two-layer fluid. *Proceedings of the Royal Society A*, *475*(2232), p. 20190331.
[24] Das, A., De, S. and Mandal, B.N., 2022. Radiation of water waves by a heaving submerged disc in a three-layer fluid. *Journal of Fluids and Structures*, *111*, p. 103575.
[25] Gayathri, R., Khan, M.B., Behera, H. and Tsai, C.C., 2022. Wave attenuation by a submerged circular porous membrane. *Journal of Ocean Engineering and Science*. DOI:10.1016/j.joes.2022.05.009.
[26] Krenk, S. and Schmidt, H., 1982. Elastic wave scattering by a circular crack. *Philosophical Transactions of the Royal Society of London. Series A, Mathematical and Physical* Sciences, 308(1502), pp. 167–198.
[27] Martin, P.A. and Farina, L., 1997. Radiation of water waves by a heaving submerged horizontal disc. *Journal of Fluid Mechanics*, *337*, pp. 365–379.
[28] Guidera, J.T. and Lardner, R.W., 1975. Penny-shaped cracks. *Journal of Elasticity*, *5*(1), pp. 59–73.
[29] Farina, L. and Martin, P.A., 1998. Scattering of water waves by a submerged disc using a hypersingular integral equation. *Applied Ocean Research*, *20*(3), pp. 121–134.
[30] Itao, K. and Crandall, S.H., 1979. Natural modes and natural frequencies of uniform, circular, free-edge plates. *Journal of Applied Mechanics*, *46*(2), pp. 448–453.
[31] Das, A., De, S. and Mandal, B.N., 2022. Radiation and scattering of flexural-gravity waves by a submerged porous disc. *Meccanica*, *57*(7), pp. 1557–1573.

7 Binary Interaction with Multiple Fluid Type Cosmology under Modified Gravity Frame

Alokananda Kar and Shouvik Sadhukhan

7.1 INTRODUCTION

Recent observational studies proved that the universe should be in the accelerated expanding stage [Capozziello et al. [17–19]]. The developments in observational studies brought several problems to the forefront of researchers, namely, the cosmological horizon problem, the magnetic monopole problem, and the vacuum energy problem. To resolve these problems, the idea of the exotic nature of cosmic matter has been introduced into physics [Kar et al. [1–10] and Sadhukhan et al. [11]]. Interacting dark matter and non-linear fluid cosmology are among the many examples of exotic matters [Chattopadhyay et al. [12–14]].

Dark matter is one candidate that discusses such an expansion scheme of the universe, but for the non-relativistic case, it cannot. The non-relativistic dark matter is considered cold dark matter and it produces zero pressure in large-scale structures [Debnath et al. [15,16]]. The warm dark matter or hot dark matter can produce large positive energy, which is again considered relativistic dark matter. Theoretically, relativistic cosmic matter can dominate the universe only after certain initial conditions or causes [Dirac et al. [20,21]]. Thus, the implementation of non-exotic matter or dark matter cannot solve the problems regarding accelerated expanding cosmology; instead, it brings different problems. Hence, the exotic type of cosmic matter or exotic dark matter has come into play [Hoyle et al. [22]].

The exotic dark matter acts like the Cardassian model, where the exotic nature comes from the self-interaction produced between dark matter particles [Zeldovich et al. [23]]. This self-interaction brings non-linearity and inhomogeneity into the equation of state of dark matter. Hence, inhomogeneous and non-linear models came into play [Banerjee et al. [24]]. These inhomogeneous and non-linear models are sometimes called fluid-type dark energy models. Another important cause behind the introduction of the inhomogeneous model is to provide a pause in cosmic inflation and initiate the reheating phase during the late-time accelerated expansion phases. Chaplygin gas and Van-Der-Waals fluid are two examples of non-linear models [Bergmann et al. [25]].

DOI: 10.1201/9781003638414-7

Modified gravity is an alternative geometry analysis mechanism that can substitute for the non-linear and inhomogeneous self-interactions to discuss the accelerated expansion of the universe. Modified gravity introduces modifications to the Einstein–Hilbert action with some higher-ordered terms. $f(R)$, $f(G)$, $f(T)$ and $f(Q)$ gravities are the common forms of modified gravity theory. In our present work, we have given the derivations of $f(R,T)$ gravity with the functional form of $f(R,T) = R + 2\lambda T$ [Nojiri et al. [26]] Introduction of a variable gravitational constant and cosmological constant, along with their coupling during the dissociation of the divergence of the energy–momentum tensor, can also be considered an alternative to the modified gravity, which can produce an accelerated expanding universe model with the variations in both G and Λ. The variable gravitational constant can result in changes to the expansion scheme during phase transitions. The variable cosmological constant varies the repulsive effect on cosmic bubbles due to Λ−type dark energies [Brevik et al. [27]].

In our present chapter, we have established an interaction between three types of fluid systems, namely, linear dark matter, inhomogeneous fluid, and a new type of three-parameter-based non-linear fluid. The three fluids discussed in the context of a binary interaction scenario. The interaction picture provides the functional variations of the gravitational constant and different cosmic phases. We have included only the mathematical analysis [Kar et al. [1–10]].

This chapter has proceeded as follows: In Section 7.2, we present the theoretical mechanism of the whole fluid dynamical process, including the interaction scenario and the analysis of the mathematical results. Finally, in Section 7.3, we have concluded this chapter.

7.2 THEORETICAL FORMALISM OF THE MODEL

The theoretical and mathematical mechanism of this chapter begins with the following action for the multiple fluid cosmologies. Here, the cosmic matter Lagrangian is a coupled form of linear dark matter, inhomogeneous fluid, and a new type of non-linear fluid system. The system works on modified gravity geometry and the presence of variable gravitational constant. Hence, the action can be written as follows.

$$S = \frac{1}{16\pi} \int d^4x \sqrt{-g} \left(\frac{f(R,T)}{G} + L_m^{\text{eff}} \right) \tag{7.1}$$

After applying the least action principle to equation (7.1), we can obtain the following dynamical equation of cosmic dynamics. The functional form we considered for modified gravity is $f(R,T) = R + 2\lambda T$ with $T = Tr\left[T_{\mu\nu}\right] = \rho_{\text{eff}} - 3p_{\text{eff}}$. Hence, we have considered $\Lambda = \lambda\left(\rho_{\text{eff}} - p_{\text{eff}}\right)$. Thus, we obtain the following equation.

$$G_{\mu\nu} = 8\pi G(t) T_{\mu\nu}^{\text{eff}} + \Lambda g_{\mu\nu} \tag{7.2}$$

Here, we can write $T_{\mu\nu}^{\text{eff}} = -\frac{2}{\sqrt{-g}} \frac{\delta\left(\sqrt{-g} L_m^{\text{eff}}\right)}{\delta g^{\mu\nu}}$ and $G_{\mu\nu} = R_{\mu\nu} - \frac{1}{2} R g_{\mu\nu}$. Now, the symmetry and other geometric modifications due to the application or coupling of cosmic matters with geometry can be discussed using the line element of the system and the evolution of the cosmic dynamical equation given in equation (7.2). Hence, we can consider the following line element.

$$ds^2 = dt^2 - a^2(t)\left(dr^2 + r^2 d\Omega^2\right) \tag{7.3}$$

We have considered the flat FRW geometry with scale factor $a(t)$ and angular coordinate $d\Omega^2 = d\theta^2 + \sin^2(\theta)d\phi^2$. By substituting this line element into the Einstein equation discussed in equation (7.3), we can derive the following two equations of cosmic dynamics, i.e., the FRW equations. Hence, we can write them as follows.

$$3H^2 = 8\pi G\rho_{\text{eff}} + \Lambda \tag{7.4a}$$

And,

$$3H^2 + 2\dot{H} = -8\pi G p_{\text{eff}} + \Lambda \tag{7.4b}$$

Now, for each cosmological model, we must nullify the divergence of the Einstein tensor as well as the energy–momentum tensor. The divergence-less condition of the energy–momentum tensor provides the idea of the conservation of energy–momentum for an effective system of cosmic fluids under cosmic horizons. By applying the divergence-less condition of Einstein tensor, as discussed in equation (7.2), we can derive the following equation.

$$8\pi\dot{G}\rho_{\text{eff}} + \dot{\Lambda} + 8\pi G\left(\dot{\rho}_{\text{eff}} + 3H\left(\rho_{\text{eff}} + p_{\text{eff}}\right)\right) = 0 \tag{7.5}$$

Now, from the divergence-less property of the energy–momentum tensor of the effective fluid system and the equation (7.5), we can derive the following two equations.

$$\dot{\rho}_{\text{eff}} + 3H\left(\rho_{\text{eff}} + p_{\text{eff}}\right) = 0 \tag{7.6a}$$

And,

$$8\pi\dot{G}\rho_{\text{eff}} + \dot{\Lambda} = 0 \tag{7.6b}$$

Hence, we can derive,

$$G(a) = G_0 - \int \frac{\dot{\Lambda}(a)}{8\pi\rho_{\text{eff}}(a)} da \tag{7.7}$$

For the cosmological constant, we assumed it to be variable like the relation $\Lambda = \alpha H^2$. Hence, we can find the following differential equation for the time variation solution of the scale factor.

$$2a(t)\ddot{a}(t) + 4\dot{a}^2(t) + 8\pi G(a)\left(p_{\text{eff}} - \rho_{\text{eff}}\right)a^2(t) - 2\alpha H_0^2 a^{2(\beta+1)} = 0 \tag{7.8}$$

In the interaction scenarios, we consider the equation of states as $p_m = \omega_m \rho_m$, $p_f = A\rho_f + BH^2$ and $p_{Nf} = C\rho_{Nf} + D\rho_{Nf}^2 - \frac{D}{\rho_{Nf}^{\alpha_1}}$. Considering this equation of states in multiple fluid cosmology, we can expand the energy-momentum conservation equation (7.6a) as follows.

$$\dot{\rho}_m + 3H\left(\rho_m + p_m\right) = -3H\rho_m\delta_1 - 3H\rho_m\delta_2 \tag{7.9a}$$

$$\dot{\rho}_f + 3H\left(\rho_f + p_f\right) = 3H\rho_m\delta_1 \tag{7.9b}$$

And,

$$\dot{\rho}_{Nf} + 3H\left(\rho_{Nf} + p_{Nf}\right) = 3H\rho_m\delta_2 \tag{7.9c}$$

The solutions for energy densities and pressures are given in Table 7.1. For the derivations of the gravitational constant, we found the same conditions as presented in Table 7.2.

The functional forms of the gravitational constant for the two secondary conditions can be given as follows, according to the order of the conditions list.

$$G(a) = G_0 - \frac{\lambda}{8\pi}\ln\left(\rho_m + \rho_f + \rho_{Nf}\right) + \frac{\lambda}{8\pi} * \text{Constant} * \ln(a) \tag{7.10a}$$

And,

$$G(a) = G_0 - \frac{\lambda}{8\pi}\ln\left(\rho_m + \rho_f + \rho_{Nf}\right) + \frac{\lambda}{8\pi} * \text{Constant}_1 * \ln(a) \tag{7.10b}$$

Here, the terms Constant and Constant_1 are as follows.

$$\text{Constant} = \frac{\begin{array}{c} 3(\delta_1 + \delta_2 - \omega_m)(1+\omega_m)\rho_{m0} - 3(1+A)A\rho_{f0} \\ + \dfrac{2(1+2\beta)(2+\beta)B}{3A+2\beta+4} - 3(1+\omega_m)(\delta_1+\delta_2)\rho_{m0} \end{array}}{\rho_{m0} + \rho_{f0} - \dfrac{3B}{3A+2\beta+4}}$$

And,

$$\text{Constant}_1 = \frac{\begin{array}{c} 3(\delta_1 + \delta_2 - \omega_m)(1+\omega_m)\rho_{m0} - 3(1+A)A\rho_{f0} + \dfrac{2(1+2\beta)(2+\beta)B}{3A+2\beta+4} \\ -3(1+\omega_m)(\delta_1+\delta_2)\rho_{m0} - \dfrac{1}{2D}\sqrt{4D^2E+1}c_1c_2 \end{array}}{\rho_{m0} + \rho_{f0} - \dfrac{3B}{3A+2\beta+4} - \dfrac{c_1}{D}\sqrt{4D^2E+1}} \tag{7.11}$$

The equation of state parameter can be written using the above relations for pressures and densities. Hence, ω_{eff} can be written as follows.

TABLE 7.1

Energy Densities and Pressures for Triple Fluid Interaction Scenario in Modified Geometry

Name of Fluid	Energy Density	Pressure
Linear dark Matter	$\rho_m = \rho_{m0} a^{-3(1+\omega_m)}$	$p_m = \left(\omega_m - \delta_1 - \delta_2\right)\rho_{m0} a^{-3(1+\omega_m)}$
Inhomogeneous fluid	$\rho_f = \rho_{f0} a^{-3(1+A)} - \dfrac{3B}{3A+2\beta+4} a^{1+2\beta}$	$p_f = \delta_1 \rho_{m0} a^{-3(1+\omega_m)} + (A)\rho_{f0} a^{-3(1+A)} + \dfrac{B(2\beta+4)}{3A+2\beta+4} a^{2\beta+1}$
New non-linear fluid	$\rho_{Nf} = \sqrt{E+\dfrac{1}{4D^2}}\left(\dfrac{\left(\left(\dfrac{\rho_{Nf0}+\dfrac{1}{2D}+\sqrt{E+\dfrac{1}{4D^2}}}{\rho_{Nf0}+\dfrac{1}{2D}-\sqrt{E+\dfrac{1}{4D^2}}}\right) a^{\frac{3}{\sqrt{4D^2E+1}}}+1\right)}{\left(\dfrac{\rho_{Nf0}+\dfrac{1}{2D}+\sqrt{E+\dfrac{1}{4D^2}}}{\rho_{Nf0}+\dfrac{1}{2D}-\sqrt{E+\dfrac{1}{4D^2}}}\right) a^{\frac{3}{\sqrt{4D^2E+1}}}-1}\right)$	$p_{Nf} = \delta_2\left(\rho_{m0} a^{-3(1+\omega_m)}\right) - \left(\sqrt{E+\dfrac{1}{4D^2}}\left(\dfrac{\left(\left(\dfrac{\rho_{Nf0}+\dfrac{1}{2D}+\sqrt{E+\dfrac{1}{4D^2}}}{\rho_{Nf0}+\dfrac{1}{2D}-\sqrt{E+\dfrac{1}{4D^2}}}\right) a^{\frac{3}{\sqrt{4D^2E+1}}}+1\right)}{\left(\dfrac{\rho_{Nf0}+\dfrac{1}{2D}+\sqrt{E+\dfrac{1}{4D^2}}}{\rho_{Nf0}+\dfrac{1}{2D}-\sqrt{E+\dfrac{1}{4D^2}}}\right) a^{\frac{3}{\sqrt{4D^2E+1}}}-1}\right) - \dfrac{1}{2D} - \dfrac{1}{D}\left(\dfrac{\left(\left(\dfrac{\rho_{Nf0}+\dfrac{1}{2D}+\sqrt{E+\dfrac{1}{4D^2}}}{\rho_{Nf0}+\dfrac{1}{2D}-\sqrt{E+\dfrac{1}{4D^2}}}\right) a^{\frac{3}{\sqrt{4D^2E+1}}}\right)}{\left(\dfrac{\rho_{Nf0}+\dfrac{1}{2D}+\sqrt{E+\dfrac{1}{4D^2}}}{\rho_{Nf0}+\dfrac{1}{2D}-\sqrt{E+\dfrac{1}{4D^2}}}\right) a^{\frac{3}{\sqrt{4D^2E+1}}}-1}\right)\right)$

$$\omega_{\text{eff}}=\frac{\left(\begin{aligned}&(\omega_m-\delta_1-\delta_2)\,\rho_{m0}a^{-3(1+\omega_m)}+\delta_1\rho_{m0}a^{-3(1+\omega_m)}+(A)\,\rho_{f0}a^{-3(1+A)}+\frac{B(2\beta+4)}{3A+2\beta+4}a^{2\beta+1}+\delta_2\left(\rho_{m0}a^{-3(1+\omega_m)}\right)-\\&\left(\sqrt{E+\frac{1}{4D^2}}\,\frac{\left(\left(\frac{\rho_{Nf0}+\frac{1}{2D}+\sqrt{E+\frac{1}{4D^2}}}{\rho_{Nf0}+\frac{1}{2D}-\sqrt{E+\frac{1}{4D^2}}}\right)a^{\frac{3}{\sqrt{4D^2E+1}}}+1\right)}{\left(\frac{\rho_{Nf0}+\frac{1}{2D}+\sqrt{E+\frac{1}{4D^2}}}{\rho_{Nf0}+\frac{1}{2D}-\sqrt{E+\frac{1}{4D^2}}}\right)a^{\frac{3}{\sqrt{4D^2E+1}}}-1}\right)-\frac{1}{2D}-\frac{1}{D}\left(\frac{\left(\left(\frac{\rho_{Nf0}+\frac{1}{2D}+\sqrt{E+\frac{1}{4D^2}}}{\rho_{Nf0}+\frac{1}{2D}-\sqrt{E+\frac{1}{4D^2}}}\right)a^{\frac{3}{\sqrt{4D^2E+1}}}\right)}{\left(\frac{\rho_{Nf0}+\frac{1}{2D}+\sqrt{E+\frac{1}{4D^2}}}{\rho_{Nf0}+\frac{1}{2D}-\sqrt{E+\frac{1}{4D^2}}}\right)a^{\frac{3}{\sqrt{4D^2E+1}}}-1}\right)^2\end{aligned}\right)}{\left(\begin{aligned}&\rho_{m0}a^{-3(1+\omega_m)}+\rho_{f0}a^{-3(1+A)}-\frac{3B}{3A+2\beta+4}a^{1+2\beta}+\\&\sqrt{E+\frac{1}{4D^2}}\left(\frac{\left(\left(\frac{\rho_{Nf0}+\frac{1}{2D}+\sqrt{E+\frac{1}{4D^2}}}{\rho_{Nf0}+\frac{1}{2D}-\sqrt{E+\frac{1}{4D^2}}}\right)a^{\frac{3}{\sqrt{4D^2E+1}}}+1\right)}{\left(\frac{\rho_{Nf0}+\frac{1}{2D}+\sqrt{E+\frac{1}{4D^2}}}{\rho_{Nf0}+\frac{1}{2D}-\sqrt{E+\frac{1}{4D^2}}}\right)a^{\frac{3}{\sqrt{4D^2E+1}}}-1}\right)\end{aligned}\right)} \tag{7.12}$$

TABLE 7.2
List of Conditions for this Model

Primary Conditions:	Dark matter dominant: $\rho_m \gg \rho_f \gg \rho_{Nf}$
	Inhomogeneous fluid dominant: $\rho_f \gg \rho_m \gg \rho_{Nf}$
	Non-linear fluid dominant: $\rho_{Nf} \gg \rho_f \gg \rho_m$
	Mixture dominant: $\rho_m \approx \rho_f \approx \rho_{Nf}$
Secondary Conditions:	Mixture dominant Case with very large Powers:
	$-3(1+\omega_m) \approx -3(1+A) \approx 1+2\beta >> a^{\sqrt[3]{4D^2E+1}} >> 1$
	Mixture dominant Case with small Powers:
	$-3(1+\omega_m) \approx -3(1+A) \approx 1+2\beta \approx a^{\sqrt[3]{4D^2E+1}} >> 1$

7.3 CONCLUSIONS

This chapter contains the mathematical analysis of multiple fluid cosmologies, where we have tried to develop different equations of state through the coupled actions for multiple fluid models with interaction scenarios. Linear dark matter, inhomogeneous fluid, and a new type of non-linear fluids have been used here under modified geometry. The idea of variable gravitational constant and cosmological constant has also been interpreted into cosmic dynamics. The equation of state contains six dynamical parameters and one power parameter, which can explain cosmological phases like the Phantom phase, Quintessence, CDM, WDM, and radiation, as well as a smooth transition between phases like Quintom, Graceful exit, etc.

7.4 FUTURE WORKS AND POSSIBILITIES

The equations of states derived here will be used in the detailed thermodynamic analysis in near future. This model will also be used to discuss the fifth force formalism as well as brane cosmic interpretations. The modified gravity to cosmological constant transition in the action used here will be used in detail to find the suitable entropies for the generalized second law of thermodynamics for the model.

REFERENCES

[1] Kar, A. and Sadhukhan, S., "Hamiltonian formalism for Bianchi type I model for perfect fluid as well as for the fluid with bulk and shearing viscosity." Basic and Applied Sciences into Next Frontiers (2021).

[2] Kar, A. and Sadhukhan, S., "Quintessence model with bulk viscosity and some predictions on the coefficient of bulk viscosity and gravitational constant." Recent Advancement of Mathematics in Science and Technology (2021): 260–276.

[3] Kar, A., Sadhukhan, S. and Chattopadhyay, S., "Energy conditions for inhomogeneous EOS and its thermodynamics analysis with the resolution on finite time future singularity problems." *International Journal of Geometric Methods in Modern Physics* 18, no. 8 (2021): 2150131.

[4] Kar, A., Sadhukhan, S. and Chattopadhyay, S., "Thermodynamics and energy condition analysis for Van-Der-Waals EOS without viscous cosmology." *Physica Scripta* (2021). https://doi.org/10.1088/1402-4896/ac2f00
[5] Kar, A., Sadhukhan, S. and Debnath, U., "Condensed body mass accretion with DBI-essence dark energy and its reconstruction with f (Q) gravity." arXiv preprint arXiv:2109.10906 (2021).
[6] Kar, A., Sadhukhan, S. and Chattopadhay, S., "Thermodynamic analysis for non-linear system (Van-der-Waals EOS) with viscous cosmology." *The European Physical Journal C* 81, no. 10 (2021): 1–21.
[7] Kar, A., Sadhukhan, S. and Debnath, U., "Reconstruction of DBI-essence dark energy with f(R) gravity and its effect on black hole and wormhole mass accretion." *Modern Physics Letters A* 36, no. 38 (2021): 2150262. https://doi.org/10.1142/S021773232150262X
[8] Kar, A., Sadhukhan, S., Malaver, M., Kasmaei, H.D and Iyer, R., "A theoretical model of dark energy stars in Einstein-Gauss-Bonnet gravity." arXiv preprint arXiv:2106.09520 (2021).
[9] Kar, A., Sadhukhan, S. and Chattopadhyay, S., "Multiple fluid theory of cosmic evolution and its thermodynamic analysis." *Indian Journal of Physics* 97, no. 3 (2023): 955–981.
[10] Kar, A., Sadhukhan, S., de la Fuente, M.M., Iyer, R., Upadhyay, S. and Gudekli, E., "Buchdahl spacetime with compact body solution of charged fluid and scalar field theory." arXiv preprint arXiv:220 (2022).
[11] Sadhukhan, S., "Quintessence model calculations for bulk viscous fluid and low value predictions of the coefficient of bulk viscosity." *International Journal of Science and Research (IJSR)* 9, no. 3 (2020):1419–1420. https://doi.org/10.21275/SR20327132301
[12] Chattopadhyay, S., "Reconstruction of f (T) gravity with interacting variable-generalised Chaplygin gas and the thermodynamics with corrected entropies." *Zeitschrift für Naturforschung A* 72, no. 3 (2017): 231–244.
[13] Chattopadhyay, S. and Debnath, U., "Interaction between phantom field and modified Chaplygin gas." *Astrophysics and Space Science* 326, no. 2 (2010): 155–158.
[14] Chattopadhyay, S. and Debnath, U., "Tachyonic field interacting with scalar (phantom) field." *Brazilian Journal of Physics* 39 (2009): 86–91.
[15] Debnath, U. and Chattopadhyay, S., "Statefinder and Om diagnostics for interacting new holographic dark energy model and generalized second law of thermodynamics." *International Journal of Theoretical Physics* 52, no. 4 (2013): 1250–1264.
[16] Debnath, U and Chakraborty, W., "Interaction between scalar field and ideal fluid with inhomogeneous equation of state." *Physics Letters B* 661, no. 1 (2008): 1–4.
[17] Capozziello, S., Cardone, V.F., Elizalde, E., Nojiri, S. and Odintsov, S.D., "Observational constraints on dark energy with generalized equations of state." *Physical Review D* 73, no. 4 (2006): 043512.
[18] Capozziello, S., De Martino, S. and Falanga, M., "Van der Waals quintessence." *Physics Letters A* 299, no. 5–6 (2002): 494–498.
[19] Capozziello, S., Carloni, S. and Troisi, A., "Quintessence without scalar fields." arXiv preprint astro-ph/0303041 (2003).
[20] Dirac, P.A.M., "The large numbers hypothesis and the Einstein theory of gravitation." *Proceedings of the Royal Society of London. A. Mathematical and Physical Sciences* 365, no. 1720 (1979): 19–30.
[21] Dirac, P.A.M., "Cosmological models and the large numbers hypothesis." *Proceedings of the Royal Society of London. A. Mathematical and Physical Sciences* 338, no. 1615 (1974): 439–446.

[22] Hoyle, F. and Narlikar, J.V., "A new theory of gravitation." *Proceedings of the Royal Society of London. Series A. Mathematical and Physical Sciences* 282 (1964): 191.

[23] Zeldovich, Y.B., *Soviet Physics JETP* 14 (1968): 1143.

[24] Banerjee, A., Duttachoudhury, S.B. and Banerjee, N., "Bianchi type-I model with cosmological constant in a generalized scalar-tensor theory of gravitation." *Physical Review D* 32 (1985): 3096.

[25] Bergmann, P.G., "Comments on the scalar-tensor theory." *International Journal of Theoretical Physics* 1 (1968): 25.

[26] Nojiri, S.I. and Odintsov, S.D., "The new form of the equation of state for dark energy fluid and accelerating universe." *Physics Letters B* 639, no. 3–4 (2006): 144–150.

[27] Brevik, I., Gorbunova, O.G. and Timoshkin, A.V., "Dark energy fluid with time-dependent, inhomogeneous equation of state." *The European Physical Journal C* 51, no. 1 (2007): 179–183.

8 Multi-Soliton, Multi-Singular Soliton and Lumps in Plasma

The Hirota Bilinear Method

Prasanta Chatterjee, Uday Narayan Ghosh, and Snehalata Nasipuri

8.1 INTRODUCTION

The integrability of partial differential equations that are non-linear has garnered significant interest from both mathematicians and physicists in recent decades. Parallel to physicists' interest in obtaining unique aspects of physical occurrences, mathematicians are motivated to discover novel approaches to problem solving. A non-linear phenomenon known as soliton has been observed in energy transport in protein molecules, as well as in water waves and optical fibres. Various mathematical strategies have been created in order to comprehend the various possible feature soliton configurations. The history of solitons begins with the colourful story of John Scott Russell, a naval architect riding on a boat moved by horses in 1844, as he was travelling from London to Edinburgh along the Union Canal. He saw a wave that moved without changing its shape. A new wave, called the 'Solitary Wave', was born, and in 1897, Korteweg and de Vries modelled the situation and derived a non-linear partial differential equation named after them. This is the famous Korteweg–de Vries equation (KdV). This is a NEE $\frac{\partial F}{\partial t} + F\frac{\partial F}{\partial x} + \frac{\partial^3 F}{\partial x^3} = 0$, where $F\frac{\partial F}{\partial x}$ is the non-linear term and $\frac{\partial^3 F}{\partial x^3}$ is the dispersion term. Although a single peak or one soliton was observed, soon scientists and mathematicians became interested in finding multi-peak solitons. These are called multi-solitons. By using the inverse scattering method, Gardner, Greene, and Kruskal et al. discovered the first multi-peak solitons of the Korteweg–de Vries (KdV) problem in 1967 [1]. Hirota created a sophisticated yet manageable method in 1971 for deriving the multi-soliton solutions of the KdV equation. It's an algebraic approach. Simple algebraic computations yield the answers with ease, but the outcomes are also far more illuminating. Integrability is the primary issue with the NEEs. Hirota did an excellent job handling this circumstance. By introducing a novel derivative called the Hirota D-operator, he was able to transform given non-linear partial differential equations (NPDEs) into the bilinear form of the Hirota D-operator [2], [3], and [4]. A bilinear equation is the outcome of

 DOI: 10.1201/9781003638414-8

converting an NPDE with a newly introduced auxiliary function $F(x,t)$ to a linear equation with $(F.F)$ as the auxiliary function by an appropriate transformation. In the Hirota sense, an equation is integrable if it can be expressed in the Hirota bilinear form. After that, the explicit formulation of the solution can be obtained using basic algebra. In the long run, when two solitons collide, their shape returns, but their phase is altered. The phase shift following a collision can also be found using Hirota's bilinear technique. This approach has been applied for many years to a broad class of NEEs, such as integrodifferential and difference-differential equations [5]. Besides completely integrable systems, this method is also being well applied to partially integrable equations. It is also shown that an integrable non-linear evolution equation can be transformed into a bilinear form. Even a bilinear form can be constructed for many non-integrable equations. Although the Inverse Scattering Technique (IST) for solving the Initial Value Problem is more general than Hirota's approach, the IST gives several types of solutions, while the latter gives only soliton solutions. However, Hirota's method has several advantages over IST. Firstly, the IST is much more complex, mathematically complicated, and time-consuming, while the Hirota method is mainly algebraic, simple, and almost straightforward. Secondly, the existence of multi-soliton solutions is not as stringent as integrability via IST. Moreover, the Hirota method is limited to obtaining N-soliton solutions; it is capable of handling a wider class of equations, including non-linear PDEs. To find out if the equation is bilinear or not, a simple algorithm is first required.

8.2 HIROTA BILINEAR METHOD

When building accurate soliton solutions for different types of non-linear PDEs, Hirota's method is among the most helpful methods. The process also determines whether or not a given equation meets the conditions needed to admit soliton solutions. Knowing the bilinear form is a disadvantage of Hirota's method. In other words, the method works with any equation that can be expressed in bilinear form, whether it is a system of coupled bilinear equations or a single bilinear equation. After the procedure is reduced to a simple algorithm, the bilinear form is achieved. Above basic algebra and calculus, little more is required. Especially for higher-order PDEs or PDEs with strongly non-linear terms, the computations can be somewhat time-consuming at times. Depending on the kind of soliton solution (two, three, four, etc.) that must be obtained, the calculations become more complex. A solution for a single soliton (1-soliton) may be computed with ease, while solutions for two and three solitons can be manually obtained with ease. However, a symbolic computation is required to obtain four solitons or more. By utilizing mathematical induction, one can acquire the general N-soliton solution of the equation by utilizing the structure of the two and three-soliton solutions, which disclose the form of higher-soliton solutions.

In its simplest form, Hirota's method calls for

i. a suitable modification of the dependent variable,
ii. Hirota differential operator D is introduced in it,
iii. solving the bilinear equation by applying a perturbation method.

8.3 STEPS OF HIROTA METHOD

There are three steps in Hirota's process, and each one needs a little care.

Step I: In order to turn the non-linear PDEs into a quadratic form for dependent variables, the dependent variable must first be suitably transformed by a rational expression (or expressions) of unknown functions. We refer to this new form as the "quadratic form". However, it's not always simple to determine such an unknown function for transformation.

Step II: The non-linearity will be eliminated in the second phase. The equation's quadratic form can be changed into a polynomial of the D-operator by introducing a particular differential operator known as the Hirota D-operator. The Hirota bilinear form is the name given to this form. For a given non-linear PDE, there is no systematic method known to create the Hirota bilinear form. However, we make the assumption that all fully integrable non-linear PDEs may be expressed in the Hirota bilinear form. As an example, the extended Kadomtsev–Petviashvili equation $\left(F_t - 6FF_x + F_{xxx}\right)_x + 3F_{yy} + aF_{tt} + bF_{ty} = 0$ is an equation that is not integrable but has the Hirota bilinear form.

Step III: Using the perturbation technique in the Hirota bilinear form is the last step. Separate analyses will be conducted for the perturbation parameter's coefficients and power. A multi-soliton solution to the problem is derived from the preceding data.

8.4 DEFINITION AND PROPERTIES OF THE HIROTA *D*-OPERATOR

By the 1960s, the KdV equation $F_t - 6FF_x + F_{xxx} = 0$ was well studied, and the solutions were obtained by the IST in the form $F = 2\partial_x^2 \log(detA)$, where A is a matrix and $detA$ is a polynomial of e^{ax+bt}. Motivated by this, Hirota (1971) introduced a new dependent variable, f by $2\partial_x^2 \log(f)$, where f is also supported by Painlevé analysis. The tau function, f, is named for its ability to provide soliton solutions and its significance in the Hirota method's theoretical framework. The D-operator introduced by Hirota is binary, as it takes a pair of functions as input and outputs a new function. These operators satisfy all the properties of bilinearity, so they are called bilinear operators. It allows us to find analytic solutions to PDEs in the form of soliton solutions. We now see the definition and some important properties as follows:

Let H be the space of all differential functions. Then the Hirota D-operator D: $\mathrm{H}\times\mathrm{H}\rightarrow\mathrm{H}$ is defined by $D_\xi(p.q) \equiv \left(\frac{\partial}{\partial \xi} - \frac{\partial}{\partial \xi_1}\right) p(\xi) q(\xi_1) |_{\xi_1=\xi}$ where $p(\xi)$ and $q(\xi)$ are called auxiliary functions, separated by the dot symbol. For higher-order derivatives the definition is

$$D_\xi^r D_\eta^s(p.q) \equiv \left(\frac{\partial}{\partial \xi} - \frac{\partial}{\partial \xi_1}\right)^r \left(\frac{\partial}{\partial \eta} - \frac{\partial}{\partial \eta_1}\right)^s p(\xi,\eta) q(\xi_1,\eta_1) |_{\xi_1=\xi,\eta_1=\eta}$$

Where r and s are non-negative integers; ξ and η are independent variables.

One can also define $D_\xi(p.q) = p_\xi q - pq_\xi$.

Note-I: Note that the Leibniz Product Rule and the Hirota derivative are the same; however, they differ in sign. It is important to remember that the conventional derivative operator acts on single spaces, while Hirota's bilinear operator operates on direct spaces.

Note-II: From the definition, it can be derived that the Hirota D-operator satisfies the following properties of bilinearity

a. $D_\xi(cp + q.h) = cD_\xi(p.h) + D_\xi(q.h)$

b. $D_\xi(p.dq + h) = dD_\xi(p.q) + D_\xi(p.h)$

That's why the Hirota D-operator is called bilinear.

Properties: If $F(D)$ is a polynomial in the Hirota D-operator, then it satisfies some important properties.

1. $F(D)(p.1) = F(\partial)p$
2. $F(D)(1.p) = F(-\partial)p$
3. $D_\xi^r(p.q) = (-1)^r D_\xi^r(q.p) \Rightarrow F(D)(p.q) = F(-D)(q.p)$
4. If$p = q$, $D_\xi^{2r-1}(p.p) = 0$, $r \geq 1$ $D_\xi^{2r}(p.p) = \sum_{i=1}^{2r}(-1)^{2r-i}\binom{2r}{i}(\partial_\xi^i p)(\partial_\xi^{2r-i}p)$, $r \geq 1$;
5. $F(D)(e^{r\xi}.e^{s\xi}) = F(r-s)e^{(r+s)\xi}$
6. $D_\xi^r D_\eta^s(e^{l_1\xi - m_1\eta}.e^{l_2\xi - m_2\eta}) = (l_1 - l_2)^r(-m_1 + m_2)^s e^{(l_1+l_2)\xi - (m_1+m_2)\eta}$

8.5 SERIES EXPANSION OF DIFFERENT SOLITON SOLUTIONS

The Hirota bilinear method is used to find soliton structures. Soliton solutions appear because of the balance between higher-order linearity and non-linearity. A soliton is a self-preserving solitary wave. Depending on the different types of non-linearity, various non-linear waves have been found. Let $G(v)$ be a given non-linear PDE. To use the Hirota method, first, we have to find an appropriate transformation, e.g., $v = H(\lambda)$ such that the given equation is transformed into a quadratic form in λ, including all partial derivatives of λ. Next, utilizing the Hirota derivative, the Hirota bilinear equation in λ has been found. Finally, perturbations on λ have to be introduced to obtain soliton profiles. Here ν is the given unknown function, and λ is the newly introduced unknown function, called the auxiliary function, also known as the tau function. Depending on various choices in the perturbation of λ, different soliton structures like regular, singular, lump, breather, etc. have been found. Let the perturbation be $\lambda = 1 + \sum_{r=1}^{\infty}\rho^r\lambda_r$, where λ_r 's are unknown functions of independent variables and ρ is the perturbation parameter.

1. **Regular Soliton Solution:** If we choose λ_r to be exponential functions in the form $e^{l\xi + m\eta + \dots}$, then regular soliton features have been observed. Here, $\xi, \eta, \dots$ are independent variables and $l, m, \dots$ are real constants.
2. **Singular Soliton Solution:** The same perturbation is used for the singular soliton as in the case of regular soliton, but there is a sign alteration in this

case, i.e., $\lambda = 1+\sum_{r=1}^{\infty}(-1)^r \rho^r \lambda_r$, where λ_r's are exponential functions in the form $e^{l\xi+m\eta+\dots}$.

3. **Lump Soliton Solution:** When λ is considered the sum of two positive quadratic functions of independent variables, lump soliton occurs. It is a rational-type solution that emphasises the appearance of a large-amplitude wave. It can be said that a lump is one type of rogue wave.
4. **Breather Solution:** A breather is an amazing soliton structure. It expresses a periodic variation in solitons's amplitude. It is obtained from soliton solutions by choosing the constants $l, m, \dots$ as complex numbers. In this case, higher-order non-linearity is required.

8.6 MULTI-SOLITON SOLUTIONS AND CONDITIONS FOR EXISTENCE OF SOLITONS

8.6.1 Explanation

The KdV equation is an established NEE. It describes weakly non-linear dispersive waves. Multiple regular soliton solutions for the KdV equation are explained here step by step through the Hirota approach. The standard KdV equation is of the form

$$\phi_\eta + 6\phi\phi_\xi + \phi_{\xi\xi\xi} = 0 \tag{8.1}$$

Consider the Cole–Hopf transformation

$$\phi(\xi,\eta) = 2\frac{\partial^2}{\partial\xi^2}(log p) \tag{8.2}$$

Get the Hirota bilinear equation

$$(D_\xi D_\eta + D_\xi^4)(p.p) = 0 \tag{8.3}$$

Then, we take a formal series expansion

$$p = 1+\sum_{r=1}^{\infty}\varepsilon^r p_r \tag{8.4}$$

We use finite perturbation expansions around $p = 1$ to obtain the multi-soliton solution of the said equation. Substituting (8.4) into (8.3), and equating the coefficients of all powers of ε to zero, we get:

$$O(\varepsilon^0): F(D)(1.1) = 0 \tag{8.5}$$

$$O(\varepsilon^1): F(D)(1.p_1 + p_1.1) = 0 \tag{8.6}$$

$$O(\varepsilon^2): F(D)(1.p_2 + p_2.1 + p_1.p_1) = 0 \tag{8.7}$$

$$O(\varepsilon^3): F(D)(1.p_3 + p_3.1 + p_1.p_2 + p_2.p_1) = 0 \tag{8.8}$$

$$O(\varepsilon^n): F(D)\left(\sum_{j=0}^{n} p_j p_{n-j}\right) = 0, \quad p_0 = 1 \tag{8.9}$$

1-Soliton Solution: For $r = 1$, we obtain a single regular soliton solution by taking one exponential $p = 1 + \varepsilon p_1$, where $p_1 = e^{\theta}$, $\theta = l\xi - m\eta + \alpha$; l, m, and α are constants. Of course, equation (8.5) is trivially satisfied. Then equation (8.6) determines the dispersion relation $m = l^3$, on the parameters l, m and equation (8.7) implies $p_2 = 0$. Therefore, $p_r = 0; r > 2$. Taking $\varepsilon = 1$, we have $p = 1 + e^{l\xi - l^3\eta + \alpha}$. Substituting p into (2), we get the 1-soliton solution of the KdV equation as:

$$\phi(\xi,\eta) = \frac{l^2}{2} Sech^2\left(\frac{l\xi - l^3\eta + \alpha}{2}\right) \tag{8.10}$$

2-Soliton Solutions: Combining two 1-soliton, one can build 2-soliton solutions by using the mechanism $p = 1 + \varepsilon p_1 + \varepsilon^2 p_2$, i.e., we take $r = 2$.

Here, $p_1 = e^{\theta_1} + e^{\theta_2}$, $\theta_r = l_r\xi - m_r\eta + \alpha_r; r = 1,2$. The dispersion relation will be $m_r = l_r^3$, for $r = 1,2$. The terms generated by $F(D)(p_1.p_1)$ justify the choice $p_2 = E_{12}e^{\theta_1+\theta_2} = E_{12}e^{(l_1+l_2)\xi-(m_1+m_2)\eta+(\alpha_1+\alpha_2)}$ and allows to calculate the constant E_{12} i.e., we get $2E_{12}F(l_1+l_2)+2F(l_1-l_2)=0$ $\Rightarrow E_{12} = -\frac{F(l_1-l_2)}{F(l_1+l_2)} = \frac{(l_1-l_2)^2}{(l_1+l_2)^2}$ using $m_r = l_r^3$, for $r = 1,2$. Subsequently, we get $p_r = 0; r \geq 3$. Hence, $p = 1 + e^{\theta_1} + e^{\theta_2} + E_{12}e^{\theta_1+\theta_2}$.

Upon substitution of p into (8.2), we obtain the 2-soliton solutions as:

$$\phi(\xi,\eta) = 2\frac{l_1^2 e^{\theta_1} + l_2^2 e^{\theta_2} + E_{12}\left(l_2^2 e^{\theta_1} + l_1^2 e^{\theta_2}\right)e^{\theta_1+\theta_2} + 2(l_1-l_2)^2 e^{\theta_1+\theta_2}}{\left(1 + e^{\theta_1} + e^{\theta_2} + E_{12}e^{\theta_1+\theta_2}\right)^2} \tag{8.11}$$

Where $\theta_r = l_r\xi - l_r^3\eta + \alpha_r; r = 1,2$ and $E_{12} = \frac{(l_1-l_2)^2}{(l_1+l_2)^2}$.

At the limiting sense, we get a nice form of (11)

$$\phi(\xi,\eta) = E_1 Sech^2\left[\frac{1}{2}(l_1\xi - m_1\eta + \alpha_1) + \Delta_1\right] + E_2 Sech^2\left[\frac{1}{2}(l_2\xi - m_2\eta + \alpha_2) + \Delta_2\right] \tag{8.12}$$

Where $E_r = \frac{1}{2}l_r^2$, $r = 1,2$; $\Delta_{1,2} = \pm\log\left|\sqrt{E_{12}}\right|$; $\sqrt{E_{12}} = \frac{l_1 - l_2}{l_1 + l_2}$, is the phase factor.

3-Soliton Solutions: We use the same mechanism, which was adopted to derive 2-soliton solutions, to obtain the 3-soliton solutions. We take $p = 1 + \varepsilon p_1 + \varepsilon^2 p_2 + \varepsilon^3 p_3$, i.e., $r = 3$. We use the form of three-terms of p_1 for 3-soliton, i.e., it is important to remember that $p_1 = e^{\theta_1} + e^{\theta_2} + e^{\theta_3}$, where $\theta_r = l_r\xi - m_r\eta + \alpha_r; r = 1,2,3$. l_r, m_r and α_r are constants. Whereas equation (8.6) determines the dispersion relation for the KdV equation as: $m_r = l_r^3$, for $r = 1,2,3$.

The term generated by $F(D)(p_1.p_1)$ justifies the choice

$$p_2 = E_{12}e^{\theta_1+\theta_2} + E_{13}e^{\theta_1+\theta_3} + E_{23}e^{\theta_2+\theta_3} = E_{12}e^{(l_1+l_2)\xi-(m_1+m_2)\eta+(\alpha_1+\alpha_2)}$$

$$+E_{13}e^{(l_1+l_3)\xi-(m_1+m_3)\eta+(\alpha_1+\alpha_3)} + E_{23}e^{(l_2+l_3)\xi-(m_2+m_3)\eta+(\alpha_2+\alpha_3)}$$

And the equation (8.7) allows the calculation of the constants E_{12}, E_{13}, and E_{23} with $m_r = l_r^3$. So, we obtain, $E_{rs} = -\frac{F(l_r - l_s)}{F(l_r + l_s)} = \frac{(l_r - l_s)^2}{(l_r + l_s)^2}$, for $r,s = 1,2,3; r < s$. Now, from $F(D)(p_1.p_2 + p_2.p_1)$, we get $p_3 = Be^{\theta_1+\theta_2+\theta_3} = Be^{(l_1+l_2+l_3)\xi-(m_1+m_2+m_3)\eta+(\alpha_1+\alpha_2+\alpha_3)}$, and $B = E_{12}E_{13}E_{23} = \frac{(l_1-l_2)^2(l_1-l_3)^2(l_2-l_3)^2}{(l_1+l_2)^2(l_1+l_3)^2(l_2+l_3)^2}$. Consequently, we have $p_r = 0; r \geq 4$. Hence, $p_3 = 1 + e^{\theta_1} + e^{\theta_2} + e^{\theta_3}$ $+E_{12}e^{\theta_1+\theta_2} + E_{13}e^{\theta_1+\theta_3} + E_{23}e^{\theta_2+\theta_3} + Be^{\theta_1+\theta_2+\theta_3}$ and the 3-soliton solutions are given by $\phi(\xi,\eta) = 2\frac{L(\xi,\eta)}{M(\xi,\eta)}$;

$$L(\xi,\eta) = \left(2(l_1-l_2)^2 + 2(l_1-l_2)^2 E_{13}E_{23}e^{2\theta_3} + E_{12}l_1^2e^{\theta_2} + E_{12}l_2^2e^{\theta_1}\right)e^{\theta_1+\theta_2}$$

$$+\left(2(l_1-l_3)^2 + 2(l_1-l_3)^2 E_{12}E_{23}e^{2\theta_2} + E_{13}l_1^2e^{\theta_3} + E_{13}l_3^2e^{\theta_1}\right)e^{\theta_1+\theta_3}$$

$$+\left(2(l_2-l_3)^2 + 2(l_2-l_3)^2 E_{12}E_{13}e^{2\theta_1} + E_{23}l_2^2e^{\theta_3} + E_{23}l_3^2e^{\theta_2}\right)\ e^{\theta_2+\theta_3}$$

$$+l_1^2e^{\theta_1} + l_2^2e^{\theta_2} + l_3^2e^{\theta_3} + Be^{\theta_1+\theta_2+\theta_3}\left(E_{12}l_3^2e^{\theta_1+\theta_2} + E_{13}l_2^2e^{\theta_1+\theta_3} + E_{23}l_1^2e^{\theta_2+\theta_3}\right)$$

$$+e^{\theta_1+\theta_2+\theta_3}(E_{12}\left(l_1^2+l_2^2+l_3^2+2l_1l_2-2l_1l_3-2l_2l_3\right) + E_{13}\left(l_1^2+l_2^2+l_3^2-2l_1l_2+2l_1l_3-2l_2l_3\right)$$

$$+E_{23}\left(l_1^2+l_2^2+l_3^2-2l_1l_2-2l_1l_3+2l_2l_3\right) + B\left(l_1^2+l_2^2+l_3^2-2l_1l_2-2l_1l_3+2l_2l_3\right))$$

and

$$M(\xi,\eta) = \left(1 + e^{\theta_1} + e^{\theta_2} + e^{\theta_3} + E_{12}e^{\theta_1+\theta_2} + E_{13}e^{\theta_1+\theta_3} + E_{23}e^{\theta_2+\theta_3} + Be^{\theta_1+\theta_2+\theta_3}\right)^2$$

For $\theta_r = l_r\xi - l_r^3\eta + \alpha_r, r = 1,2,3;$ $E_{rs} = \frac{(l_r - l_s)^2}{(l_r + l_s)^2}$; $r,s = 1,2,3; r < s$ and $B = E_{12}E_{13}E_{23}$.

Note: A non-linear PDE is said to be Hirota integrable if it can be written in Hirota bilinear form and multi-soliton solutions can be expressed as a combination of a finite number of 1-soliton solutions.

8.7 CONDITIONS

When Hirota analysed the multi-soliton solutions for the KdV equation, he noticed that multi-soliton solutions occur under certain situations. Now, we will explain those situations.

Let $G(\phi,\phi_\xi,\phi_\eta,\phi_{\xi\eta},\ldots)=0$ be a given non-linear partial differential equation. Here, ϕ is the wave's amplitude (unknown function) and $\xi,\eta,\ldots$ are independent variables. Now, we want to obtain its Hirota bilinear form, so we are considering a transformation on ϕ,

Let $\phi=W\left[p(\xi,\eta,\ldots,)\right]$, W is a function of $p(\xi,\eta,\ldots,)$, and let it get a polynomial in the Hirota D-operator $F(D)$, which operates on a pair of functions $(p.p)$. The $F(D)(p.p)$ is the Hirota bilinear form of the given PDE. If the Hirota bilinear equation $F(D)(p.p)=0$ satisfies the condition $F(\xi,\eta,\ldots)=F(s)=0$, then a 1-soliton solution of the given PDE exists and is given by $\phi=W\left[p(\xi,\eta,\ldots)\right]=W\left[1+e^{\theta}\right]$. Here, θ is a polynomial in the independent variables, i.e., $\theta=l\xi-m\eta+\ldots+c$; $l,m,\ldots,c$ are constants. Hence, 1-soliton solution is restricted only by the dispersion relation, which is derived from the equation $F(\xi,\eta,\ldots)=0$.

Now, let's move on to the 2-soliton solutions. The 2-soliton of the given PDE exists if two conditions are fulfilled: (i) $F(\xi_r,\eta_r,\ldots)=F(s_r)=0$ (which gives the dispersion relation) and (ii) The phase factor (E_{12}) must be of the form $E_{12}=-\dfrac{F(s_1-s_2)}{F(s_1+s_2)}$.

Then, the expression of the 2-soliton solutions will be $\phi=W\left[p(\xi,\eta,\ldots)\right]=W\left[1+e^{\theta_1}+e^{\theta_2}+E_{12}e^{\theta_1+\theta_2}\right]$, where $\theta_r=l_r\xi-m_r\eta+\ldots+c_r$; $l_r,m_r,\ldots,c_r$ $(r=1,2.)$ are constants. Therefore, 2-soliton is restricted by the dispersion relation, and phase factor should be of the form $E_{12}=-\dfrac{F(s_1-s_2)}{F(s_1+s_2)}=-\dfrac{F(l_1-l_2,-m_1+m_2)}{F(l_1+l_2,-m_1-m_2)}$.

To obtain 3-soliton solutions, there is an additional condition

$$\sum_{\Upsilon_r=\pm1}F(\Upsilon_1s_1+\Upsilon_2s_2+\Upsilon_3s_3)F(\Upsilon_1s_1-\Upsilon_2s_2)F(\Upsilon_2s_2-\Upsilon_3s_3)F(\Upsilon_3s_3-\Upsilon_1s_1)=0, \tag{8.13}$$

along with the conditions for the 2-soliton solutions. Therefore, the 3-soliton solutions are restricted by the dispersion relation, given by $F(s_r)=0, r=1,2,3,$ the phase factor $E_{rt}=-\dfrac{F(s_r-s_t)}{F(s_r+s_t)}$; $r,t=1,2,3,\ r<t$, and the additional condition given in (8.13). Hence, the expression of 3-soliton solution will be

$$\phi=W\begin{bmatrix}p(\xi,\eta,\ldots)]=W[1+e^{\delta_1}+e^{\delta_2}+e^{\delta_3}+E_{12}e^{\delta_1+\delta_2}\\+E_{13}e^{\delta_1+\delta_3}+E_{23}e^{\delta_2+\delta_3}+Be^{\delta_1+\delta_1+\delta_3}\end{bmatrix},$$

where $\delta_r=l_r\xi-m_r\eta+\ldots+c_r$; $l_r,m_r,\ldots,c_r$ $(r=1,2,3.)$ are constants and $B=E_{12}E_{13}E_{23}$.

This condition can be extended for higher-order soliton solutions. Given as:

$$V[X,m] = \sum_{\Upsilon_r=\pm 1} \begin{bmatrix} X\left(\sum_{r=1}^{m}\Upsilon_r l_r - \sum_{r=1}^{m}\Upsilon_r m_r\right) \\ \prod_{r<t}^{(m)} X\left(\Upsilon_r l_r - \Upsilon_t l_t, -\Upsilon_r m_r + \Upsilon_t m_t\right)\Upsilon_r\Upsilon_t \end{bmatrix} = 0$$

Consequently, p is left with no remaining freedom; the parameters are further limited by the phase factors E_{rt} and the dispersion relation. Therefore, the presence of three or more solitons does not occur by itself; rather, the polynomial X must satisfy strict conditions.

8.8 2-SOLITON INTERACTION OF THE KORTEWEG–DE VRIES EQUATION (KDV)

Now, we study the regular soliton interaction of electron acoustic solitary waves based on the foundation of the basic hydrodynamical equations. This is a very well-known set of governing equations that we have taken here. The variables and parameters associated with this set of equations bear standard meaning (for details see ref [6])

$$\frac{\partial \eta_e}{\partial t} + \frac{\partial(\eta_e \upsilon_e)}{\partial x} = 0 \tag{8.14}$$

$$\frac{\partial \upsilon_e}{\partial t} + \upsilon_e \frac{\partial \upsilon_e}{\partial x} = \frac{\partial \Phi}{\partial x} \tag{8.15}$$

$$\frac{\partial^2 \Phi}{\partial x^2} = \eta_e - \eta_{il} - \eta_{ih} \tag{8.16}$$

Corresponding ions are distributed as:

$$\eta_{il} = \mu\left[1 - A_1 s\Phi + B_1 s^2\Phi^2 - C_1 s^3\Phi^3 + D_1 s^4\Phi^4 + \ldots\right]$$

$$\eta_{ih} = \nu\left[1 - A_1 s\sigma\Phi + B_1 s^2\sigma^2\Phi^2 - C_1 s^3\sigma^3\Phi^3 + D_1 s^4\sigma^4\Phi^4 + \ldots\right],$$

Where $A_1 = G\dfrac{G_{21}G_{31}}{2G_{11}}$; $B_1 = H\dfrac{G_{22}G_{32}}{8G_{11}}$; $C_1 = I\dfrac{G_{23}G_{33}}{16G_{11}}$; $D_1 = J\dfrac{G_{24}G_{34}}{128G_{11}}$;

$$G = (q-1)^{\frac{-1}{2(r+1)}}, H = -(q-1)^{\frac{-2}{(r+1)}}, I = (q-1)^{\frac{-3}{(r+1)}}, J = -5(q-1)^{\frac{-4}{(r+1)}},$$

$$G_{11} = \Gamma\left[1 + \frac{1}{2(r+1)}\right]\Gamma\left[q - \frac{1}{2(r+1)}\right], G_{21} = \Gamma\left[q + \frac{1}{2(r+1)}\right],$$

$$G_{22} = \Gamma\left[q + \frac{3}{2(r+1)}\right],$$

$$G_{23}=\Gamma\left[q+\frac{5}{2(r+1)}\right],\ G_{24}=\Gamma\left[q+\frac{7}{2(r+1)}\right],G_{31}=\Gamma\left[1-\frac{1}{2(r+1)}\right],$$

$$G_{32}=\Gamma\left[1-\frac{3}{2(r+1)}\right],\ G_{33}=\Gamma\left[1-\frac{5}{2(r+1)}\right],\ G_{34}=\Gamma\left[1-\frac{7}{2(r+1)}\right].$$

8.8.1 Derivation of the KdV Equation

The perturbation method (PM) is no longer able to produce prominent features when non-linearity is incorporated into the evolution equations. Since nature is commonly non-linear, non-linearity must occur in the evolution equations. The Standard PM (SPM) cannot handle the non-linear term, and it produces results by chopping the non-linear effect rather than incorporating it. To get the effect of non-linearity in the subsequently derived evolution equation, a special limiting process of SPM has been chosen so that the non-linear effect will be carried forward to the evolution equations. This is known as the reductive perturbation technique (RPT), and the non-linear effect has been extracted using this technique. RPT is very effective in producing simple canonical equations from systems with non-linearity. The stretching is taken as

$$\xi_1=\varepsilon^{\frac{1}{2}}(x-\lambda t),\ \tau_1=\varepsilon^{\frac{3}{2}}t, \tag{8.17}$$

where ε measures the non-linearity strength, and the phase velocity is represented by λ. The expansion of dependent variables is prescribed by the dispersion relation, which is taken as

$$\eta_e=1+\varepsilon\eta^{(1)}+\varepsilon^2\eta^{(2)}+\varepsilon^3\eta^{(3)}+\varepsilon^4\eta^{(4)}+\ldots \tag{8.18}$$

$$\upsilon_e=\varepsilon\upsilon^{(1)}+\varepsilon^2\upsilon^{(2)}+\varepsilon\upsilon^{(3)}+\varepsilon^4\upsilon^{(4)}+\ldots \tag{8.19}$$

$$\Phi=\varepsilon\Phi_1+\varepsilon^2\Phi_2+\varepsilon^3\Phi_3+\varepsilon^4\Phi_4+\ldots \tag{8.20}$$

Stretched coordinates (8.17) and the expression of dependent variables (8.18)–(8.20) have been substituted into the basic normalized equations (8.14)–(8.16). By equating different powers of ε up to the second order, a system of equations containing a second-order dependent variable is obtained. These relations lead to the KdV equation

$$\frac{\partial\Phi_1}{\partial\tau_1}+A\Phi_1\frac{\partial\Phi_1}{\partial\xi_1}+B\frac{\partial^3\Phi_1}{\partial\xi_1{}^3}=0, \tag{8.21}$$

Where $A=B\left[2B_1\dfrac{\mu+\nu\sigma^2}{(\mu+\nu\sigma)^2}-\dfrac{3}{\lambda^4}\right]$, $B=\dfrac{\lambda^3}{2}$ and phase velocity $\lambda^2=\dfrac{1}{A_1}$, and A_1,B_1 are the coefficients of the R-Q distribution. Let us transform ξ_1 by $\xi B^{1/3}$, τ_1 by τ and $\Phi_1$by $6\Phi A^{-1}B^{1/3}$. Finally, the standard KdV equation is obtained as follows:

$$\frac{\partial\Phi}{\partial\tau}+6\Phi\frac{\partial\Phi}{\partial\xi}+B\frac{\partial^3\Phi}{\partial\xi^3}=0 \tag{8.22}$$

Solving (8.22) using the Hirota bilinear method, the 2-soliton solutions of the KdV equation are obtained, and the expression is given as follows:

$$\Phi(\xi,\tau)=\alpha_1 Sech^2\left[\left(\frac{\zeta_1}{2}\right)+\delta_1\right]+\alpha_2 Sech^2\left[\left(\frac{\zeta_2}{2}\right)+\delta_2\right]$$

Where $\zeta_{1,2}=k_{1,2}\xi-k_{1,2}^3\tau;\ \alpha_{1,2}=\frac{-1}{2}k_{1,2}^2;\ \delta_{1,2}=\pm\log|E_{12}|;\ E_{12}=\frac{k_1-k_2}{k_1+k_2}$ (phase factor).

Using the inverse of the transformation applied to the independent variables, we get the 2-soliton solutions of the originally derived KdV equation

$$\Phi_1(\xi_1,\tau_1)=\frac{6B^{1/3}}{A}\alpha_1 Sech^2\left[\frac{k_1}{2B^{\frac{1}{3}}}\left(\xi_1-B^{\frac{1}{3}}k_1^2\tau_1+\frac{2B^{\frac{1}{3}}}{k_1}\log\left|\frac{k_1-k_2}{k_1+k_2}\right|\right)\right]$$

$$+\frac{6B^{1/3}}{A}\alpha_2 Sech^2\left[\frac{k_2}{2B^{\frac{1}{3}}}\left(\xi_1-B^{\frac{1}{3}}k_2^2\tau_1+\frac{2B^{\frac{1}{3}}}{k_2}\log\left|\frac{k_1-k_2}{k_1+k_2}\right|\right)\right]$$

8.8.2 Result and Discussion

Large-amplitude solitary waves travel faster than smaller ones when they propagate in the same direction. Therefore, larger waves surpass smaller waves. This overtaking collision is studied using the Hirota bilinear method (HBM). Using RPT, we derive the KdV equation from the governing equations. Next, we use HBM to extract the soliton solution from the resulting KdV equation. For different times, we have plotted the same soliton solution and shown the overtaking interaction between solitary waves of different amplitudes. For $q=2, r=0,\ \mu=0.25,\ \sigma=0.5$, we have plotted Figure 8.1 at different time. This figure set depicts the overtaking interaction. Initially, the waves are rather far away from one another. As time progresses, they get closer, interact with each other, and then they eventually separate. Ultimately, the large-amplitude solitary wave overtakes the small amplitude wave.

8.9 MULTIPLE SINGULAR SOLITON SOLUTIONS

A singular soliton emerges from the soliton solution that carries singularities [7]. The solutions with singularities are obtained similarly by HBM. We have observed that regular solitons appear in the form of $Sech^2$, $Sech$, etc. The process by which we extract the singular soliton solution using HBM for the modified Korteweg-de Vries (mKdV) equation is now explained. We consider the mKdV equation as:

$$\phi_\tau-6\phi^2\phi_\xi+\phi_{\xi\xi\xi}=0 \tag{8.23}$$

Using the transformation $\phi(\xi,\tau)=\frac{\partial}{\partial\xi}\left[log\frac{m(\xi,\tau)}{n(\xi,\tau)}\right]=\frac{nm_\xi-mn_\xi}{nm}$, we get the following Hirota bilinear form for the mKdV equation

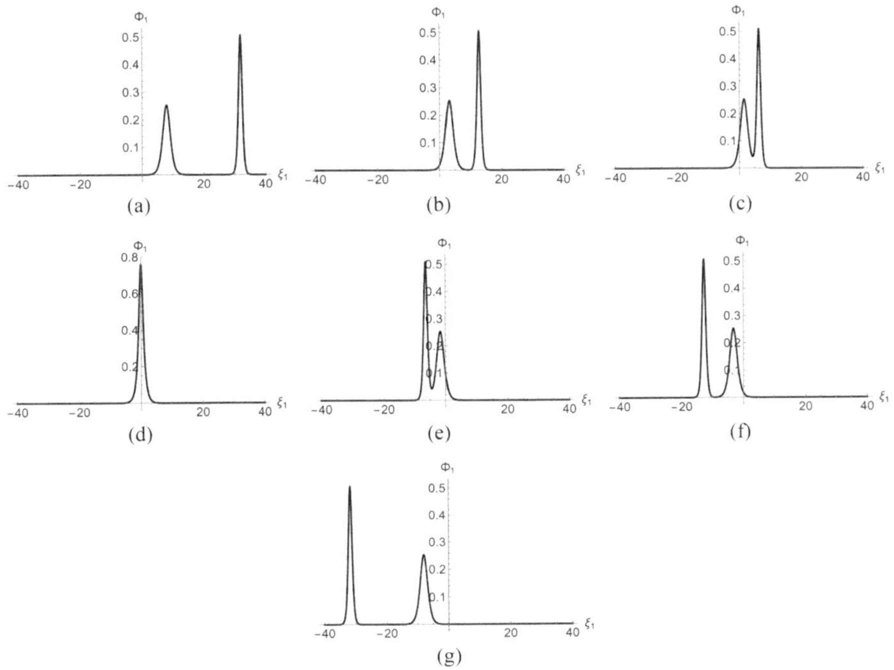

FIGURE 8.1 Interaction between the large-amplitude soliton and the small amplitude soliton, for $q=2, r=0, \mu=0.25, \sigma=0.5$ at time $\tau_1=5, \tau_1=2, \tau_1=1, \tau_1=0, \tau_1=-1, \tau_1=-2, \tau_1=-5$ respectively.

$$S_1(D)(m.n)=0 \Rightarrow \left(D_\tau + D_\xi^3\right)(m.n)=0$$

$$S_2(D)(m.m+n.n)=0 \Rightarrow D_\xi^2(m.m+n.n)=0$$

Where S_1 is the odd-order polynomial in Hirota derivative with respect to ξ, τ, and S_2 is the even order polynomial in Hirota derivative with respect to ξ, not including constant term.

To determine the singular 2-soliton solutions, we take

$$m(\xi,\tau)=1+e^{\delta_1}+e^{\delta_2}+E_{12}e^{\delta_1+\delta_2}$$

$$n(\xi,\tau)=1-e^{\delta_1}-e^{\delta_2}+F_{12}e^{\delta_1+\delta_2}$$

Where $\delta_r = k_r\xi - \rho_r\tau + a_r$, $r=1,2$. The coefficient parameters satisfying the relation $S_1(k_r, -\rho_r)=0 \Rightarrow \rho_r = k_r^3$, $r=1,2$, are known as the dispersion relation.

Further calculation gives us the expressions of the phase factors as follows:

$$E_{12}=F_{12}=\frac{S_2(k_1-k_2,\ -\rho_1+\rho_2)}{S_2(k_1+k_2,\ -\rho_1-\rho_2)}=\frac{(k_1-k_2)^2}{(k_1+k_2)^2}$$

Therefore, the singular 2-soliton solution is

$$\phi(\xi,\tau)=\frac{\partial}{\partial\xi}\left[log\frac{m(\xi,\tau)}{n(\xi,\tau)}\right]$$

Where

$$m(\xi,\tau)=1+\exp\left[k_1\xi-k_1^3\tau+a_1\right]+\exp\left[k_2\xi-k_2^3\tau+a_2\right]+\frac{(k_1-k_2)^2}{(k_1+k_2)^2}$$
$$\exp\left[(k_1+k_2)\xi-\left(k_1^3+k_2^3\right)\tau+(a_1+a_2)\right];$$

$$n(\xi,\tau)=1-\exp\left[k_1\xi-k_1^3\tau+a_1\right]-\exp\left[k_2\xi-k_2^3\tau+a_2\right]+\frac{(k_1-k_2)^2}{(k_1+k_2)^2}$$
$$\exp\left[(k_1+k_2)\xi-\left(k_1^3+k_2^3\right)\tau+(a_1+a_2)\right].$$

Note: If we use the transformation

$$\phi(\xi,\tau)=\frac{\partial}{\partial\xi}\left[\log\left[m(\xi,\tau)\right]\right]$$

Then the general form of $m(\xi,\tau)$ for singular N-soliton solutions for the mKdV equation will be: $m(\xi,\tau)=\sum_{\mu_r\in\{0,1\}}(-1)^{\sum_{r=1}^{N}\mu_r}\left(\exp\left[\sum_{r<s}^{(N)}b_{rs}\mu_r\mu_s+\sum_{r=1}^{N}\mu_r\delta_r\right]\right)$

$\delta_r=k_r\xi-\rho_r\tau+a_r$, $r=1,2,\ldots,N$. The dispersion relation will be the same as the previous $\rho_r=k_r^3$ and $E_{rs}=\exp\left[b_{rs}\right]=\frac{(k_r-k_s)^2}{(k_r+k_s)^2}$, $1\le r<s\le N$.

8.9.1 Electron Acoustic Singular Solitons Interaction of the Modified Korteweg-de Vries (mKdV) Equation

Considering the same set of governing equations used in the previous section, we obtain the parameter values q = 2.428475; r = 0.01; μ = 0.5; σ = 0.28 which leads $A\to 0$ of the equation (8.21). To study the interaction between singular solitons around that critical parameter region, we stretch the independent variables [8,9] as

$$\xi_2=\varepsilon(x-\lambda_1 t);\ \tau_2=\varepsilon^3 t,$$

and the same dependent variable expansion is used. The RPT leads to the mKdV equation

$$\frac{\partial\varphi_1}{\partial\tau_2}+A_2\varphi_1^2\frac{\partial\varphi_1}{\partial\xi_2}+B_2\frac{\partial^3\varphi_1}{\partial\xi_2^3}=0 \tag{8.24}$$

Where $A_2 = B_2\left[\frac{15}{\lambda_1^6} - 3C_1 s^3\left(\mu + \nu\sigma^3\right)\right]$, $B_2 = \frac{\lambda_1^3}{2}$, $\lambda_1 = A_1^{\frac{-1}{2}}$; A_1 and C_1 are the coefficient of the R-Q distribution. Transforming ξ_2 by $\eta B_2^{1/3}$, τ_2 by ϱ and φ_1 by $\sqrt{6}\psi A_2^{-1/2} B_2^{1/6}$, the extracted mKdV equation (8.24) is converted to the standard mKdV equation

$$\frac{\partial \Psi}{\partial \varrho} + 6\Psi^2 \frac{\partial \Psi}{\partial \eta} + \frac{\partial^3 \Psi}{\partial \eta^3} = 0 \tag{8.25}$$

We use the Hirota bilinear approach to the problem (8.25), yielding multiple singular soliton solutions. Then, the singular 2-soliton solutions of the mKdV equation are obtained via simple algebraic scheming, and the expression is as follows:

$$\psi(\eta, \varrho) = \beta_1 Cosech\left[\frac{\delta_1}{2} + \Theta_1\right] + \beta_2 Cosech\left[\frac{\delta_2}{2} + \Theta_2\right] \tag{8.26}$$

Where $\delta_{1,2} = k_{1,2}\eta - k_{1,2}^3\,\varrho$; $\beta_{1,2} = \frac{-1}{2} k_{1,2}^2$; $\Theta_{1,2} = \pm\log|a_{12}|$; $a_{12} = \frac{k_1 - k_2}{k_1 + k_2}$ (phase factor).

By taking the reverse transformation, the singular 2-soliton solutions of the originally derived mKdV equation are obtained as

$$\begin{aligned}\varphi_1\ (\xi_2, \tau_2) = {} & \frac{\sqrt{6}B_2^{1/6}}{A_2^{1/2}}\beta_1 Cosech\left[\frac{k_1}{2B_2^{1/3}}\left(\xi_2 - B_2^{1/3}k_1^2\tau_2 + \frac{2B_2^{1/3}}{k_1}\log\left|\frac{k_1 - k_2}{k_1 + k_2}\right|\right)\right] \\ & + \frac{\sqrt{6}B_2^{1/6}}{A_2^{1/2}}\beta_2 Cosech\left[\frac{k_2}{2B_2^{1/3}}\left(\xi_2 - B_2^{1/3}k_2^2\tau_2 + \frac{2B_2^{1/3}}{k_2}\log\left|\frac{k_1 - k_2}{k_1 + k_2}\right|\right)\right]\end{aligned}$$

8.9.2 Result and Discussion

The parameter values $q = 2.428475$; $r = 0.01$; $\mu = 0.5$; $\sigma = 0.28$ lead to the quadratic non-linear coefficient $A \to 0$ of the equation (8.21). So, the KdV equation is no longer valid for that parameter set. To study the non-linear features of this parameter set, we have derived the mKdV equation. Two singular soliton solutions have been obtained by HBM. Plotting Figure 8.2 for a different times, we have depicted the overtaking interaction of the two singular solutions of the mKdV equation. We have presented the fat singular soliton on the smooth line and the thin singular soliton on the dotted line. As observed in the case of the 2-soliton interaction with the KdV equation, the same phenomenon occurs here. Initially, the dotted soliton moves behind the smooth soliton, but as time progresses, they get closer and interact with one another. After the interaction, the thin singular soliton overtakes and moves faster than the fat singular soliton.

8.10 LUMP SOLITON STRUCTURES IN MAGNETIZED AURORAL PLASMA

Lump Soliton is a unique rational function solution that has localised wave structures in all spatial directions. A lump soliton is one type of rogue wave, characterized by a large- amplitude wave, usually more than twice the normal amplitude of the

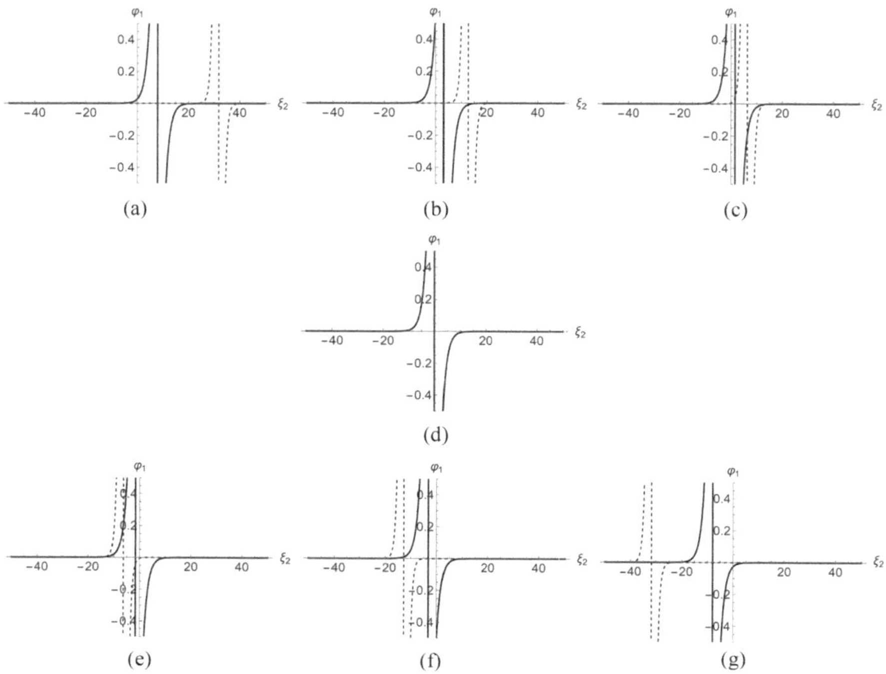

FIGURE 8.2 Interaction between the fat singular soliton (smooth line) and the thin singular soliton (dotted line), for $q = 2.428475$; $r = 0.01$; $\mu = 0.5$; $\sigma = 0.28$ at time $\tau_2 = 5, \tau_2 = 2, \tau_2 = 1, \tau_2 = 0, \tau_2 = -1, \tau_2 = -2, \tau_2 = -5$ respectively.

surrounding waves, with unexpected appearance features and disappearance without a trace. The remarkable non-linear properties of lump solitons have drawn the attention of numerous researchers. Based on the basic hydrodynamical equations, this is a very well-known set of governing equations that we have used here. The variables and parameters associated with this set of equations bear their standard meanings (for details, see ref [10]).

The wave is thought to be travelling in the $X - Z$ plane. Next, the set of normalized equations can be expressed as

$$\frac{\partial N_r}{\partial T} + \frac{\partial (N_r U_{rX})}{\partial X} + \frac{\partial (N_r U_{rZ})}{\partial Z} = 0 \tag{8.27}$$

$$\frac{\partial U_{rX}}{\partial T} + \left(U_{rX} \frac{\partial}{\partial X} + U_{rZ} \frac{\partial}{\partial Z} \right) U_{rX} = -\frac{\partial \Phi}{\partial X} + U_{rY} \tag{8.28}$$

$$\frac{\partial U_{rY}}{\partial T} + \left(U_{rX} \frac{\partial}{\partial X} + U_{rZ} \frac{\partial}{\partial Z} \right) U_{rY} = -U_{rX} \tag{8.29}$$

$$\frac{\partial U_{rZ}}{\partial T} + \left(U_{rX} \frac{\partial}{\partial X} + U_{rZ} \frac{\partial}{\partial Z} \right) U_{rZ} = -\frac{\partial \Phi}{\partial Z} \tag{8.30}$$

$$\Omega\left(\frac{\partial^2}{\partial X^2}+\frac{\partial^2}{\partial Z^2}\right)\Phi = N_c - N_h - N_r \tag{8.31}$$

The Lorentzian kappa distribution is followed by hot electrons and the Maxwellian distribution by cold electrons as

$$N_c = Fe^{(\alpha_c\Phi)} \tag{8.32}$$

$$N_h = (1-F)\left(1-\frac{\alpha_h\Phi}{\kappa-\frac{3}{2}}\right)^{\left(-\kappa+\frac{1}{2}\right)} \tag{8.33}$$

where $a_1 = -\dfrac{\alpha_h(1-F)\left(-\kappa+\frac{1}{2}\right)}{\left(\kappa-\frac{3}{2}\right)}$; $b_1 = \dfrac{\alpha_h^{\,2}(1-F)\left(-\kappa+\frac{1}{2}\right)\left(-\kappa-\frac{1}{2}\right)}{2\left(\kappa-\frac{3}{2}\right)^2}$;

$c_1 = \dfrac{\alpha_h^{\,3}(1-F)\left(-\kappa+\frac{1}{2}\right)\left(-\kappa-\frac{1}{2}\right)\left(-\kappa-\frac{3}{2}\right)}{6\left(\kappa-\frac{3}{2}\right)^3}$.

Now, we derive the KP equation from this set of governing equation using RPT.

8.10.1 Perturbation Technique to Derive Kadomtsev–Petviashvili (KP) Equation

We will now examine new stretched coordinates γ, λ, and μ such that

$$\gamma = \epsilon^2 X,\ \lambda = \epsilon(Z-VT),\ \mu = \epsilon^3 T, \tag{8.34}$$

where ϵ is a small parameter that represents the strongest point of the non-linearity and V is the phase velocity of the ion-acoustic wave. The $X-Z$ plane is where the waves are said to be propagating. The stretched coordinates are not coincidentally selected. This relation for the independent variable and the expansion of the dependent variable [8] are provided by the dispersion relation. Then,

$$\frac{\partial}{\partial X} = \epsilon^2\frac{\partial}{\partial \gamma},\ \frac{\partial}{\partial Z} = \epsilon\frac{\partial}{\partial \lambda},\ \frac{\partial}{\partial T} = -\epsilon V\frac{\partial}{\partial \lambda}+\epsilon^3\frac{\partial}{\partial \mu}$$

Dependent variables should be enlarged as follows:

$$N_r = 1+\epsilon^2 N_{r1}+\epsilon^4 N_{r2}+\ldots, \tag{8.35}$$

$$U_{rX,rY} = \epsilon^3 U_{rX_1,rY_1}+\epsilon^4 U_{rX_2,rY_2}+\ldots, \tag{8.36}$$

$$U_{rZ} = \epsilon^2 U_{rZ_1}+\epsilon^4 U_{rZ_2}+\ldots, \tag{8.37}$$

$$\Phi = \epsilon^2\Phi_1 + \epsilon^4\Phi_2 + \ldots, \tag{8.38}$$

We obtain a set of equations by substituting these expressions (8.34)–(8.38) into the equations (8.27)–(8.33) and taking various powers of ϵ into consideration. Using back substitutions to express all dependent variables in terms of Φ_1, we obtain the KP equation

$$\frac{\partial}{\partial\lambda}\left[\frac{\partial\Phi_1}{\partial\mu} + A\Phi_1\frac{\partial\Phi_1}{\partial\lambda} + B\frac{\partial^3\Phi_1}{\partial\lambda^3}\right] - C\frac{\partial^2\Phi_1}{\partial\gamma^2} = 0 \tag{8.39}$$

where $A = \frac{V^2}{2}\left\{\frac{3}{V^3} - \frac{V}{2}\left(\frac{\alpha_c^2 F}{2} + b_1\right)\right\}$, $B = \frac{V^3}{2}$, $C = -\frac{V}{2}$ and $V^2 = \frac{1}{\alpha_c F + a_1}$ (phase velocity).

8.10.2 KP Equation in Standard Form

We turn the resulting KP equation (8.39) into the well-known form of the KP equation by applying an appropriate transformation. Let's look at λ by $\lambda_1 A^{m_1} B^{n_1} C^{p_1}$, Φ_1 by

$$6\varphi_1 A^{m_2} B^{n_2} C^{p_2},\ \gamma \text{ by } \frac{1}{\sqrt{v}}\gamma_1 A^{m_3} B^{n_3} C^{p_3}, \text{ and } \mu \text{ by } \mu_1.$$

This gives us now

$$\frac{\partial}{\partial\lambda_1}\left[\frac{\partial\varphi_1}{\partial\mu_1} + 6\varphi_1\frac{\partial\varphi_1}{\partial\lambda_1} + \frac{\partial^3\varphi_1}{\partial\lambda_1^3}\right] - v\frac{\partial^2\varphi_1}{\partial\gamma_1^2} = 0 \tag{8.40}$$

8.10.3 Lump Soliton of the KP Equation

The lump solutions to the KP equation will be shown here. Equation (8.40) is referred to as the KP-I equation if $v = 1$, and the KP-II equation if $v = -1$. We now consider the transition

$$\varphi_1(\lambda_1,\gamma_1,\mu_1) = 2\,\frac{\partial^2}{\partial\lambda_1^2}\left(\log\left[G(\lambda_1,\gamma_1,\mu_1)\right]\right) \Rightarrow 2\left(\frac{G_{\lambda_1\lambda_1}G - G_{\lambda_1}^2}{G^2}\right) \tag{8.41}$$

However, as equation (8.41) results from a correct guess regarding the transformation of the dependent variables when using the Hirota bilinear approach, an explanation of this expression is not available. The Hirota bilinear method requires the setting of an appropriate transformation in order to convert non-linear partial differential equations that are in quadratic form in dependent variables into bilinear form. Setting the transformation to obtain the soliton solution of a non-linear evolution equation using the Hirota bilinear technique is the most difficult step. By forming the auxiliary function independently, it would be possible to use HBM to find the lump soliton solution to the derived KP equation (8.39) instead of the standard KP equation (8.40). It transforms into the bilinear equation shown below:

$$(G_{\lambda_1\mu_1}G-G_{\lambda_1}G_{\mu_1})-\nu\left(G_{\gamma_1\gamma_1}G-G_{\gamma_1}^2\right)+\left(G_{\lambda_1\lambda_1\lambda_1\lambda_1}G-4G_{\lambda_1\lambda_1\lambda_1}G_{\lambda_1}+3G_{\lambda_1\lambda_1}^2\right)=0 \quad (8.42)$$

The Hirota bilinear form of (42) is

$$\left(D_{\lambda_1}D_{\mu_1}+D_{\lambda_1}^4\right)(G.G)-\nu D_{\gamma_1}^2(G.G)=0;\ \nu=\pm1 \quad (8.43)$$

Now, we take $\nu=1$, so the quadratic form of the KP-I equation will be

$$(G_{\lambda_1\mu_1}G-G_{\lambda_1}G_{\mu_1})-\left(G_{\gamma_1\gamma_1}G-G_{\gamma_1}^2\right)+\left(G_{\lambda_1\lambda_1\lambda_1\lambda_1}G-4G_{\lambda_1\lambda_1\lambda_1}G_{\lambda_1}+3G_{\lambda_1\lambda_1}^2\right)=0 \quad (8.44)$$

To construct lump solutions, let's start with

$$G=P^2+Q^2+l_9;P=l_1\lambda_1+l_2\gamma_1+l_3\mu_1+l_4,\ Q=l_5\lambda_1+l_6\gamma_1+l_7\mu_1+l_8 \quad (8.45)$$

Next, we will evaluate the values of the real parameters l_r, $1\le r\le 9$. By substituting the value of G, which we received from (8.45) and its partial derivatives into (8.44), we derive the expression of l_3,l_7 and l_9, taking l_1,l_2,l_4,l_5, l_6, l_8 as free variables.

$$\left\{l_3=\frac{l_1l_2^2-l_1l_6^2+2l_2l_5l_6}{l_1^2+l_5^2},\ l_7=\frac{l_5l_6^2-l_5l_2^2+2l_1l_2l_6}{l_1^2+l_5^2},\ l_9=\frac{3\left(l_1^2+l_5^2\right)^3}{(l_1l_6-l_2l_5)^2}\right\} \quad (8.46)$$

In addition, the restriction

$$l_1l_6-l_2l_5=\begin{vmatrix} l_1 & l_2 \\ l_5 & l_6 \end{vmatrix}\neq 0 \quad (8.47)$$

should be satisfied.

This set (8.45) produces a family of lump soliton to the KP-I equation (8.44), given by

$$\varphi_1(\lambda_1,\gamma_1,\mu_1)=\frac{4\left(l_1^2+l_5^2\right)G-8(l_1P+l_5Q)^2}{G^2} \quad (8.48)$$

Where

$$G=\left[l_1\lambda_1+l_2\gamma_1+\frac{l_1l_2^2-l_1l_6^2+2l_2l_5l_6}{l_1^2+l_5^2}\mu_1+l_4\right]^2$$
$$+\left[l_5\lambda_1+l_6\gamma_1+\frac{l_5l_6^2-l_5l_2^2+2l_1l_2l_6}{l_1^2+l_5^2}\mu_1+l_8\right]^2+\frac{3\left(l_1^2+l_5^2\right)^3}{(l_1l_6-l_2l_5)^2},$$

$$P=l_1\lambda_1+l_2\gamma_1+\frac{l_1l_2^2-l_1l_6^2+2l_2l_5l_6}{l_1^2+l_5^2}\mu_1+l_4,\ Q=l_5\lambda_1+l_6\gamma_1+\frac{l_5l_6^2-l_5l_2^2+2l_1l_2l_6}{l_1^2+l_5^2}\mu_1+l_8. \quad (8.49)$$

All six of the relevant parameters in l_1,l_2,l_4,l_5,l_6, and l_8 are arbitrary in this class of lump solutions, as long as the solutions are well-defined, that is, as long as equation (8.47) is satisfied. The exact meaning of the determinant requirement is that two directions in the λ_1-γ_1 plane that are not parallel are (l_1, l_2), and (l_5, l_6). Consequently, the family of lump soliton to the KP-I equation (8.39) can be obtained by reversing the coefficient. Hence, the expression of solution will be

$$\Phi_1(\lambda,\gamma,\mu) = \left(\frac{24B^{\frac{1}{3}}}{A}\right)\left[\frac{\left(l_1^2+l_5^2\right)G - 2\left(l_1P + l_5Q\right)^2}{G^2}\right]; \tag{8.50}$$

$$P = \frac{1}{B^{\frac{1}{3}}C^{\frac{1}{2}}}\left[C^{\frac{1}{2}}l_1\lambda + B^{\frac{1}{6}}l_2\gamma\right] + \frac{1}{B^{\frac{1}{3}}C^{\frac{1}{2}}}\left[B^{\frac{1}{3}}C^{\frac{1}{2}}l_3\mu + B^{\frac{1}{3}}C^{\frac{1}{2}}l_4\right],$$

and

$$Q = \frac{1}{B^{\frac{1}{3}}C^{\frac{1}{2}}}\left[C^{\frac{1}{2}}l_5\lambda + B^{\frac{1}{6}}l_6\gamma\right] + \frac{1}{B^{\frac{1}{3}}C^{\frac{1}{2}}}\left[B^{\frac{1}{3}}C^{\frac{1}{2}}l_7\mu + B^{\frac{1}{3}}C^{\frac{1}{2}}l_8\right],$$

Where

$$\left\{ l_3 = \frac{l_1l_2^2 - l_1l_6^2 + 2l_2l_5l_6}{l_1^2+l_5^2},\; l_7 = \frac{l_5l_6^2 - l_5l_2^2 + 2l_1l_2l_6}{l_1^2+l_5^2},\; l_9 = \frac{3\left(l_1^2+l_5^2\right)^3}{\left(l_1l_6 - l_2l_5\right)^2} \right\}$$

If we choose the parameters as: $l_1 = 1, l_2 = c, l_4 = 0, l_5 = 0, l_6 = d, l_8 = 0$, then the expression of the lump soliton is as follows

$$\Phi_1(\lambda,\gamma,\mu) = \left(\frac{24B^{\frac{1}{3}}}{A}\right)\frac{S_1}{S_2} \tag{8.51}$$

$$S_1 = -\left[\frac{\lambda}{B^{\frac{1}{3}}} + \frac{c\gamma}{B^{\frac{1}{6}}C^{\frac{1}{2}}} + \left(c^2-d^2\right)\mu\right]^2 + d^2\left[\frac{\gamma}{B^{\frac{1}{6}}C^{\frac{1}{2}}} + 2c\mu\right]^2 + \frac{3}{d^2};$$

$$\text{and } S_2 = \left[\left[\frac{\lambda}{B^{\frac{1}{3}}} + \frac{c\gamma}{B^{\frac{1}{6}}C^{\frac{1}{2}}} + \left(c^2-d^2\right)\mu\right]^2 + d^2\left[\frac{\gamma}{B^{\frac{1}{6}}C^{\frac{1}{2}}} + 2c\mu\right]^2 + \frac{3}{d^2}\right]^2.$$

8.10.4 Result and Discussion

We have obtained the KP equation from the hydrodynamical governing equations in the auroral plasma by using the extended RPT. The generated KP equation from the basic governing equations has been inserted into the standard KP equation using the appropriate transformation. By applying the Hirota bilinear technique, lump soliton solutions have been obtained. By acquiring alternative sets of actual parameters, one can produce different lump soliton solutions. We now display our results for several fundamental variables of the system and examine the impact of those parameters on

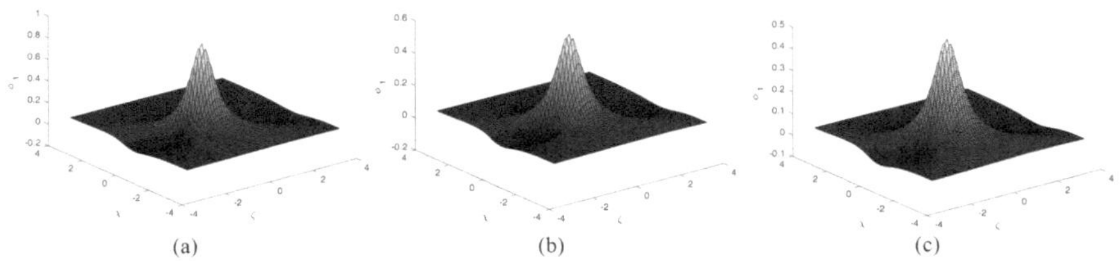

FIGURE 8.3 Lump structures for the parameter values $F = 0.1$; $N_c = 0.2\ \text{cm}^{-3}$; $N_h = 1.8\ \text{cm}^{-3}$; $T_c = 1$ eV; $T_h = 26$ eV; $T_{\text{eff}} = 7$ eV; $\kappa = 2.0$; $\kappa = 2.5$; $\kappa = 3.0$ at time $\mu = 0$ respectively.

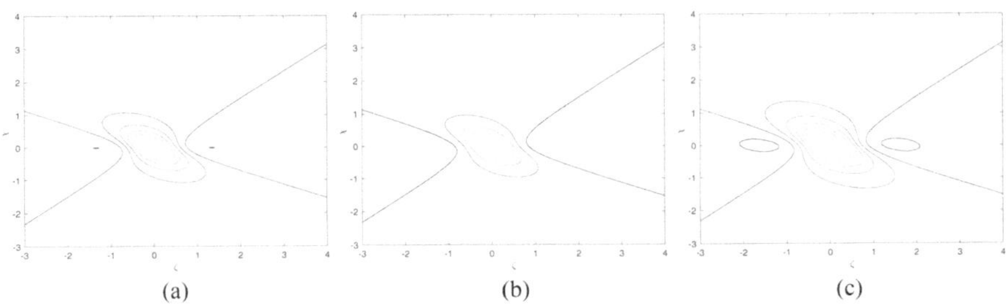

FIGURE 8.4 Contour profiles of lump solitons for the parameter values $F = 0.1$; $N_c = 0.2\ \text{cm}^{-3}$; $N_h = 1.8\ \text{cm}^{-3}$; $T_c = 1$ eV; $T_h = 26$ eV; $T_{\text{eff}} = 7$ eV; $\kappa = 2.0$; $\kappa = 2.5$; $\kappa = 3.0$ at time $\mu = 0$ respectively.

the results. We are plotting a single lump soliton solution and illustrating its associated properties. We can observe the lump soliton features for a variety of parameter values in Figure 8.3. Here, we have plotted the electrostatic potential profile at a given time for various values of the system plasma parameter while maintaining the same values for the other parameters. The solution features of the phase diagram are shown in Figure 8.4, Figure 8.3 shows the lump soliton profile. The images shown in Figure 8.3 demonstrate how the lump solitons' amplitude gradually reduces as the spectral index increases.

Figure 8.4 shows the closeness and conservative regions of the system for the same parameter regime. To facilitate the analysis of the lump soliton structures, we map them in a single dimension. Plots in Figure 8.5 illustrate how the behaviour of solitons varies in one-dimension when $\kappa = 3$. The solid line, dashed line, and dotted line represent the three frames where the soliton structure varies. In Figure 8.5a, where $\lambda = 1$ (dotted line), $\lambda = 1.4$ (dashed line), and $\lambda = 2$ (solid line), we show the soliton structure is shown in reference to the coordinate axis γ. Figure 8.5b displays the identical behaviour for three distinct frames: $\gamma = 1$ (dotted line), $\gamma = 1.4$ (dashed line), and $\gamma = 1.6$ (solid line) and the coordinate axis λ is in line with the soliton structures. The distribution still exhibits dispatch from thermal equilibrium when it is in region $\kappa \approx 2-3$, but it behaves more like a Maxwellian distribution when $\kappa \to 3$. A change

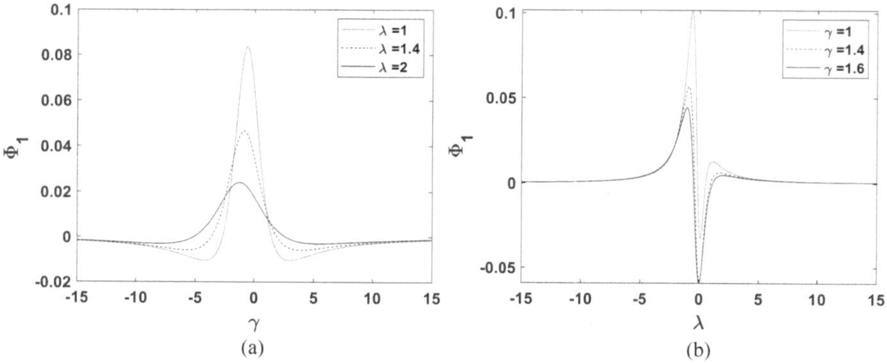

FIGURE 8.5 2-D profiles of lump solitons for the parameter values $F = 0.1$; $N_c = 0.2\ \text{cm}^{-3}$; $N_h = 1.8\ \text{cm}^{-3}$; $T_c = 1$ eV; $T_h = 26$ eV; $T_{\text{eff}} = 7$ eV; $\kappa = 3.0$ at time $\mu = 0$. (a) represents Φ_1 vs γ diagram and (b) is Φ_1 vs λ diagram.

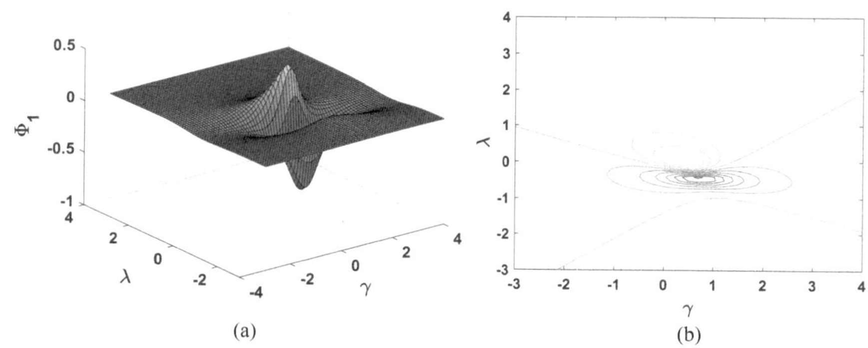

FIGURE 8.6 (a) represents the 3-D structure of lump soliton and (b) is its contour plot for the parameter values $F = 0.1$; $N_c = 0.2\ \text{cm}^{-3}$; $N_h = 1.8\ \text{cm}^{-3}$; $T_c = 1$ eV; $T_h = 26$ eV; $T_{\text{eff}} = 7$ eV; $\kappa = 2.6$ at time $\mu = 0.3$.

in the lump structures is observed when we depart from the original time by a small amount. This variance is shown in Figure 8.6a. Plots in Figures 8.6 and 8.7 are for the spectral index $\kappa = 2.6$ and time $\mu = 0.3$. The contour plot in Figure 8.6a is shown in Figure 8.6b. The potential profile (Φ_1) is plotted in relation to the coordinate axes γ and λ in Figure 8.7a and b, for $\kappa = 2.6$, at time $\mu = 0.3$. At time $\mu = 0.3$, the dotted line represents $\lambda = 0.2$, the dashed line represents $\lambda = 0.4$, and the solid line represents $\lambda = 0.6$ in Figure 8.7a. Figure 8.7b shows the dotted line, dashed line, and solid line, for $\gamma = 1$, $\gamma = 1.2$, $\gamma = 1.4$, respectively. In Figure 8.7a compared to Figure 8.5a, the height of the peak increases threefold, and the depth of the valley doubles if we decrease the value of λ and slightly increase the duration. When comparing Figure 8.5b and 8.7b, we can see that the left peak's height falls up to seven times and turns negative, while the right peak's height increases up to four times and turns positive as time increases.

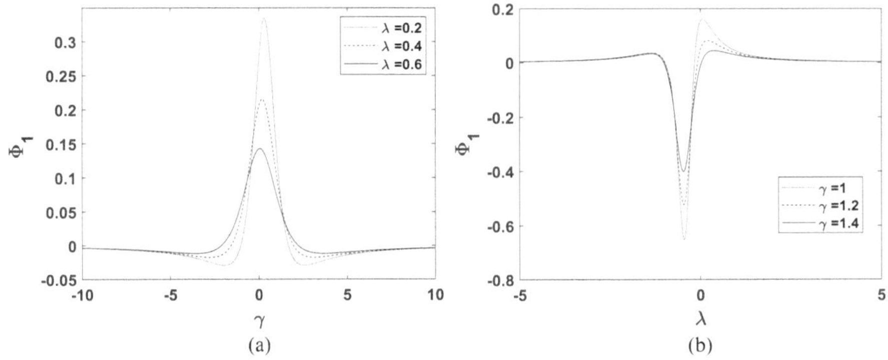

FIGURE 8.7 2-D profiles of lump soliton. (a) is Φ_1 vs γ diagram and (b) Φ_1 vs λ diagram for the parameter values $F = 0.1$; $N_c = 0.2\ \text{cm}^{-3}$; $N_h = 1.8\ \text{cm}^{-3}$; $T_c = 1$ eV; $T_h = 26$ eV; $T_{\text{eff}} = 7$ eV; $\kappa = 2.6$ at time $\mu = 0.3$.

8.11 CONCLUSION

A detailed explanation of the Hirota bilinear method is contained in this chapter. Basically, it is very difficult to solve a non-linear partial differential equation and investigate its physical phenomena. So, we have studied such a prominent method, i.e., the Hirota method, to get various non-linear features. The motivation for this study is explained first in the introduction. Then, the story behind the Hirota method and the discovery of the Hirota derivative are explained sequentially. Depending on the expansion in the perturbation step of the Hirota method, we get various non-linear wave structures, which are discussed here. The conditions under which multi-soliton solutions occur are also described in this chapter. Some beautiful non-linear structures, viz., multi-soliton, multiple singular solitons, and lump soliton, are observed in various plasma fields. Considering a fundamental hydrodynamical model, the interaction of 2-soliton solutions of the KdV and the interaction of singular 2-soliton solutions of the mKdV equations have been shown here consecutively. Lastly, the lump profiles of the KP equation in magnetized auroral plasma have been studied here.

ACKNOWLEDGMENT

Dr. Uday Narayan Ghosh and Prof. Prasanta Chatterjee gratefully acknowledge the CSIR-HRDG project grant 25/0320/23/EMR-II, letter number 03/1479/23/EMR-II, dated 10.08.2023. Ms. Snehalata Nasipuri is thankful to the Council of Scientific and Industrial Research (CSIR), India, for providing financial support under the Senior Research Fellowship (SRF) program (File No: 09/202(0120)/2021-EMR-I).

REFERENCES

[1] Clifford, S., Gardner, C. S., Greene, J. M., Kruskal, M. D., & Miura, R. M. (1967). Method for solving the Korteweg-de Vries equation. *Physical Review Letters*, 19 (19), 1095.

[2] Hirota, R. (1973). Exact N-soliton solutions of the wave equation of long waves in shallow water and in nonlinear lattices. *Journal of Mathematical Physics*, 14 (7), 810–814.

[3] Hirota, R., & Ito, M. (1983). Resonance of solitons in one dimension. *Journal of the Physical Society of Japan*, 52 (3), 744–748.

[4] Hirota, R., & Satsuma, J. (1981). Soliton solutions of a coupled Korteweg-de Vries equation. *Physics Letters A*, 85 (8–9), 407–408.

[5] Wazwaz, A. M. (2008). Solitons and singular solitons for the Gardner–KP equation. *Applied Mathematics and Computation*, 204 (1), 162–169.

[6] Ghosh, U. N. (2022). Electron acoustic singular solitons interaction in the Earth's magnetotail region. *Advances in Space Research*, 70 (7), 2070–2080.

[7] Nakamura, A. (1982). Simple multiple explode-decay mode solutions of a two-dimensional nonlinear Schrodinger equation. *Journal of Mathematical Physics*, 23 (3), 417–420.

[8] Drazin, P. G., & Johnson, R. S. (1989). *Solitons: An Introduction*. Cambridge University Press, p. 2.

[9] Davidson, R. C. (1972). *Methods in Nonlinear Plasma Theory*. Academic.

[10] Ghosh, U. N., Nasipuri, S., & Chatterjee, P. (2022). Study of lump soliton structures in quantum electron-positron-ion magnetoplasma. *Contribution to Plasma Physics*, 62 (8), e202200094.

9 Numerical Simulation of Indian Ocean Dipole and Its Impact on the Bay of Bengal

Saikat Pramanik and Sourav Sil

9.1 INTRODUCTION

The ocean currents impact the climate by transporting warm and cold water between the polar and tropical regions. The global oceanic circulation pattern makes significant contributions to the energy, heat, and salt balances. Understanding and monitoring the ocean circulation will help in several economic activities, including transportation, fishing, cyclone and weather forecasting, coastal and urban maintenance, civil management, and many more. However, the available satellite and sparse in-situ observations are limited in their ability to allow a thorough examination and understanding of the surface and subsurface oceanic processes and associated variability. To overcome this limitation, developing a multiscale ocean state hindcast system or ocean model simulation is necessary.

The Indian Ocean (IO) is different from the other two major oceans, the Atlantic Ocean and the Pacific Ocean, due to its geographical location as its northern boundaries are closed by the Asian landmass and the associated wind system. Southeast trade winds are relatively steady over the southern IO (south of 10°S). However, the wind over the northern IO reverses twice every year. During June–September, winds blow from the southwest, known as summer monsoonal winds, and during December–February, they blow from the northeast, known as winter monsoonal winds. During the transition months, March–May and October–November, winds are very weak. The seasonally reversing monsoonal winds create a distinct circulation in the northern IO. During fall (October–November), the southward-propagating East India Coastal Current (EICC) along the east coast of India transports fresh water from the Ganges Brahmaputra delta toward the south. The winter monsoon current (WMC) during December–February propagates from east to west along 8–10°N. One part of this current propagates northward along the West coast of India as a West India Coastal Current (WICC), while another part moves southward as the Somalia Current (SC) after hitting the Somalia coast [1]. The Southwest Monsoon Current (SMC) during June–September flows eastward at around 6°N and carries nutrient-rich water from the southwest coast of India to the Bay of Bengal (BoB) region [2–4].

DOI: 10.1201/9781003638414-9

The equatorial and coastal Kelvin waves and reflected Rossby waves indirectly affect the oceanic processes [5,6]. The interannual climatic event Indian Ocean Dipole (IOD, [7]) modulates the oceanic processes remotely through circulation. The IOD comprises an aperiodic oscillation of sea surface temperatures (SST) and sea surface pressure (SSP) between its negative and positive phases. A positive phase of the IOD (positive IOD, denoted as PIOD) results in lower (higher) SSP, more (less) precipitation, and a positive (negative) SST anomaly in the western (eastern) IO. The reverse phenomenon occurs during negative IOD events (NIOD) [5].

In the current study, we develop a multiscale ocean model with interannual surface and boundary forcing and river influxes in the form of point sources along the Indian coastline in the IO region. The rest of the article is structured as follows, an explanation of the governing model equations and observational datasets is given in Section 9.2. Section 9.3 covers the seasonal variability of SST, SSHA, and surface currents from the model simulation and available observations. It also describes the surface and subsurface oceanic variability during IOD events. Finally, Section 9.4 presents the conclusions and future directions.

9.2 MODEL CONFIGURATION AND DATA

We have chosen the Regional Ocean Modeling System (ROMS) as it is a primitive-equations-based, sigma-coordinate model that is extensively utilized by the research community for various applications such as physical [5], biophysical [3], biogeochemical [8], bio-optical [9], tides [10], sediment transport [11], sea ice [12], etc. The details of the computational algorithm for the ROMS model are provided in Shchepetkin and McWilliams [13]. The governing momentum equations are solved using the leapfrog scheme in the computational environment (https://www.myroms.org/). The used variables are shown in Table 9.1.

The momentum equations in the x- and y-directions are

$$\frac{\partial u}{\partial t} + \vec{v}.\nabla u - fv = -\frac{\partial \varnothing}{\partial x} - \frac{\partial}{\partial z}\left(\overline{u'w'} - v\frac{\partial u}{\partial z}\right) + F_u + D_u \tag{9.1}$$

TABLE 9.1
List of the Used Variables

D_u, D_v, D_C	Diffusive terms
F_u, F_v, F_C	Forcing terms
v_θ, v	Diffusivity and molecular viscosity
x, y, z, t	Zonal, meridional, vertical, time coordinates
$f(x, y)$	Coriolis parameter
u, v, w	Velocity ($\vec{v}$) components
g	Acceleration of gravity
P	Total pressure $P \approx -\rho_0 gz$
$\varnothing$	Dynamic pressure
$\rho_0 + \rho$	Total in-situ density

$$\frac{\partial v}{\partial t}+\vec{v}.\nabla v+fu=-\frac{\partial \varnothing}{\partial y}-\frac{\partial}{\partial z}\left(\overline{v'w'}-v\frac{\partial v}{\partial z}\right)+F_v+D_v \tag{9.2}$$

A scalar field $C(x,y,z,t)$, such as temperature, salinity, chlorophyll, etc., is evaluated by the advective-diffusive equation

$$\frac{\partial C}{\partial t}+\vec{v}.\nabla C=-\frac{\partial}{\partial z}\left(\overline{C'w'}-v_\theta\frac{\partial C}{\partial z}\right)+F_C+D_C \tag{9.3}$$

An equation of state is given by

$$\rho=\rho(T,S,P) \tag{9.4}$$

The density variation is neglected in the horizontal momentum equations, known as the Boussinesq approximation. The hydrostatic approximation is used in the model, which balances the vertical pressure gradient and the buoyancy force,

$$\frac{\partial \varnothing}{\partial z}=-\frac{\rho g}{\rho_0} \tag{9.5}$$

The fluid is considered incompressible, and the mass conservation equation

$$\frac{\partial u}{\partial x}+\frac{\partial v}{\partial y}+\frac{\partial w}{\partial z}=0 \tag{9.6}$$

Here, the surface net heat is computed from SST, sea-level-specific humidity, sea surface air temperature, air density, and wind speed. The surface forcings given in Table 9.2, are used from the TropFlux datasets [14] during 1988–2015.

TABLE 9.2
List of Parameters in the Forcing Fields

Forcing Types	Parameters
Winds forcing	Zonal winds
	Meridional winds
Heat fluxes	Shortwave radiation
	Longwave radiation
	Latent heat flux
	Sensible heat flux
	Sea surface temperature
	Sea-level air temperature
	Specific humidity
	Air density
Freshwater flux	Sea surface salinity
	Evaporation minus precipitation
	River influxes

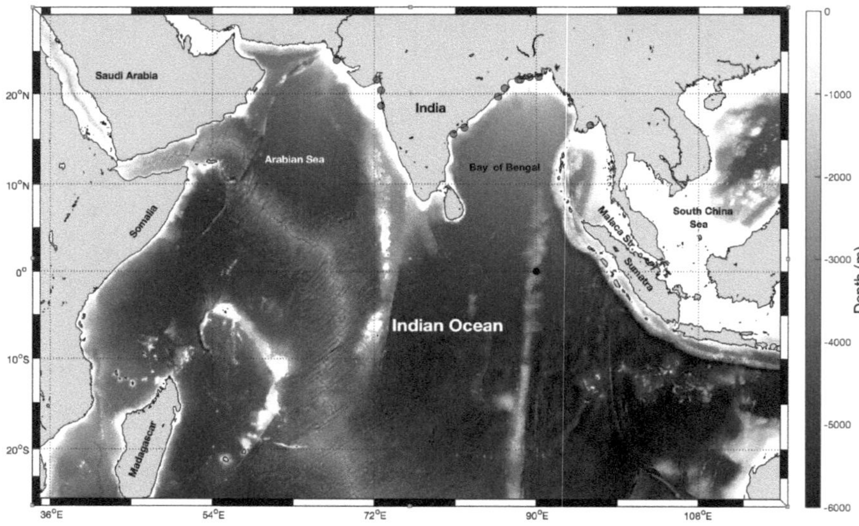

FIGURE 9.1 The model domain with topography (m, shaded) from ETopo2 datasets. Red points on the coastline indicate the river mouths along the Indian subcontinent. Equatorial Research Moored Array for African-Asian-Australian Monsoon Analysis and Prediction (RAMA) location at 90°E has been denoted by a black dot, where the zonal current is validated.

TropFlux data has been able to produce seasonal and intraseasonal variabilities better than other available flux datasets over the northern IO region [14–16]. The governing model equations are solved using open oceanic boundary conditions on the eastern and southern sides, with initial conditions from Simple Ocean Data Assimilation (SODA, [17]). These equations are also solved on a 25 × 25 km mesh grid with 42 sigma levels following the domain topography from the 2-min resolution Earth Topography (ETopo2) datasets over the IO region (34°E–118°E, 26°S–29°N, Figure 9.1). The vertical structure of the model grid along 12°N and the equator is shown in Figure 9.2a-b. Vertical levels are closer to each other in the shallow region, with 14 levels in the upper 100 m, where the depth is maximum. Additionally, the Arabian Sea is deeper than the Bay of Bengal along the 12°N channel (Figure 9.2a). The model configuration details are given by Pramanik et al. [5]; Pramanik and Sil [18].

For validation of the model results and analysis, we have used the World Ocean Atlas 2018 (WOA2018, Source: https://www.ncei.noaa.gov/access/world-ocean-atlas-2018/) monthly climatological temperature, Archiving Validation and Interpretation of Satellite Oceanographic (AVISO, Source: https://www.aviso.altimetry.fr/en/data/data-access.html) monthly SSHA fields and Ocean Surface Current Analysis Real-time (OSCAR, Source: https://podaac.jpl.nasa.gov/dataset/OSCAR_L4_OC_third-deg) 5-day current fields over the study domain.

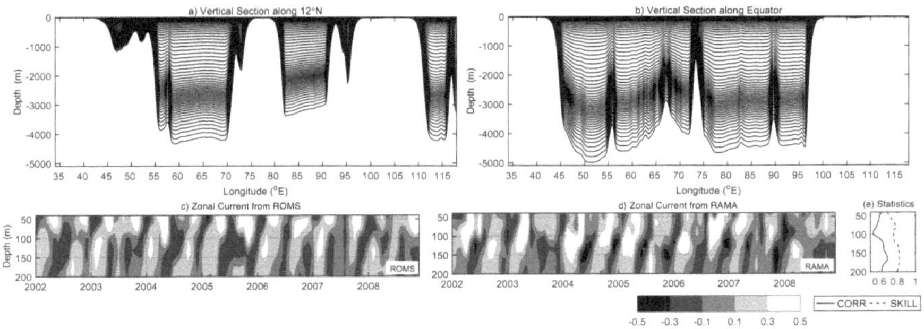

FIGURE 9.2 Longitudinal section of the domain topography along (a) 12°N and (b) equator. Time-depth variation of equatorial zonal currents (m/s) at 90°E (please see the location in Figure 9.1) from (c) ROMS simulation and (d) RAMA buoy. The light shade represents westward currents, and the deep shade represents eastward currents. (e) Depth-wise correlation and skill score between the ROMS and RAMA are shown in solid and dashed lines, respectively.

The correlation coefficient (R) and skill score (S) between the model and observation have been calculated for each depth

$$R = \frac{\sum (x_i - \bar{x})(y_i - \bar{y})}{\sqrt{\sum (x_i - \bar{x})^2 \sum (y_i - \bar{y})^2}} \tag{9.7}$$

$$S = \frac{4\,(1+R)}{\left(r + \frac{1}{r}\right)^2 (1+R_0)} \tag{9.8}$$

where x_i and y_i denote the sample points from the model and observation. r is the standard deviation ratio between the model and observation, with means $\bar{x}$ and $\bar{y}$, respectively. The composite of multiple PIOD years (1994, 1997, 2006, 2012, 2015) and NIOD years (1992, 1996, 1998, 2005, 2010) is used to show the comparative analysis of ocean temperature during different seasons [5].

9.3 RESULTS AND DISCUSSIONS

The results of the current article are discussed in the following sections. Section 9.3.1 provides a comparison of the zonal current with Research Moored Array for African-Asian-Australian Monsoon Analysis and Prediction (RAMA) observations and statistics. A seasonal comparison of ROMS-simulated SST with WOA2018 is shown in Section 9.3.2. Section 9.3.3 discusses the seasonality of SSHA with the surface current. The capability of ROMS to capture the surface temperature dipole is discussed in Section 9.3.4. Finally, the influence of IOD on the BoB SST and surface current is analyzed in Section 9.3.5.

9.3.1 Equatorial Zonal Current Variability

The ROMS-simulated time-depth structure of the equatorial zonal current at 90°E has been compared with the RAMA observations during 2002–2008 (Figure 9.2c–e). The figure shows westward currents during the summer and winter monsoons and eastward currents during the transition months (April–June and October–November) in the upper layers up to depths of 100 m (Figure 9.2c-d). The westward propagation during the monsoon period is mainly due to the easterly winds over the equator, while the eastward propagation during the transition period is attributed to the effect of the Wyrtki jet [19]. Along with semi-annual variations, a smaller scale oscillation in the upper 60 m has been observed in both ROMS and RAMA observations, particularly during 2006 and 2008, driven by the atmospheric intraseasonal oscillation [20]. Additionally, we have calculated the depth-wise correlation and the model's skill score. The correlation and skill scores are greater than 0.7 and 0.8 throughout the depth, except near 100 m depth (Figure 9.2e). Hence, the configured ROMS model captured the seasonality and interannual variability of equatorial currents accurately, with good statistics to further analyze the model results.

9.3.2 Seasonal SST Variability

In this section, we analyze the model's performance in capturing the seasonal SST variability over the Indian Ocean. We compare the ROMS simulated seasonal SST with the WOA2018 datasets during (i) Dec–Feb (winter), (ii) Mar–May (spring), (iii) Jun–Sep (summer), and (iv) Oct–Nov (fall). In the tropical IO, SST is higher than 28°C during all seasons of the year, while in the southern IO, SST is less than 25°C, especially during the summer and fall seasons (Figure 9.3a–d). The ROMS model accurately captures the spatial distribution of SST over the IO, except in some regions.

ROMS SST shows a cooling pattern (~1°C–1.5°C) over the northern Bay of Bengal region during winter (Figure 9.3a and e), the northern Arabian Sea region during winter and spring (Figure 9.3a–b and 9.3e–f), and western IO (near the Oman and Somalia coasts) during spring and fall (Figure 9.3c–d and 9.3g–h). There are

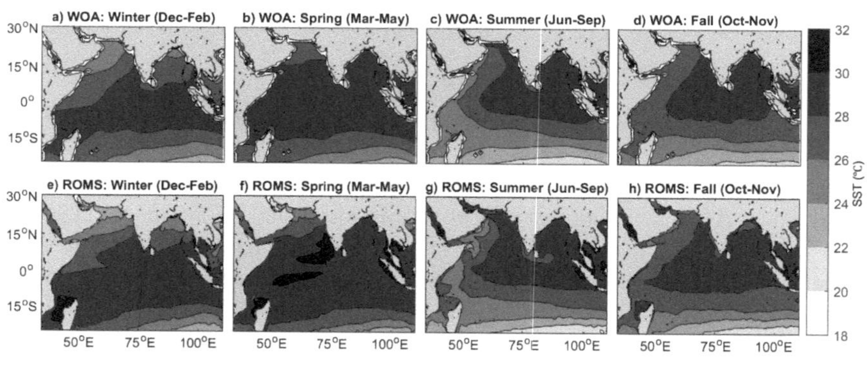

FIGURE 9.3 Comparison of ROMS simulated SST with WOA18 during Dec–Feb (winter), Mar–May (spring), Jun–Sep (summer), and Oct–Nov (fall) seasons.

overestimates (~0.2°C–0.6°C) of the SST pattern south of Somalia during winter (Figure 9.3a and e) and over the equatorial and north equatorial region (south of 12°N) during spring (Figure 9.3b and f).

9.3.3 Seasonal SSHA and Current Variability

We compared the ROMS simulated seasonal average SSHA with AVISO SSHA, and surface currents with OSCAR observations during 1993–2016 (Figure 9.4). The western IO is dominated by low SSHA, whereas the north-eastern IO and the BoB are mainly dominated by positive sea surface height anomalies. The current system in the IO is well captured by the ROMS model. During winter, the westward-propagating North Equatorial Current (NEC) along ~5°N, and South Equatorial Current (SEC) south of ~9°S are well captured by the ROMS (Figure 9.4a and e). This NEC during winter helps to generate the southward-propagating Somali current (SC).

The Equatorial Counter Current (ECC) persists almost every season and propagates eastward along the equator, which is also well captured by ROMS (Figure 9.4). Additionally, ROMS simulates the anticyclonic circulation induced by the positive SSHA and northward-propagating SC during summertime very well (Figure 9.3c and g). The BoB, which is mainly dominated by positive SSHA most of the time, particularly during winter and spring, is well captured by ROMS. The northward-propagating Western Boundary Current (WBC) during spring, the northeastward propagating SMC during summer, and the southward-propagating EICC during fall are well captured by ROMS in the BoB.

The eastward follows SMC bifurcates into two parts near southern Sri Lanka, one part propagates northeastward and enters the bay through the Sri Lanka Dome region, while the other part propagates eastward. Additionally, during this time, the

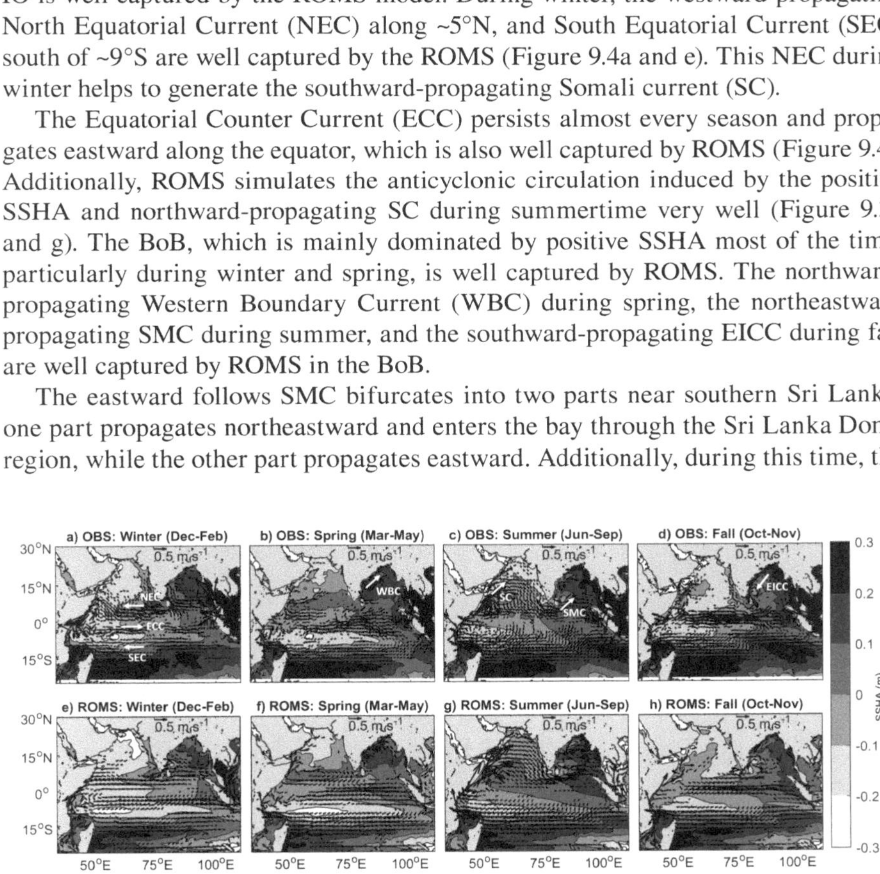

FIGURE 9.4 Spatial comparison of ROMS (lower panel) simulated surface currents (m/s, vectors) and SSHA (m, shaded) with observations (upper panel) during Dec–Feb (winter), Mar–May (spring), Jun–Sep (summer), and Oct–Nov (fall) seasons. Arrows represent the propagation of currents 1. North Equatorial Current (NEC), 2. South Equatorial Current (SEC), 3. Equatorial Counter Current (ECC), 4. Somalia Current (SC), 5. Western Boundary Current (WBC), 6. Southwest Monsoon Current (SMC), 7. East India Coastal Current (EICC), 8. Northward West India Coastal Current (WICC), 9. Winter Monsoon Current (WMC) and 10. Southward West India Coastal Current (WICC).

WICC propagates southward along the west coast of India and the EICC propagates northward along the east coast of India. Hence, the seasonal variation of SSHA, and surface currents is well captured by ROMS as compared with the observations to further analyze the thermal structure in the BoB. Different factors are influence the surface and subsurface processes. Local forcing, such as winds, thermal fluxes, and freshwater influxes through precipitation over the ocean and river influxes, directly affects the circulation and thermal structure of the water column at the surface.

9.3.4 Equatorial Temperature during IOD Events

This section discusses the surface and subsurface (100 m) temperature variabilities during positive and negative IOD events. We then discuss the vertical structure of temperature along the equatorial IO region. The seasonal variations of 100 m temperature anomalies with SST anomalies (in contours) during composite PIOD and NIOD events show that the IOD effects are well captured by the ROMS model (Figure 9.5). It also indicates that these effects are more significant in the subsurface

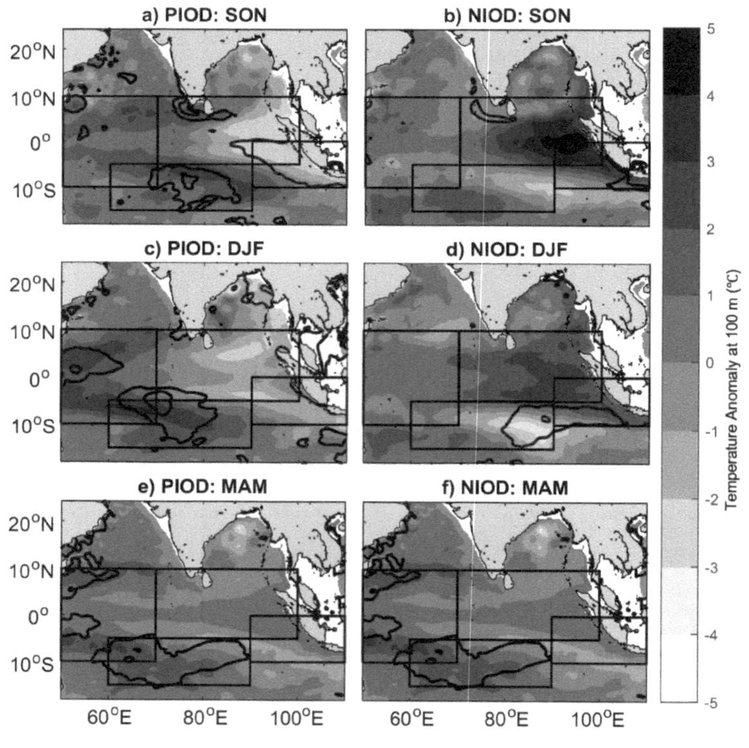

FIGURE 9.5 Spatial distribution of anomalous 100 m temperature (°C, shaded) and SST (°C, contours) from ROMS simulation during SON (Sep-Nov), DJF (Dec-Feb), and MAM (Mar-May) in positive and negative IOD composite years. Solid (dashed) contours represent the positive (negative) SST anomalies. Black boxes represent the eastern and western IO to calculate the dipole mode index, and green boxes represent the northern and southern IO to calculate the subsurface dipole index.

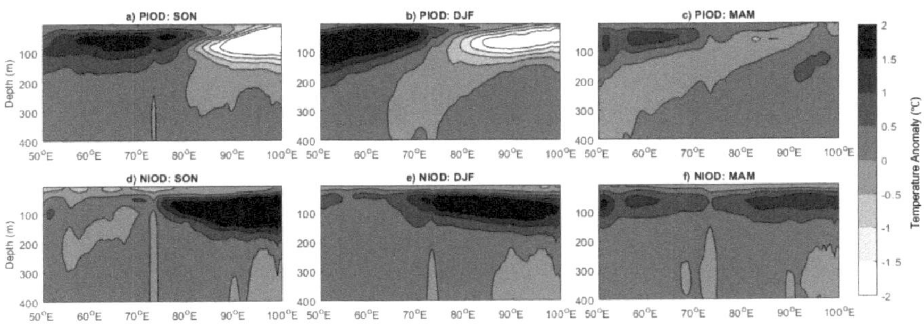

FIGURE 9.6 Vertical distribution of temperature anomalies (°C, shaded) over equatorial IO (averaged over 5°N to 5°S) from the IOM during SON (Sep-Nov), DJF (Dec-Feb), and MAM (Mar-May) in positive and negative IOD composite years.

than at the surface. The ROMS simulated SST displayed the well-known east-west temperature gradient over the equatorial IO. The subsurface north-south dipole [21] is also captured by ROMS during PIOD composites, but it is more prominent during NIOD composites.

The peak east-west temperature gradient in PIOD years is observed in SON (Sep–Nov), and it persists until DJF (Dec–Feb). In NIOD years, the peak north–south temperature gradient is observed in DJF and persists until MAM (Mar–May). Additionally, the SST variations (contours) suggest only the east-west temperature gradient during SON. The depth–longitude variation of subsurface temperature anomalies is investigated over the equator (average from 10°S to the equator) during three seasons (fall, winter, and spring) of IOD events (Figure 9.6). In PIOD events, strong upwelling along the Sumatra coast contributes to positive (negative) temperature anomalies over the western (eastern) equatorial IO [7].

These negative temperature anomalies in eastern IO are strong in SON, spread in DJF, and become weaker in MAM. It is important to note that, the IOD effects extend up to 200 m depths. During this time, strong anomalous easterlies are observed over the equatorial IO, which significantly impacts the upwelling in the eastern IO. Conversely, during NIOD years, easterly winds are weak, and sometimes, westerlies replace them. As a result, downwelling with positive temperature anomalies occurs in the eastern IO and the opposite phenomenon occurs in the western IO with negative temperature anomalies. Due to the weakening of easterlies, the second upwelling Kelvin wave in the equatorial IO is weak, and the second downwelling Kelvin waves started to propagate much earlier than in normal years [22]. This may be the possible reason for the negative temperature anomalies during NIOD composite years. These analyses confirm the capability of the ROMS model to simulate the IOD events.

9.3.5 Bay of Bengal SST and Surface Currents during IOD Events

The seasonal variations of SST anomalies (shaded) over the BoB during IOD composite events are shown in Figure 9.7. Anomalous negative (positive) surface temperatures during PIOD events and anomalous positive (negative) surface temperatures

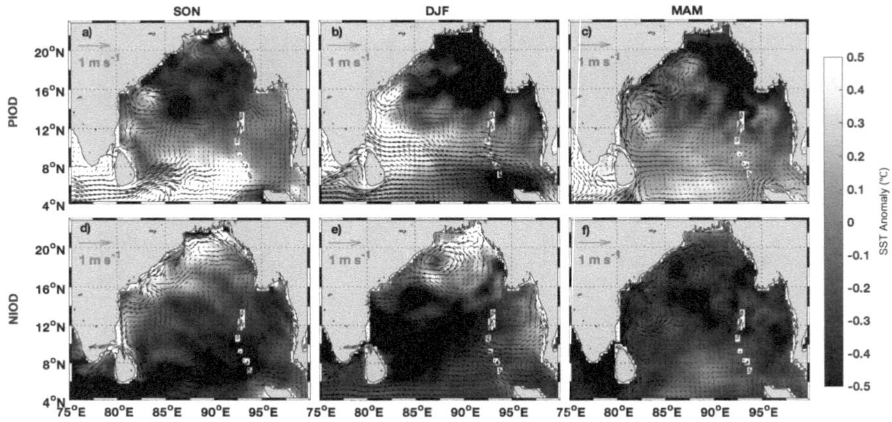

FIGURE 9.7 Model simulated currents (vectors, m/s) with sea surface temperature (SST, °C) anomaly (shaded) during SON (Sep-Nov), DJF (Dec-Feb), and MAM (Mar-May) in positive and negative IOD composite years.

during NIOD events are observed in the northern (southwestern) BoB (Figure 9.7a–c). Conversely, in NIOD composites, positive (negative) SST anomalies are observed in the northern (southwestern) BoB (Figure 9.7d–f). This temperature gradient in the BoB, similar to the subsurface dipole in the IO, begins to form during SON, reaches its maximum in DJF, and persists until MAM [21].

The reversal boundary currents on the western side of the BoB, the northward-propagating springtime WBC and the southward-propagating autumn time EICC, are well simulated by the ROMS model. In the fall (SON), a stronger EICC is observed during NIOD years, but the EICC is weak during PIOD years (Figure 9.7a and c). In addition to these comparisons, the observational studies mentioned earlier support the results obtained from the model simulation. The North Equatorial Counter Current (NECC, [23]) becomes weaker during the fall each year except during NIOD. The NECC is stronger in NIOD years, and reaches up to the northern tip of Sumatra (Figure 9.7d). In NIOD years, strong anomalous westerlies over the BoB make the NECC stronger during this event. However, easterly winds over the BoB during PIOD and EN composite years make the NECC very weak at this time [22].

During winter, a well-defined north equatorial current propagates westward from the Malacca Strait, and it does not vary much according to different events (Figure 9.7b and e). In NIOD composites, it propagates northwestward due to the influence of the cyclonic eddy at the northern tip of Sri Lanka (Figure 9.7e), but in PIOD composites, it reflects off the east coast of Sri Lanka at around 6°N and propagates along the west coast of the basin (Figure 9.7b). During spring, two clockwise eddies in the western BoB and the northward-propagating Western Boundary Current (WBC) are present in all events (Figure 9.7 and f). Among these two eddies, the northwestern eddies during PIOD composites are stronger compared to NIOD composites.

9.4 CONCLUSIONS

The ocean circulation in the Indian Ocean is unique due to the seasonal reversal of winds and events like the Indian Ocean Dipole. This article presents the configuration of the regional ocean model designed to simulate the Indian Ocean Dipole and associated processes at both the surface and subsurface. The analysis is carried out with special emphasis on the Bay of Bengal region. The work provides a state-of-the-art modeling framework and multiscale validations of the regional model.

The conclusions of the current study are as follows:

1. The validations with the seasonal surface (WOA) and interannual subsurface (RAMA) data show that the present modeling framework can capture the seasonal and interannual ocean features. The analysis indicates the semi-annual variation in the upper 100 m, driven by the semi- annual Wyrtki jets in the equatorial IO.
2. The analysis reveals that ROMS can capture the seasonal variability of upper oceanic parameters like SST, SSHA, and surface currents very well as compared with WOA2018 climatological temperature, satellite altimetry sea surface height, and OSCAR current observations. The configured ROMS model realistically captures known surface currents such as the western boundary current, east India coastal current, southwest monsoon current over the BoB, and north and south equatorial currents along with the equatorial counter current over the equatorial IO.
3. The ROMS model in the IO produces surface and subsurface temperature variability during IOD events. The study analyzed the conventional east-west temperature gradient over the equatorial IO along with a newly identified north–south dipole in the subsurface over the IO. The east–west temperature gradient during IOD events peaks in SON and persists up to DJF, whereas the north-south temperature gradient peaks in DJF and continues up to MAM. Interestingly, the north-south temperature gradient is more prominent during NIOD composite years, while the east–west temperature gradient is weaker during these events. The vertical structure of temperature anomalies over the IO shows that IOD events extend up to 200 m in depth.
4. The study also analyses the surface temperature and current variations over the BoB region during both phases of the IOD using the ROMS model. During PIOD events, negative (positive) temperature anomalies are observed over the northern (western and southwestern) BoB. Conversely, during NIOD events, the BoB is dominated by negative temperature anomalies except in the western BoB region. The study indicates that the springtime WBC (fall time EICC) is stronger during PIOD (NIOD) events.

The impact of the IOD on the eddy in the BoB will be studied. In addition, the thermal inertia during the IOD influences the monsoon and cyclones in the BoB. Therefore, the present modeling framework will be used to quantify the role of the ocean in atmospheric variability.

ACKNOWLEDGMENTS

The authors thank the Science and Engineering Research Board (SERB), Government of India (Grant: CRG/2019/005842) for the financial support. The figures are produced using MATLAB. The authors thank the Editor and reviewers for valuable suggestions.

REFERENCES

[1] Shankar, D., Vinayachandran, P. N., & Unnikrishnan, A. S. (2002). The monsoon currents in the north Indian Ocean. *Progress in Oceanography*, *52*(1), 63–120. https://doi.org/10.1016/S0079-6611(02)00024-1

[2] Burns, J. M., Subrahmanyam, B., & Murty, V. S. N. (2017). On the dynamics of the Sri Lanka Dome in the Bay of Bengal. *Journal of Geophysical Research: Oceans*. https://doi.org/10.1002/2017JC012986

[3] Pramanik, S., Sil, S., Gangopadhyay, A., Singh, M. K., & Behera, N. (2020). Interannual variability of the Chlorophyll-a concentration over Sri Lankan Dome in the Bay of Bengal. *International Journal of Remote Sensing*, *41*(15), 5974–5991. https://doi.org/10.1080/01431161.2020.1727057

[4] Vinayachandran, P. N., Shankar, D., Vernekar, S., Sandeep, K. K., Amol, P., Neema, C. P., & Chatterjee, A. (2013). A summer monsoon pump to keep the Bay of Bengal salty. *Geophysical Research Letters*, *40*(9), 1777–1782. https://doi.org/10.1002/grl.50274

[5] Pramanik, S., Sil, S., Mandal, S., Dey, D., & Shee, A. (2019). Role of interannual equatorial forcing on the subsurface temperature dipole in the Bay of Bengal during IOD and ENSO events. *Ocean Dynamics*, *69*(11–12). https://doi.org/10.1007/s10236-019-01303-0

[6] Rao, R. R., Girish Kumar, M. S., Ravichandran, M., Rao, A. R., Gopalakrishna, V. V., & Thadathil, P. (2010). Interannual variability of Kelvin wave propagation in the wave guides of the equatorial Indian Ocean, the coastal Bay of Bengal and the southeastern Arabian Sea during 1993–2006. *Deep-Sea Research Part I: Oceanographic Research Papers*, *57*(1), 1–13. https://doi.org/10.1016/j.dsr.2009.10.008

[7] Saji, N. H., Goswami, B. N., Vinayachandran, P. N., & Yamagata, T. (1999). A dipole mode in the tropical Indian Ocean. *Nature*, *401*(6751), 360–363. https://doi.org/10.1038/43854

[8] Fennel, K. J., Wilkin, J., Levin, W., Moisan, J., O'Reilly, J., & Haidvogel, D. (2006). Nitrogen cycling in the Middle Atlantic Bight: Results from a three-dimensional model and implications for the North Atlantic nitrogen budget. *Global Biogeochemical Cycles*, *20*, GB3007. https://doi.org/10.1029/2005GB002456

[9] Bissett, W. P., Walsh, J. J., Dieterle, D. A., & Carder, K. L. (1999). Carbon cycling in the upper waters of the Sargasso Sea: I. Numerical simulation of differential carbon and nitrogen fluxes. *Deep-Sea Research Part I: Oceanographic Research Papers*. https://doi.org/10.1016/S0967-0637(98)00062-4

[10] Pramanik, S., Mandal, S., Shee, A., Halder, S., & Sil, S. (2019a). Tidal circulation studies using regional model in the Bay of Bengal. In *Proceedings of the Fourth International Conference in Ocean Engineering (ICOE2018)* (pp. 829–836). Springer, Singapore.

[11] Warner, J. C., Sherwood, C. R., Arango, H. G., & Signell, R. P. (2005). Performance of four turbulence closure models implemented using a generic length scale method. *Ocean Modelling*. https://doi.org/10.1016/j.ocemod.2003.12.003

[12] Budgell, W. P. (2005). Numerical simulation of ice-ocean variability in the Barents Sea region. *Ocean Dynamics*. https://doi.org/10.1007/s10236-005-0008-3

[13] Shchepetkin, A. F., & McWilliams, J. C. (2005). The regional oceanic modeling system (ROMS): A split-explicit, free-surface, topography-following-coordinate oceanic model. *Ocean Modelling*. https://doi.org/10.1016/j.ocemod.2004.08.002

[14] Kumar, B. P., Vialard, J., Lengaigne, M., Murty, V. S. N., & McPhaden, M. J. (2012). TropFlux: Air-sea fluxes for the global tropical oceans-description and evaluation. *Climate Dynamics*, *38*(7–8), 1521–1543. https://doi.org/10.1007/s00382-011-1115-0

[15] Dey, D., Sil, S., Jana, S., Pramanik, S., & Pandey, P. C. (2017). An assessment of TropFlux and NCEP air-sea fluxes on ROMS simulations over the Bay of Bengal region. *Dynamics of Atmospheres and Oceans*, *80*, 47–61. https://doi.org/10.1016/j.dynatmoce.2017.09.002

[16] Pramanik, S., & Sil, S. (2021). Assessment of SCATSat-1 scatterometer winds on the upper ocean simulations in the North Indian Ocean. *Journal of Geophysical Research: Oceans*. https://doi.org/10.1029/2020JC016677

[17] Carton, J. A., & Giese, B. S. (2008). A reanalysis of ocean climate using Simple Ocean Data Assimilation (SODA). *Monthly Weather Review*, *136*(8), 2999–3017. https://doi.org/10.1175/2007MWR1978.1

[18] Pramanik, S. and Sil, S. (2021a). Investigation of Surface and Sub-Surface Processes in the Bay of Bengal using Biophysical Regional Ocean Modeling System (ROMS). Ph. D. Thesis. IIT Bhubaneswar.

[19] Wyrtki, K. (1973). An Equatorial Jet in the Indian Ocean. In *Science* (Vol. 181, No. 4096, pp. 262–264). https://doi.org/10.1126/science.181.4096.262

[20] Masumoto, Y., Hase, H., Kuroda, Y., Matsuura, H., & Takeuchi, K. (2005). Intraseasonal variability in the upper layer currents observed in the eastern equatorial Indian Ocean. *Geophysical Research Letters*, *32*(2), 1–4. https://doi.org/10.1029/2004GL021896

[21] Sayantani, O., & Gnanaseelan, C. (2015). Tropical Indian Ocean subsurface temperature variability and the forcing mechanisms. *Climate Dynamics*. https://doi.org/10.1007/s00382-014-2379-y

[22] Sreenivas, P., Gnanaseelan, C., & Prasad, K. V. S. R. (2012). Influence of El Nino and Indian Ocean Dipole on sea level variability in the Bay of Bengal. *Global and Planetary Change*, *80–81*, 215–225. https://doi.org/10.1016/j.gloplacha.2011.11.001

[23] Hacker, P., Firing, E., Hummon, J., Gordon, A. L., & Kindle, J. C. (1998). Bay of Bengal currents during the northeast monsoon. *Geophysical Research Letters*, *25*(15), 2769–2772. https://doi.org/10.1029/98GL52115

10 Role of Stent Embedment and Release Kinetics on Two-Phase Binding of Drug in a Three-Layered Arterial Tissue

Akash Pradip Mandal and Prashanta Kumar Mandal

10.1 INTRODUCTION

Today, bare-metal stents (BMSs) are far less effective and safer than drug-eluting stents (DESs), which has revolutionised the treatment of coronary artery disease (CAD). One of the most difficult clinical problems in the stenting process is the endeavour to introduce local drugs into sick regions in order to stop smooth muscle cells created by the implantation of DES [1]. To investigate the local pharmacokinetics of stent-based drug administration, several theoretical and experimental investigations have been established [2–7].

To increase the effectiveness of the device, various problems must be resolved. To investigate the mechanisms underlying drug release from DES, early studies used a straightforward one-dimensional model [8,9], and later, two-dimensional [10–12], and finally, complex three-dimensional [13,14] models to investigate drug release mechanisms from DES. Studies have shown that the drug transport phenomena are remarkably influenced by several clinical key factors, such as strut design, choice of drug coated on the stent, local fluid dynamics, etc. [3,12,15,16]. Escuer et al. [17] explored the impact of vascular curvature on drug release kinetics recently, and they noted that after two hours of stenting, channel curvature significantly influences drug transport. However, Chen et al. provide a detailed demonstration of the impacts of various bending angles in the drug transport phenomenon [18]. The significance of luminal flow on the drug delivery mechanism was highlighted by Kolachalama et al. [16] in their use of a square-shaped strut. In their opinion, both luminal flow and strut design significantly affect the drug transport phenomenon. To understand various biomechanical impacts on drug delivery mechanisms from DESs, such as coating of therapeutic drugs, anisotropic diffusion, tissue porosity, etc., experiments on circular-shaped struts were undertaken by Vairo et al. [19]. However, some publications fail to account for the filtration of plasma and protein-binding effects when modelling local haemodynamics [20–25]. It is necessary to discuss the ideas of drug

 DOI: 10.1201/9781003638414-10

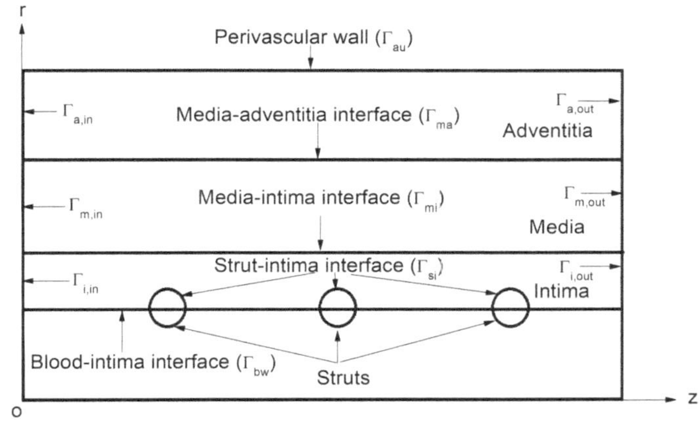

FIGURE 10.1 Geometrical view of the stented artery.

binding and free drug dispersion to adequately define drug delivery systems. Many academics utilised the partition coefficient to study drug binding mechanisms [8]. However, a two-species drug delivery model is adapted by Horner et al. to elucidate the binding mechanism of drug [26]. Chemical reactions are also used to characterise the drug binding mechanism, and an equilibrium constant is used to connect association and dissociation coefficients. A nonlinear reaction model is used by Groh et al. [27] to find how chemotherapeutic agents interact with the cell microenvironment during drug delivery to tumours. However, McGinty and Pontrelli [28] present many binding models, such as nonlinear saturable irreversible binding, nonlinear saturable reversible binding, linear irreversible binding, and linear reversible binding to examine drug binding kinetics in the artery wall. Researchers now discover that, in addition to drug binding to the specific receptor (SR), non-specific binding can also occur as a result of drug grabbing in the extracellular matrix [29]. Two distinct drug binding equations – one for non-specific binding to the general extracellular matrix (ECM) and the other for specific binding to receptors – were examined by Tzafriri et al. [30]. This chapter illustrates the specific and non-specific drug binding phases in the artery wall during stent-based drug delivery in the three states of unstable state, quasi-steady state, and steady state. In our model, a straightforward Cardio Coil stent with three partially embedded struts is taken into account. The target domain is a three-layered artery wall, which is shown in Figure 10.1, and the possible values of the involved parameters are derived from our most recent study [31].

10.2 UNSTEADY DRUG TRANSPORT

10.2.1 Governing Equations

The unsteady transportation of free drug is governed by the convection-diffusion-reaction process and the binding of the drug through a nonlinear saturable reversible chemical reaction. Here, the normalised concentration of free drug is denoted by c_f, the normalised specific bound (SR bound) drug concentration by c_s and non-specific bound (ECM bound) drug concentration by c_{ns}. The governing equations representing the transport of free drug, non-specific bound drug and specific bound drug may now be written as

$$\frac{\partial c_{f_j}}{\partial t} + Pe\left[\frac{1-\sigma_j}{\epsilon}\right]\frac{\partial c_{f_j}}{\partial r} = \frac{\theta_j}{\epsilon^2}\left[\frac{\partial^2 c_{f_j}}{\partial r^2} + \frac{1}{r}\frac{\partial c_{f_j}}{\partial r} + \epsilon^2\frac{\partial^2 c_{f_j}}{\partial z^2}\right] - \left[\frac{\partial c_{s_j}}{\partial t} + \frac{\partial c_{\mathrm{ns}_j}}{\partial t}\right], \quad (10.1)$$

$$\frac{\partial c_{s_j}}{\partial t} = Da_s\left[c_{f_j}\left(c_s^{\max} - c_{s_j}\right) - \frac{c_{s_j}}{K_s}\right], \quad (10.2)$$

$$\frac{\partial c_{\mathrm{ns}_j}}{\partial t} = Da_{ns}\left[c_{f_j}\left(c_{\mathrm{ns}}^{\max} - c_{\mathrm{ns}_j}\right) - \frac{c_{\mathrm{ns}_j}}{K_{\mathrm{ns}}}\right]. \quad (10.3)$$

In this case, the dimensionless radial and axial coordinates, r and z are normalised with respect to the strut diameter (d) and the artery wall thickness (δ_W), respectively. Scaled by the initial strut drug concentration (c_0), c_{f_j}, c_{s_j} and c_{ns_j} represent the dimensionless concentrations of the free drug, specific drug, and non-specific bound drug, respectively. The dimensionless local densities for specific and non-specific bound drugs, denoted as, $c_s^{\max}$ and $c_{\mathrm{ns}}^{\max}$, are also normalised with respect to the initial strut drug concentration (c_0). Here, dimensionless time (t), Damköhler number (Da_s, Da_{ns}), Peclet number (Pe), equilibrium constants (K_s, K_{ns}) and scaling parameters (ϵ, θ_j) are as follows:

$$t\left(=\frac{\bar{t}D_m}{d^2}\right),\ Da_s = \frac{k_{f_s}c_0 d^2}{D_m},\ Da_{\mathrm{ns}} = \frac{k_{f_{\mathrm{ns}}}c_0 d^2}{D_m},\ Pe = \frac{d\,V_{\mathrm{filt}}}{D_m},$$

$$K_s = \frac{c_0 k_{f_s}}{k_{r_s}},\ K_{\mathrm{ns}} = \frac{c_0 k_{f_{\mathrm{ns}}}}{k_{r_{\mathrm{ns}}}},\ \epsilon = \frac{\delta_W}{d},\ \theta_j = \frac{D_j}{D_m}.$$

Here, where k_{f_s} and $k_{f_{\mathrm{ns}}}$ are the association and dissociation rate constants, respectively, V_{filt} is the transmural filtration velocity of plasma flow from lumen into the tissue through the blood-tissue interface. D_j denotes the drug diffusivity, σ_j is the Staverman reflection coefficient [32]. Here, j represents layer [$j = i$ (the intima layer), m (the media layer) and a (the adventitia layer)]. Here, and in the sequel, s and ns refer to specific and the non-specific binding. Again, $k_{r_s}, k_{r_{\mathrm{ns}}}$ are the dissociation rate constants for the specific and non-specific bindings, respectively.

10.2.2 Boundary and Initial Conditions

Impermeability boundary conditions are assumed along the proximal $\left(\Gamma_{j.\mathrm{in}}, j = i,m,a,\right)$ and distal $\left(\Gamma_{j.\mathrm{out}}, j = i,m,a\right)$ boundaries [15,33]:

$$\frac{\partial c_{f_j}}{\partial z} = 0 \quad \text{on} \quad \Gamma_{j,\mathrm{in}} \text{ and } \Gamma_{j,\mathrm{out}}. \quad (10.4)$$

At the perivascular wall (Γ_{au}), a zero concentration condition (sink condition) is applied [34,35]:

$$c_{f_a} = 0 \text{ at } \Gamma_{\mathrm{au}}. \quad (10.5)$$

At the strut surface, two different drug release mechanisms are used, namely, a constant and a time-dependency system [22,36]:

$$c_{f_i} = 1 \text{ or } e^{-\phi t} \text{ on } \Gamma_{\text{si}}, \tag{10.6}$$

where $\phi = \dfrac{\lambda\, d^2}{D_m}$, and λ is the drug release rate.

10.2.3 Interface Conditions

At the blood-tissue interface (Γ_{bw}), a no-flux boundary condition is taken into account [15,34]:

$$\frac{\partial c_{f_i}}{\partial r} = 0 \text{ on } \Gamma_{\text{bw}}. \tag{10.7}$$

Due to membrane rejection between different layers, a jump interfacial condition is imposed at both media-intima and media-adventitia interfaces [37–39]. Therefore, the media-intima (Γ_{mi}) interface condition may be written as:

$$\left[(1-\sigma_i)\epsilon Pe\, c_{f_i} - \theta_i \frac{\partial c_{f_i}}{\partial r}\right] = \left[(1-\sigma_m)\epsilon Pe\, c_{f_m} - \frac{\partial c_{f_m}}{\partial r}\right] \text{ on } \Gamma_{\text{mi}}. \tag{10.8}$$

Similarly, media-adventitia interface (Γ_{ma}) condition may also be written as

$$\left[(1-\sigma_m)\epsilon Pe\, c_{f_m} - \frac{\partial c_{f_m}}{\partial r}\right] = \left[(1-\sigma_a)\epsilon Pe\, c_{f_a} - \theta_a \frac{\partial c_{f_a}}{\partial r}\right] \text{ on } \Gamma_{\text{ma}}. \tag{10.9}$$

10.2.4 Solution Procedure

To solve the above governing equations along with boundary, interface and initial conditions, we first transform the irregular adapted domain into a simple rectangular one by utilising the following transformations to avoid the discretised error:

$$\alpha = 1 + \frac{r - \Gamma_{\text{fl}}}{R_i}, \beta = 2 + \frac{r - \Gamma_{\text{mi}}}{R_m}, \gamma = 3 + \frac{r - \Gamma_{\text{ma}}}{R_a} \tag{10.10}$$

where $\Gamma_{\text{fl}} = \Gamma_{\text{bw}} \cup \Gamma_{\text{si}}$, $R_i = \Gamma_{\text{mi}} - \Gamma_{\text{fl}}$, $R_m = \Gamma_{\text{ma}} - \Gamma_{\text{mi}}$ and $R_a = \Gamma_{\text{au}} - \Gamma_{\text{ma}}$ such that the intima region is converted to $[0, L] \times [1, 2]$, the media to $[0, L] \times [2, 3]$ and the adventitia to $[0, L] \times [3, 4]$; L being the axial length of the arterial segment. Now, all the converted governing equations along with the set of boundary, initial and interface conditions are discretised using forward time, centred space discretisation technique and computed by applying finite difference method in colocated grid. The details discretisation techniques have already been documented in our previous work [39].

10.3 QUASI STEADY-STATE DRUG TRANSPORT

10.3.1 Governing Equations and Boundary Conditions

Since drug transport usually occurs within the cell region, its reaction rate is significantly faster than the diffusion's time scale [40]. Therefore, we are able to reduce the nonlinear reversible saturable binding model by combining equations (10.1)–(10.3) into an equation for the total drug concentration $\left(c_{t_j} = c_{f_j} + c_{s_j} + c_{\text{ns}_j}\right)$ and a simple algebraic expression for the two types of drug binding. Now, the dimensionless form of the quasi steady-state transport of the drug is given by:

$$\frac{\partial c_{t_j}}{\partial t} = \frac{dc_{f_j}}{dc_{t_j}}\left\{-Pe\left[\frac{1-\sigma_j}{\epsilon}\right]\frac{\partial c_{t_j}}{\partial r} + \frac{\theta_j}{\epsilon^2}\left[\frac{\partial^2 c_{t_j}}{\partial r^2} + \frac{1}{r}\frac{\partial c_{t_j}}{\partial r} + \epsilon^2\frac{\partial^2 c_{t_j}}{\partial z^2}\right]\right\}, \tag{10.11}$$

$$c_{s_j} = \frac{c_{f_j} c_s^{\max}}{K_s^{-1} + c_{f_j}}, \tag{10.12}$$

$$c_{\text{ns}_j} = \frac{c_{f_j} c_{\text{ns}}^{\max}}{K_{\text{ns}}^{-1} + c_{f_j}}, \tag{10.13}$$

where

$$\sqrt{\frac{dc_{f_j}}{dc_{t_j}}} = \frac{\left(K_s^{-1} + c_{f_j}\right)\left(K_{\text{ns}}^{-1} + c_{f_j}\right)}{\left(K_s^{-1} + c_{f_j}\right)\left(K_{\text{ns}}^{-1} + c_{f_j}\right) + \sqrt{K_s^{-1} c_s^{\max}}\left(K_{\text{ns}}^{-1} + c_{f_j}\right) + \sqrt{K_{\text{ns}}^{-1} c_{\text{ns}}^{\max}}\left(K_s^{-1} + c_{f_j}\right)}.$$

Using the boundary, initial and interface conditions for free drug c_{f_j} already discussed in Sections 10.2.2 and 10.2.3, the same can be obtained for total drug concentration.

10.3.2 Solution Procedure

Now, using the transformations (10.10), the stented domain has been transformed into a rectangular domain, as already discussed in Section 10.2.4. All the transformed equations for the total drug along with initial, boundary and interface conditions are discretised using the forward time, centred space discretisation technique and solved numerically using the finite difference method. This gives the values of the normalised total drug concentration $\left(c_{t_j}\right)$ at each grid point. Again, at quasi steady-state, the expression for total drug concentration $\left(c_{t_j}\right)$ can be written as $c_{t_j} = c_{f_j} + c_{s_j} + c_{\text{ns}_j}$, and using expressions (10.12) and (10.13), we get a cubic equation in c_{f_j}, which can be written as:

$$c_{f_j}^3 + A_j c_{f_j}^2 + B_j c_{f_j} + C_j = 0, \tag{10.14}$$

where $A_j = c_s^{\max} + c_{\text{ns}}^{\max} + K_s^{-1} + K_{\text{ns}}^{-1} - c_{t_j}$, $B_j = K_s^{-1} c_s^{\max} + K_{\text{ns}}^{-1} c_{\text{ns}}^{\max} + K_{\text{ns}}^{-1} + K_s^{-1} + K_s^{-1} K_{\text{ns}}^{-1} - c_{t_j}$, $C_j = -c_{t_j} K_s^{-1} K_{\text{ns}}^{-1}$. In the limit of rapid binding, the unsteady coupled drug transport equations (equations 10.1–10.3) may be reduced to a single partial differential equation (equation 10.11) together with simple algebraic expressions (equations 10.12 and 10.13) for both specific and non-specific bound drugs. For physical meaningful result, the mean concentration of free drug $\left(c_{f_j}\right)$ is always ≥ 0, so we consider only the real positive root of equation (10.14), which can be solved using Cardan's method. After obtaining the values of c_{f_j}, we can easily compute c_{s_j} and c_{ns_j} from the simple algebraic expressions (10.12 and 10.13).

10.4 STEADY-STATE DRUG TRANSPORT

10.4.1 Governing Equations and Boundary Conditions

The steady-state distribution of free drug and two-phase binding of the drug, corresponding to time-dependent unsteady equations (equations 10.1–10.3) may be written as:

$$Pe\left[\frac{1-\sigma_j}{\epsilon}\right]\frac{\partial c_{f_j}}{\partial r} = \frac{\theta_j}{\epsilon^2}\left[\frac{\partial^2 c_{f_j}}{\partial r^2} + \frac{1}{r}\frac{\partial c_{f_j}}{\partial r} + \epsilon^2\frac{\partial^2 c_{f_j}}{\partial z^2}\right], \tag{10.15}$$

$$c_{f_j}\left(c_s^{\max} - c_{s_j}\right) - \frac{c_{s_j}}{K_s} = 0, \tag{10.16}$$

$$c_{f_j}\left(c_{\text{ns}}^{\max} - c_{\text{ns}_j}\right) - \frac{c_{\text{ns}_j}}{K_{\text{ns}}} = 0, \tag{10.17}$$

Where the initial, boundary and interface conditions are maintained as for the unsteady situation (Sections 10.2.2 and 10.2.3) and initially, a constant drug release mechanism is imposed at the strut surface.

10.4.2 Solution Procedure

Now, all the drug transport equations along with boundary, interface and initial conditions are converted to a rectangular domain by utilising the transformation (10.10) and the centred space discretisation approach is used to discretise the transformed drug transport equations. The discretised version of these equations forms a system of algebraic equations, which is solved numerically using Successive-Over-Relaxation (SOR) method with over-relaxation parameter 1.25.

10.5 RESULTS AND DISCUSSION

Although several experimental and numerical drug delivery models have been created recently, the drug release process still requires further studies. One of the most crucial concerns for further developing the model in terms of safety and efficacy is the drug release profile. This section presents some preliminary findings from an investigation into the effects of various therapeutically difficult factors on three-state drug delivery models, including strut embedment, strut diameter, strut radius and drug diffusivity.

The impact of strut embedment on drug distribution in the context of pharmacokinetics is an intriguing subject that continues to surround the stenting operation. Figure 10.2a–c show the variation of normalised mean free drug concentrations and

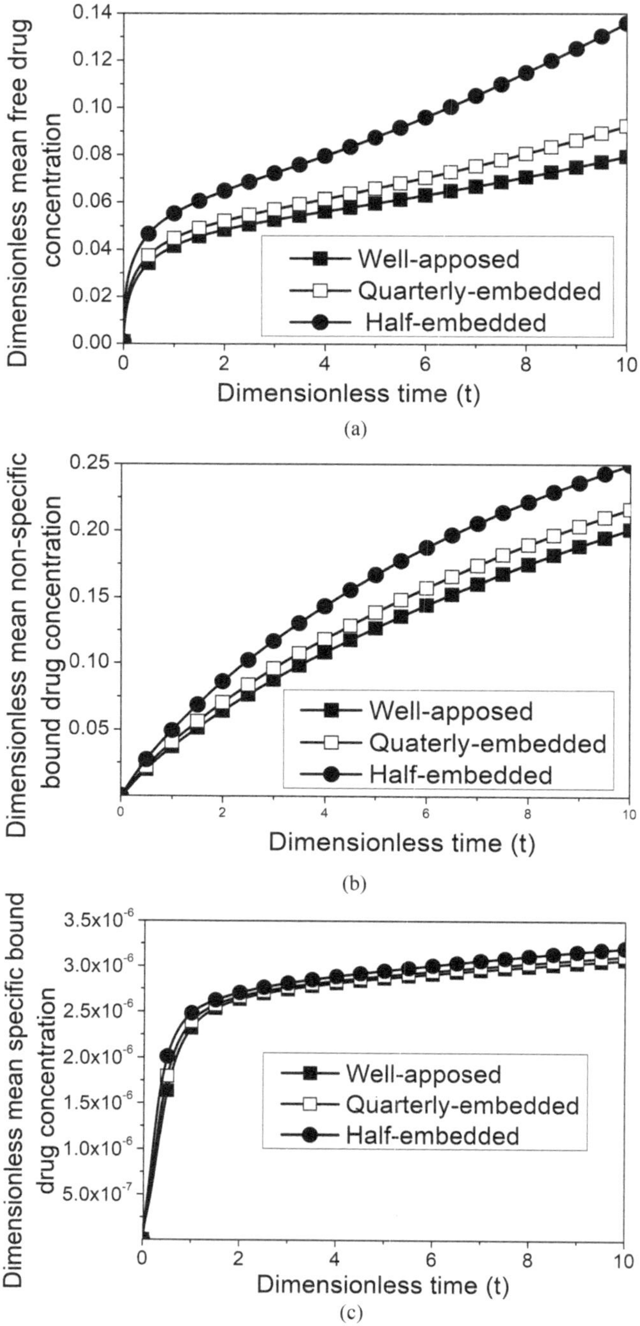

FIGURE 10.2 Time-varying dimensionless mean drug concentration for different strut embedment levels, (a) Free drug, (b) Non-specific bound drug, (c) Specific bound drug.

two distinct phases of normalised mean bound drug concentrations (specific and non-specific) over time for different degrees of strut embedment. According to the data, half-embedded struts had higher free drug and two-phase bound drug concentrations than quarter-embedded and well-apposed struts. It is significant to remember that the half-embedded stent increases the amount of drug that enters into the arterial wall by creating a larger contact surface. Consequently, a higher concentration profile of the free drug is produced as the degree of stent embedment increases, causing the binding site to fill more quickly.

Utilising our quasi steady-state model from Section 10.3, Figure 10.3a and b demonstrate how interstrut distance affects drug binding in two distinct stages. With a rising interstrut distance value, the mean concentration of the bound drug in the two phases rises. The blood-intima contact (Γ_{bw}) should be sufficiently permeable to

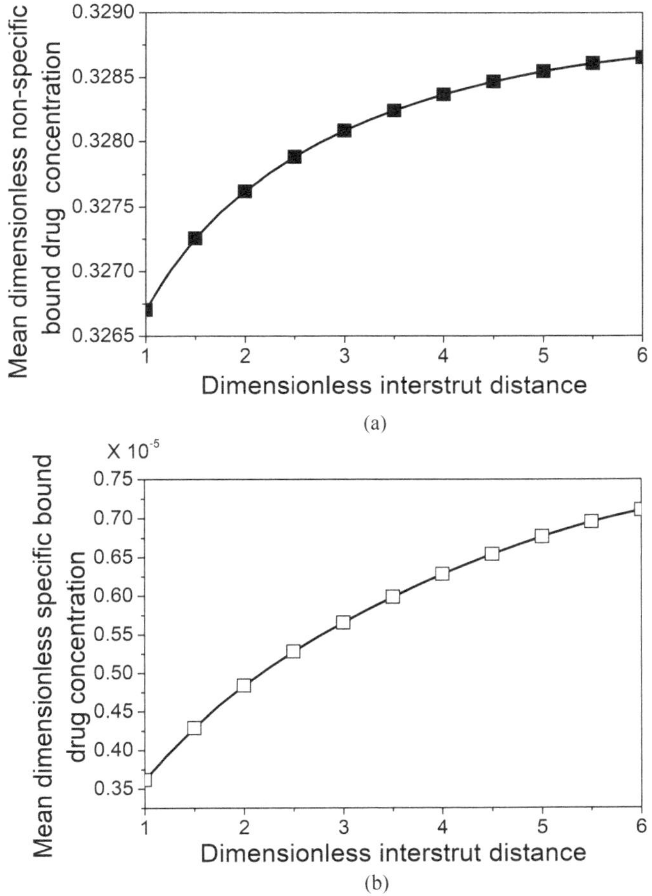

FIGURE 10.3 Variation of dimensionless mean bound drug concentration with dimensionless interstrut distance at quasi steady-state, (a) Non-specific bound drug, (b) Specific bound drug.

allow an acceptable amount of plasma to pass through, according to one significant conclusion of our research.

Figure 10.4a and b shows the effect of the dimensionless scaling parameter ϵ $(= \frac{\delta_l}{d})$, where δ_l is the radius of the lumen, on the drug delivery mechanism in relation to our steady-state drug delivery model discussed in Section 10.4. Since the scaling parameter is the ratio of lumen radius to strut diameter, the parameter ϵ has important significance in drug delivery. The mean concentrations of specific and non-specific bound drug increase with the increasing values of ϵ. These findings suggest that for a thinner strut, bound drugs are rapidly saturated to the binding sites. This occurs because a thinner strut works better than a thicker one in stent-based delivery.

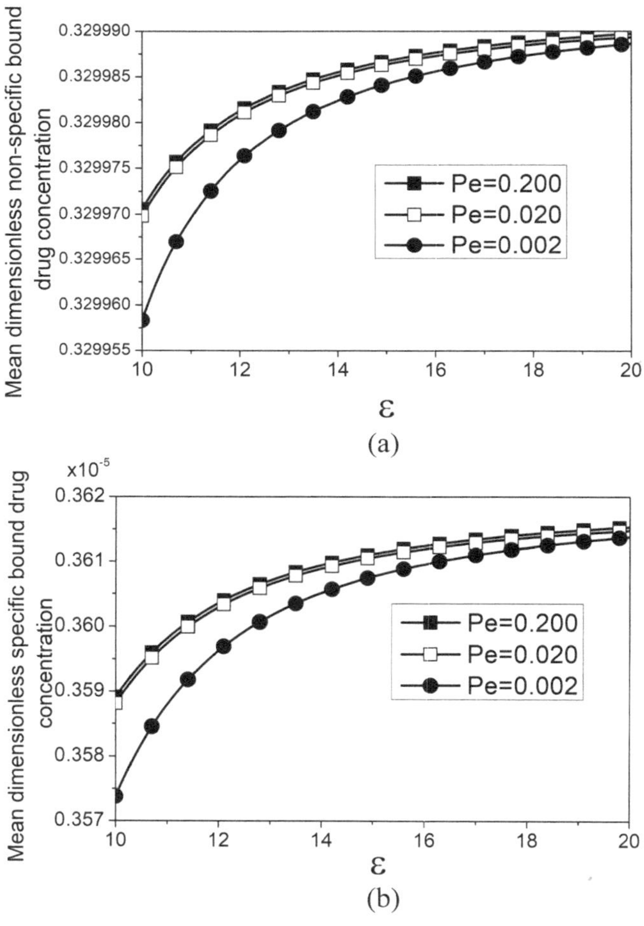

FIGURE 10.4 Variation of normalised mean concentrations of bound drug with dimensionless parameter ϵ for different Peclet numbers at steady-state, (a) Non-specific bound drug, (b) Specific bound drug.

As a result, our findings demonstrate that the stent's dimension has a significant impact on steady-state concentrations and a thinner strut may have better therapeutic efficacy than a thicker strut. Once more, the choice of drug and its diffusivity have a potential impact on drug delivery models. Figure 10.4a and b also shows how the mean drug concentration in two phases changes with Peclet numbers (Pe). The figures indicate that a decrease in drug diffusivity or an increase in Peclet number leads to higher mean concentration profiles for each bound drug form.

10.6 CONCLUSION

In this chapter, a two-dimensional axi-symmetric model of drug delivery in a three-layered arterial wall is considered. The goal of the work is to investigate drug binding mechanisms in three different states. At a quasi steady-state, we can use a single partial differential equation for total drug and simple algebraic expressions for specific and non-specific bound drug. The study reveals that non-specific bound drug concentration is always higher compared to specific bound drug concentration, and the total drug concentration inside an arterial wall is equal to the sum of free drug, specific bound drug and non-specific bound drug. Another important implication is that a thinner strut may have higher efficacy in stent-based delivery.

The salient outcomes of the present chapter are listed below:

- Comparing half-embedded stents to well-apposed and quarterly-embedded stents, higher free drug concentration profiles and faster drug binding saturation can be attained.
- Quasi steady-state drug transport is modelled by a single unsteady partial differential equation of total drug together with two simple algebraic equations for two types of bound drug.
- Since, convection is greatly accelerated by the transmural plasma flow, the interstrut distance should be optimised so that a reasonable amount of plasma can filtrate through the arterial wall.
- The saturation rate of both types of bound drug is inversely proportional to drug diffusivity.
- The total amount of drug in a target domain is the sum of the free drug, the non-specific bound drug, and the specific bound drug.
- A thinner strut is more efficacious compare to a thicker one in stent-based drug delivery.
- In realistic stent-based drug delivery model, two types of bound drug should be considered together with free drug.

REFERENCES

[1] Schwartz, R. S., Edelman, E. R., Carter, A., Chronos, N., Rogers, C., Robinson, K. A., & Waksman, R. (2002). Drug-Eluting Stents in Preclinical Studies: Recommended Evaluation from a Consensus Group. *Circulation* 106 (14): 1867–1873.

[2] Frank, A. O., Walsh, P. W., & Moore, J. E. (2002). Computational Fluid Dynamics and Stent Design. *Artificial Organs* 26 (7): 614–621.

[3] Mandal, A. P., Sarifuddin, & Mandal, P. K. (2015). An Unsteady Analysis of Arterial Drug Transport from Half-Embedded Drug-Eluting Stent. *Applied Mathematics and Computation* 266: 968–981.
[4] Li, F., Gu, Y., Hua, R., Ni, Z., & Zhao, G. (2018). In Vitro Release Study of Sirolimus from a Pdlla Matrix on a Bioresorbable Drug-Eluting Stent. *Journal of Drug Delivery Science and Technology* 48: 88–95.
[5] Song, J., Kouidri, S., & Bakir, F. (2021). Review on the Numerical Investigations of Mass Transfer from Drug Eluting Stent. *Biocybernetics and Biomedical Engineering* 41 (3): 1057–1070.
[6] Koch, T., Lenz, T., Joner, M., Xhepa, E., Koppara, T., Wiebe, J., Coughlan, J., et al. (2021). Ten-Year Clinical Outcomes of Polymer-Free Versus Durable Polymer New-Generation Drug-Eluting Stent in Patients with Coronary Artery Disease with and Without Diabetes Mellitus. *Clinical Research in Cardiology* 110: 1586–1598.
[7] Hanus, J., & Zahora, J. (2005). Measurement and Comparison of Mechanical Properties of Nitinol Stents. *Physica Scripta* 2005 (T118): 264.
[8] Migliavacca, F., Gervaso, F., Prosi, M., Zunino, P., Minisini, S., Formaggia, L., & Dubini, G. (2007). Expansion and Drug Elution Model of a Coronary Stent. *Computer Methods in Biomechanics and Biomedical Engineering* 10 (1): 63–73.
[9] Pontrelli, G., & Monte, F. D. (2014). A Two-Phase Two-Layer Model for Transdermal Drug Delivery and Percutaneous Absorption. *Mathematical Biosciences* 257: 96–103.
[10] Grassi, M., Pontrelli, G., Teresi, L., Grassi, G., Comel, L., Ferluga, A., & Galasso, L. (2009). Novel Design of Drug Delivery in Stented Arteries: A Numerical Comparative Study. *Mathematical Biosciences & Engineering* 6 (3): 493–508.
[11] Hossainy, S., & Prabhu, S. (2008). A Mathematical Model for Predicting Drug Release from a Biodurable Drug-Eluting Stent Coating. *Journal of Biomedical Materials Research Part A: An Official Journal of the Society for Biomaterials, the Japanese Society for Biomaterials, and the Australian Society for Biomaterials and the Korean Society for Biomaterials* 87 (2): 487–493.
[12] Mandal, A. P., & Mandal, P. K. (2016). Drug Elution Model of Coronary Stent: Effects of Stent Embedment and Binding of Drug. *International Journal of Biomedical Engineering and Technology* 20 (2): 150–165.
[13] Cutri, E., Zunino, P., Morlacchi, S., Chiastra, C., & Migliavacca, F. (2013). Drug Delivery Patterns for Different Stenting Techniques in Coronary Bifurcations: A Comparative Computational Study. *Biomechanics and Modeling in Mechanobiology* 12 (4): 657–669.
[14] Hose, D. R., Narracott, A. J., Griffiths, B., Mahmood, S., Gunn, J., Sweeney, D., & Lawford, P. V. (2004). A Thermal Analogy for Modelling Drug Elution from Cardiovascular Stents. *Computer Methods in Biomechanics and Biomedical Engineering* 7 (5): 257–264.
[15] Sarifuddin, R. S., and Mandal, P. K. (2020). Computational Model of Stent-Based Delivery from a Half-Embedded Two-Layered Coating. *Computer Methods in Biomechanics and Biomedical Engineering* 23 (12): 815–831.
[16] Kolachalama, V. B., Tzafriri, A. R., Arifin, D. A., & Edelman, E. R. (2009). Luminal Flow Patterns Dictate Arterial Drug Deposition in Stent-Based Delivery. *Journal of Controlled Release* 133 (1): 24–30.
[17] Escuer, J., Aznar, I., McCormick, C., Pena, E., McGinty, S., & Martinez, M. A. (2021). Influence of Vessel Curvature and Plaque Composition on Drug Transport in the Arterial Wall Following Drug-Eluting Stent Implantation. *Biomechanics and Modeling in Mechanobiology* 20 (2): 767–786.
[18] Chen, Y., Xiong, Y., Jiang, W., Wong, M. S., Yan, F., Wang, Q., & Fan, Y. (2016). Numerical Simulation on the Effects of Drug-Eluting Stents with Different Bending Angles on Hemodynamics and Drug Distribution. *Medical & Biological Engineering & Computing* 54 (12): 1859–1870.

[19] Vairo, G., Cioffi, N., Cottone, R., Dubini, G., & Migliavacca, F. (2010). Drug Release from Coronary Eluting Stents: A Multidomain Approach. *Journal of Biomechanics* 43 (8): 1580–1589.

[20] Balakrishnan, B., Tzafriri, A. R., Seifert, P., Groothuis, A., Rogers, C., & Edelman, E. R. (2005). Strut Position, Blood Flow, and Drug Deposition: Implications for Single and Overlapping Drug-Eluting Stents. *Circulation* 111 (22): 2958–2965.

[21] Balakrishnan, B., Dooley, J. F., Kopia, G., & Edelman, E. R. (2007). Intravascular Drug Release Kinetics Dictate Arterial Drug Deposition, Retention, and Distribution. *Journal of Controlled Release* 123 (2): 100–108.

[22] Balakrishnan, B., Dooley, J., Kopia, G., & Edelman, E. R. (2008). Thrombus Causes Fluctuations in Arterial Drug Delivery from Intravascular Stents. *Journal of Controlled Release* 131 (3): 173–180.

[23] Mongrain, R., Leask, R., Brunette, J., Faik, I., Bulman-Feleming, N., & Nguyen, T. (2005). Numerical Modeling of Coronary Drug Eluting Stents. *Studies in Health Technology and Informatics* 113: 443–458.

[24] Mongrain, R., Faik, I., Leask, R. L., Rodés-Cabau, J., Larose, E., & Bertrand, O. F. 2007. Effects of Diffusion Coefficients and Struts Apposition Using Numerical Simulations for Drug Eluting Coronary Stents. *Journal of Biomedical Engineering* 129 (5): 733–742.

[25] Borghi, A., Foa, E., Balossino, R., Migliavacca, F., & Dubini, G. (2008). Modelling Drug Elution from Stents: Effects of Reversible Binding in the Vascular Wall and Degradable Polymeric Matrix. *Computer Methods in Biomechanics and Biomedical Engineering* 11 (4): 367–377.

[26] Horner, M., Joshi, S., Dhruva, V., Sett, S., Stewart, S. F. C. (2010). A Two-Species Drug Delivery Model Is Required to Predict Deposition from Drug-Eluting Stents. *Cardiovascular Engineering and Technology* 1 (3): 225–234.

[27] Groh, C. M., Hubbard, M. E., Jones, P. F., Loadman, P. M, Periasamy, N., Sleeman, B. D., Smye, S. W., Twelves, C. J., & Phillips, R. M. (2014). Mathematical and Computational Models of Drug Transport in Tumours. *Journal of the Royal Society Interface* 11 (94): 20131173.

[28] McGinty, S., & Pontrelli, G. (2015). A General Model of Coupled Drug Release and Tissue Absorption for Drug Delivery Devices. *Journal of Controlled Release* 217: 327–336.

[29] Lauffenburger, D. A., & Linderman, J. (1996). *Receptors: Models for Binding, Trafficking, and Signaling*. Oxford University Press: New York.

[30] Tzafriri, A. R., Groothuis, A., Price, G. S., & Edelman, E. R. (2012). Stent Elution Rate Determines Drug Deposition and Receptor-Mediated Effects. *Journal of Controlled Release* 161 (3): 918–926.

[31] Mandal, A. P., & Mandal, P. K. (2022b). Two-Phase Binding of Drug in a Three-Layered Arterial Wall Following Stent Implantation. *Physica Scripta* 97 (4): 045004.

[32] Staverman, A. J. (1951). The Theory of Measurement of Osmotic Pressure. *Recueil Des Travaux Chimiques Des Pays-Bas* 70 (4): 344–352.

[33] Khakpour, M., & Vafai, K. (2008). Critical Assessment of Arterial Transport Models. *International Journal of Heat and Mass Transfer* 51 (3–4): 807–822.

[34] Kolandaivelu, K., O'Brien, C. C, Shazly, T., Edelman, E. R., & Kolachalama, V. B. (2015). Enhancing Physiologic Simulations Using Supervised Learning on Coarse Mesh Solutions. *Journal of the Royal Society Interface* 12 (104): 20141073.

[35] Mandal, A. P., & Mandal, P. K. (2017). Computational Modelling of Three-Phase Stent-Based Delivery. *Journal of Exploratory Research in Pharmacology* 2 (2): 31–40.

[36] McGinty, S., Wheel, M., McKee, S., & McCormick, C. (2015). Does Anisotropy Promote Spatial Uniformity of Stent-Delivered Drug Distribution in Arterial Tissue? *International Journal of Heat and Mass Transfer* 90: 266–279.

[37] Iasiello, M., Vafai, K., Andreozzi, A., Bianco, N., & Tavakkoli, F. (2015). Effects of External and Internal Hyperthermia on LDL Transport and Accumulation within an Arterial Wall in the Presence of a Stenosis. *Annals of Biomedical Engineering* 43 (7): 1585–1599.

[38] Yang, N., & Vafai, K. (2008). Low-Density Lipoprotein (LDL) Transport in an Artery–A Simplified Analytical Solution. *International Journal of Heat and Mass Transfer* 51 (3–4): 497–505.

[39] Mandal, A. P., & Mandal, P. K. (2022a). Specific and Nonspecific Binding of Drug Eluted from a Half-Embedded Stent in Presence of Atherosclerotic Plaque. *Computer Methods in Biomechanics and Biomedical Engineering* 25 (8): 922–935.

[40] Tzafriri, A. R., Levin, A. D., & Edelman, E. R. (2009). Diffusion-Limited Binding Explains Binary Dose Response for Local Arterial and Tumour Drug Delivery. *Cell Proliferation* 42 (3): 348–363.

11 Impact of Rumor Credibility and User Maturity on Rumor Propagation

Moumita Ghosh, Arijit Malakar, Ritwick Banerjee, and Pritha Das

11.1 INTRODUCTION

Rapid development of technology and the emergence of the Internet have accelerated the spread of information worldwide through various channels. The advent of Social networking sites (SNS) offers platforms where numerous users can easily share their opinions, propagate information among acquaintances, and result in collective discussions on various topic. As each user of SNS can share and also be a recipient of information, SNS are becoming the breeding ground of rumors due to their large number of users and increasing popularity. Rumors are often created to target insecurities and uncertainties in people's life by distorting or improvising real data. According to reference [1], a rumor can be defined as "circulating story of questionable veracity, which is apparently credible but hard to verify, and produces sufficient skepticism and/or anxiety so as to motivate finding out the actual truth". Sometimes, the spread of rumors can lead to very serious consequences [2,3]. The study in reference [4] presents a theory that a rumor influenced the result of the US election in 2016. These kinds of events encourage investigation into the dynamics of rumor propagation on complex social networks. In 1960 Daley and Kendall formulated the first model, DK (Daley-Kendall) [5], on rumor propagation, considering three groups of people: ignorant people, who do not know about rumor; spreaders, who spend the rumor; and stiflers, who know the rumor but do not spread it. Then Maki and Thompson revised this work and developed a new model, considering that interaction between two spreaders makes them stiflers [6]. Some researchers have found that the dynamical behavior of rumor propagation has similarities to the nature of disease transmission. Many research articles have been published based on Susceptible-Infected-Recovered (SIR) and Susceptible-Exposed-Infected-Recovered (SEIR) models [7,8]. In studies [7 and 9], the authors introduced forgetting, hesitation mechanisms in rumor spread dynamics. Some researchers have studied the effect of delay on the propagation of rumors [7,10]. As we have already discussed, the spread of rumors can sometimes have serious consequences after going viral (when the rumor reaches a large number of users).

DOI: 10.1201/9781003638414-11

Studies in [11,12] investigated optimal control theory to eradicate the negative impacts of rumor using media. Some researchers considered network model to study the effect of rumors on people [13,14]. The authors in reference [15] considered heterogeneous network model to study the complex dynamics of rumor propagation. In reference [16], the authors compared the dynamics of rumor propagation on heterogeneous network with homogeneous network models.

Very few of the above studies have discussed the persuasive factor of rumor to netizens. A rumor that is more relevant to people's lives and more trending, is more credible to users on SNS. A rumor with higher credibility can easily influence its recipients to spread it further. The consideration of the credibility of rumors is relatively new in rumor spread dynamics. Moreover, here we divide the newcomer class based on their maturity. Mature ignorants are more efficient at discriminating between rumor/misinformation and other information. When a mature ignorant comes into contact with any information spreader, he/she does not spread it. They hesitate and search for an authentic source of the information and joining the latent group. In contrast, people in the immature ignorant class are prone to believe any information they come across and pass it to others. Next, we derive the threshold of the influence of a rumor, which lays out the requirements for rumors to continue spreading on social networks. It also gives the condition under which the rumor eventually disappears from social networking sites.

The remainder of the study is organized as follows: In Section 11.2, we formulate a rumor propagation model and prove that the solution space is positively invariant on a compact set. Next, in Section 11.3, we find the steady-states and the conditions for stability of the system around the equilibria. In this section we also derive the threshold value of the influence of the rumor. In Section 11.4, we present some numerical plots to validate our analytical results. Finally, we conclude with some important remarks in Section 11.5.

11.2 MODEL FORMULATION

In this article, we propose a rumor propagation model that includes rumor credibility and user-maturity. User maturity refers to a user's ability to discriminate a rumor from other information based on their logical thinking, knowledge, intuition, and past experiences. Moreover, users on SNS respond differently after hearing a rumor. Some spread the rumor, some hesitate to spread it, and some spreaders stop spreading after losing interest in that particular rumor or getting involved in another one. Considering the hesitating and forgetting mechanisms, along with other characteristics of users on SNS, we categorize users into five groups according to their attitudes toward rumor as follows:

1. **Immature Newcomers** (I_U): Individuals in this class do not know about the rumor, and after hearing it, they usually believe it and spread it to others without any contemplation.
2. **Mature Newcomers** (I_E): Individuals in this class do not know about the rumor, and after hearing it, they usually prefer to think about its authentication and seek confirmation before spreading it.

3. **Latent Class** (L): Individuals in this class know about the rumor but do not spread it out of hesitation.
4. **Spreader** (S): People in this class spread rumor.
5. **Aware Class** (A): People who know the rumor and have a logical explanation and authentic information. Therefore, they do not spread rumor and also try to convince others not to spread it.

Based on our discussions, we propose our model as follows:

$$\begin{aligned}
\frac{dI_U}{dt} &= (1-p)\Lambda - \frac{\beta cSI_U}{N} - \mu I_U \\
\frac{dI_E}{dt} &= p\Lambda - \frac{\beta cSI_E(1-c)}{N} - \mu I_E \\
\frac{dL}{dt} &= \frac{\beta cSI_E(1-c)}{N} - \phi_1 LA - \eta_1 L - \mu L \\
\frac{dS}{dt} &= \frac{\beta cSI_U}{N} + \eta_1 L - \eta_2 S - \mu S \\
\frac{dA}{dt} &= \phi_1 LA + \eta_2 S - \mu A
\end{aligned} \tag{11.1}$$

Here, $N(t) = I_U(t) + I_E(t) + L(t) + S(t) + A(t)$ is the total number of active user accounts on social media at time t. We have considered a constant recruitment rate Λ in the total newcomer section. As Λ shows the number of newcomers to the whole group, it is a non-negative integer. When a rumor breaks out on the internet, individuals in the spreader class start to spread it over various social media at a rate β. Different individuals in the newcomer class respond differently based on their maturity level. Mature individuals do not spread it immediately, they join the latent class L with proportion $\frac{\beta c(1-c)SI_E}{N}$, where c is the credibility of the rumor. c is strictly less than 1, as full credibility of a rumor does not occur in reality. The higher the value of c, the more credible the rumor is to netizens. Immature ignorants switch to the spreader class with a proportion $\frac{c\beta ScI_U}{N}$ after hearing the rumor. However, individuals in I_E are more mature than in I_U, they do not become spreader immediately after they get to know the rumor. Rather they rely on their intuition and past experiences and join the latent class due to hesitation. A rumor become more suspicious when $c = 0.5$, as when the rumor is fully credible, people in I_E usually ignore it. When $c = 0$, rumor is completely unreliable, no one spread it. Therefore, it is realistic to consider the term $c(1-c)$ in the transmission from mature individuals I_E to the latent class L. Moreover, individuals in the latent class join with spreaders at a rate η_1, and some spreaders stop spreading the rumor and join the aware class at a rate η_2 after forgetting that particular rumor. When people in the latent class comes into contact with aware people, they receive a logical explanation of the rumor and switch to the aware class at a rate ϕ_1. The propagation process is also shown by the schematic diagram in

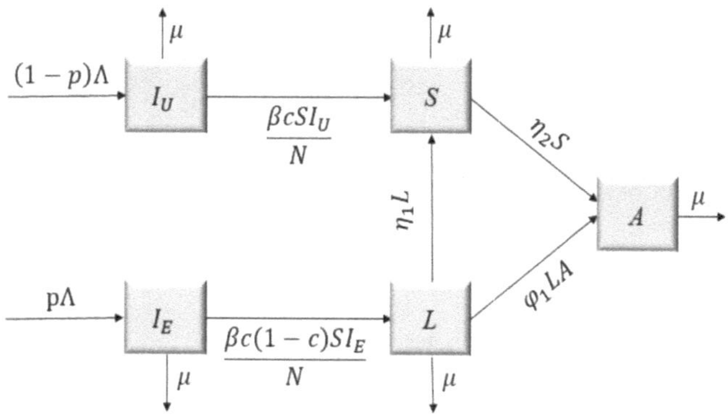

FIGURE 11.1 Schematic diagram showing transmission process of rumor.

TABLE 11.1
Presents the Definition of Each Parameter

Parameter	Definition
p	Probability that a newcomer is mature enough to not believe in the rumors
b	Number of new accounts on SNS per unit time
c	Credibility of rumor
β	Spreading rate of rumors
μ	The rate of deactivation of social media accounts the particular event
η_1	The rate at which individuals in exposed class switch to spreader class.
η_2	Forgetting rate of rumor
ϕ_1	The rate with which individuals in exposed class become aware of the vulnerability of spreading and switch to aware class

Figure 11.1. The parameters p, Λ, β, ϕ_1, η_1, η_2, μ, c are positive. Table 11.1 presents the definition of each parameter.

Theorem 11.2.1

The system's solution space (11.2) is positively invariant in the compact set $\Gamma = \left\{ (I_U, I_E, L, S, A) \in R^5 \mid 0 \le N(t) \le \frac{\Lambda}{\mu} \right\}$.

Proof Considering $N(t) = I_U(t) + I_E(t) + L(t) + S(t) + A(t)$ as the total number of active users on SNS, adding all the equations in system (11.1) gives us,

$$\frac{dN}{dt} = \Lambda - \mu N$$

Now, integrating, we get

$$N(t)=\left(N(0)-\frac{\Lambda}{\mu}\right)e^{-\mu t}+\frac{\Lambda}{\mu}. \tag{11.2}$$

Therefore, the positive invariance set for system (11.1) is given by $\Gamma=\left\{(I_U,\ I_E,\ L,\ S,\ A)\in\ R^5 \mid 0\le N(t)\le\frac{\Lambda}{\mu}\right\}$.

11.3 DYNAMICAL NATURE OF THE MODEL NEAR EQUILIBRIA

11.3.1 Rumor Free Equilibrium and the Threshold Value of Influence

The system always has a rumor-free equilibrium (RFE) $\underline{E}=\left(\underline{I_U},\ \underline{I_E},\ 0,\ 0,\ 0\right)$, where $\underline{I_U}=\frac{(1-p)\Lambda}{\mu}$ and $\underline{I_E}=\frac{p\Lambda}{\mu}$, and $\underline{L}=\underline{S}=\underline{A}=0$. As we discussed earlier, the rumor spreading model has similarities with the disease transmission model. The 'Threshold value of influence', R_0, denotes the expected number of secondary cases by a single spreader in a completely ignorant host, similar to the basic reproduction number for the disease model. R_0 plays a key role in the propagation of the rumor. $R_0>1$ implies the rumor prevails in SNS, while $R_0<1$ implies that the rumor will eventually disappear. First, we normalize the system by taking $I_U(t)+I_E(t)+L(t)+S(t)+A(t)=1$. Then, we calculate the value of $R_0=FV^{-1}$ with the help of reference [19]. Suppose $w=y$, then from model (11.1)

$$\dot{w}=F-V$$

where

$$F=\begin{pmatrix}\dfrac{\dfrac{\beta SI_E c(1-c)}{N}-\eta_1 L}{\dfrac{\beta SI_U c}{N}+\eta_1 L}\end{pmatrix} V=\begin{pmatrix}\dfrac{\phi_1 LA+\mu L}{\eta_2 S+\mu S}\end{pmatrix}$$

Next, calculating the Jacobians F, V of F, V at the rumor free equilibrium, $\underline{E}=\left(\frac{(1-p)\Lambda}{\mu},\frac{p\Lambda}{\mu},\ 0,\ 0,\ 0\right)$, we obtain

$$F=\begin{pmatrix}-\eta_1 & \eta_1 & \dfrac{\beta I_E c(1-c)}{N} & \dfrac{\beta I_U c}{N}\end{pmatrix} \tag{11.3}$$

$$V=(\mu\ 0\ 0\ \eta_2+\mu\ ) \tag{11.4}$$

Where F gives the matrix of the rate of secondary spreader growth. We then obtain the next-generation matrix FV^{-1}.

Then the threshold value of influence is defined as $R_0=\rho\left(FV^{-1}\right)$, the largest eigenvalue of FV^{-1}. We then obtain the final parametric expression of threshold value of influence for our model with $N=1$.

$$R_0 = \frac{\beta(1-p)\Lambda c - \eta_1(\eta_2+\mu)N + \sqrt{\left(\eta_1 N(\eta_2+\mu)+\beta(1-p)\Lambda c\right)^2 + 4\eta_1\beta cp\Lambda N(1-c)(\eta_2+\mu)}}{2\mu(\eta_2+\mu)N} \tag{11.5}$$

The basic reproduction number, R_0 is of great importance in contagious disease models. If $R_0 > 1$, then the disease invades and persists, otherwise, the disease eventually wipes out. Similarly, here, we can utilize R_0 as the threshold value of influence to eradicate the rumor from SNS. Now, as R_0 depends on a large number of parameters $\beta,\ \Lambda, \mu,\ \eta_1,\ \eta_2,\ p,\ c$.

11.3.2 Existence of Rumor Prevailing Equilibrium

Here, we find the endemic state, Rumor Prevailing Equilibrium (RPE) of the system (11.1). At RPE $\left(\widehat{I_U},\ \widehat{I_E},\ \hat{L},\ \hat{S},\ \hat{A}\right)$, with $\hat{S} \neq 0$, we have

$$\begin{aligned}
&(1-p)\Lambda - \frac{\beta cSI_U}{N} - \mu I_U = 0\\
&p\Lambda - \frac{\beta cSI_E(1-c)}{N} - \mu I_E = 0\\
&\frac{\beta cSI_E(1-c)}{N} - \phi_1 LA - \eta_1 L - \mu L = 0\\
&\frac{\beta cSI_U}{N} + \eta_1 L - \eta_2 S - \mu S = 0\\
&\phi_{1LA} + \eta_2 S - \mu A = 0
\end{aligned} \tag{11.6}$$

Solving first and second equations of (11.6), we get:

$$\begin{aligned}
\hat{I}_U &= \frac{(1-p)\Lambda}{\frac{\beta Sc}{N}+\mu}\\
\hat{I}_E &= \frac{p\Lambda}{\frac{\beta Sc(1-c)}{N}+\mu}
\end{aligned} \tag{11.7}$$

Substituting the values of (11.7) into the fourth equation of the system (11.6), we get:

$$L = \left[\eta_2 + \mu - \frac{(1-p)\Lambda}{S+\frac{\mu N}{\beta c}}\right]\frac{S}{\eta_1} \tag{11.8}$$

Substituting the value of (11.8) into the fifth equation of the system (11.6), we get:

$$A = \frac{\eta_1\eta_2 S}{\mu\eta_1 - \phi_1\left(\eta_2 + \mu - \dfrac{(1-p)\Lambda}{S + \dfrac{\mu N}{\beta c}}\right)S} \tag{11.9}$$

Substituting the values of (11.7), (11.8) and (11.9), into the third equation of the system (11.6), we get:

$$a_4 S^4 + a_3 S^3 + a_2 S^2 + a_1 S - a_0 = 0 \tag{11.10}$$

where

$$a_0 = \frac{\mu^3\eta_1 N^2}{\beta^2 c^2 \phi_1(\eta_2+\mu)}\left[p\Lambda\eta_1 + \frac{(\eta_1+\mu)\left[(1-p)\Lambda-(\eta_2+\mu)\right]}{1-c}\right] a_1$$

$$= \frac{\mu N}{\beta c(1-c)}\left(\eta_1\left[\frac{\mu N}{\beta c} - \frac{(1-p)\Lambda}{\eta_2+\mu}\right]\left[p\Lambda(1-c)+\eta_2\frac{\mu N}{\beta c}\right]\right.$$

$$-(\eta_1+\mu)\left[(\eta_2+\mu)\frac{\mu^2N^2}{\beta^2c^2} - \frac{(1-p)^2\Lambda^2}{\eta_2+\mu}\right]$$

$$\left.-\frac{(2-c)\mu\eta_1}{\phi_1}\left[\frac{2p\Lambda N}{\eta_2+\mu} + (\eta_1+\mu)\left[\frac{(1-p)\Lambda}{\eta_2+\mu} - 1\right]\right]\right) a_2$$

$$= p\Lambda\eta_1\left[\frac{2\mu N}{\beta c} - \frac{(1-p)\Lambda c}{\eta_2+\mu} - \frac{\mu\eta_1}{\phi_1(\eta_2+\mu)}\right]$$

$$+\eta_1\eta_2\frac{\mu N}{\beta c(1-c)}\left[\frac{\mu N}{\beta c}(3-2c) + (2-c)\left[1 - \frac{(1-p)\Lambda}{\eta_2+\mu}\right]\right]$$

$$-\left[(\eta_2+\mu)\frac{\mu^2N^2}{\beta^2c^2} + \frac{(1-p)^2\Lambda^2}{\eta_2+\mu}\right] - \frac{\mu N}{\beta c}\frac{(\eta_1+\mu)}{(1-c)}\left[\frac{\mu N}{\beta c} - (1-p)\Lambda\right]$$

$$+\mu(\eta_1+\mu)\frac{\eta_1}{\phi_1}\left[1 - \frac{(1-p)\Lambda}{\eta_2+\mu}\right] a_3$$

$$= \eta_1\left[p\Lambda + \eta_2\frac{\mu N}{\beta c}\left(\frac{3-2c}{1-c}\right)\right] + 2(\eta_1+\mu)\left[(1-p)\Lambda - \frac{\mu N}{\beta c}\left(\frac{2-c}{1-c}\right)\right] a_4$$

$$= \eta_1\eta_2 - (\eta_2+\mu)$$

If all a_i, $i = 1,2,3,4$ are positive, then by Descarte's rule of signs, there exists a positive root of equation (11.10) (say $\hat{S}$).

Substituting the positive root $\hat{S}$ of the equation (11.10) into (11.8) and (11.9), we can claim the existence of a positive value of L (say $\hat{L}$) and A (say $\hat{A}$) such that:

$$\hat{L} = \left[\eta_2 + \mu - \frac{(1-p)\Lambda}{\hat{S} + \frac{\mu N}{\beta c}} \right] \frac{\hat{S}}{\eta_1}$$

$$\hat{A} = \frac{\eta_1 \eta_2 \hat{S}}{\mu \eta_1 - \phi_1 \left(\eta_2 + \mu - \frac{(1-p)\Lambda}{\hat{S} + \frac{\mu N}{\beta c}} \right) \hat{S}} \tag{11.11}$$

Therefore, under certain conditions, we can guarantee the existence of at least one rumor prevailing equilibrium, $\hat{E}$.

11.3.3 Local Stability Analysis around Equilibria

In this subsection, we investigate the dynamical nature of the system near the equilibria. The Jacobian matrix for the system (11.1) at the co-existing equilibrium point is given by

$$\hat{J} = (-\beta \hat{S} c - \mu \ 0 \ 0 \ -\beta c \hat{I}_U \ 0 \ 0 \ -\beta \hat{S} c(1-c) - \mu \ 0 \ -\beta \hat{I}_E c(1-c) \ 0 \ 0 \ \beta \hat{S} c(1-c)$$
$$-\phi_1 \hat{A} - \eta_1 - \mu \ \beta \hat{I}_E c(1-c) - \phi_1 \hat{L} \ \beta \hat{S} c \ 0 \ \eta_1 \ \beta c \hat{I}_U - \eta_2 - \mu \ 0 \ 0 \ 0 \ \phi_1 \hat{A} \ \eta_2 \ -\phi_1 \hat{L} - \mu) \tag{11.12}$$

Now, we obtain the Jacobian matrix J at the RFE:

$$J_0 = \begin{pmatrix} -\mu \ 0 \ 0 - \frac{\beta c(1-p)}{\mu} \ 0 \ 0 - \mu \ 0 - \frac{\beta c(1-c)p\Lambda}{\mu} \ 0 \ 0 \ 0 - \eta_1 \\ -\mu \ \frac{\beta c(1-c)p\Lambda}{\mu} \ 0 \ 0 \ 0 \ \eta_1 \ \frac{\beta(1-p)\Lambda}{\mu} - \eta_2 - \mu \ 0 \ 0 \ 0 \ 0 \ \eta_2 \ -\mu \end{pmatrix} \tag{11.13}$$

The characteristic polynomial at the RFE is given by:

$$\lambda^5 + b_4\lambda^4 + b_3\lambda^3 + b_2\lambda^2 + b_1\lambda + b_0 = 0 \tag{11.14}$$

where

$$b_0 = \mu^2\left[\mu^3 + \mu^2(\eta_1+\eta_2) + \mu(\eta_1\eta_2 - \beta(1-p)\Lambda) - \beta\eta_1\Lambda[c(1-c)p+(1-p)]\right]b_1$$

$$= \mu^2\left[5\mu^2 + 4\mu(\eta_1+\eta_2) - 4\beta(1-p)\Lambda + 3\eta_1\left[\eta_2 - \frac{\beta\Lambda}{\mu}[c(1-c)p+(1-p)]\right]\right]b_2$$

$$= \mu\left[11\mu^2 + 6\mu(\eta_1+\eta_2) - 6\beta(1-p)\Lambda + 3\eta_1\left[\eta_2 - \frac{\beta\Lambda}{\mu}[c(1-c)p+(1-p)]\right]\right]b_3$$

$$= 10\mu^2 + 4\mu(\eta_1+\eta_2) - 4\beta(1-p)\Lambda + \eta_1\left[\eta_2 - \frac{\beta\Lambda}{\mu}[c(1-c)p+(1-p)]\right]b_4$$

$$= 5\mu + \eta_1 + \eta_2 - \frac{\beta(1-p)\Lambda}{\mu}$$

Therefore, $\lambda = -\mu$ is a root of equation (11.14) with multiplicity 3. Therefore, to make the RFE stable, the roots of the following equation need to be stable.

$$\lambda^2 + \lambda\left(2\mu + \eta_1 + \eta_2 - \frac{\beta(1-p)\Lambda}{\mu}\right)$$

$$+\left[\mu^2 + \mu(\eta_1+\eta_2) + \eta_1\left[\eta_2 - \frac{\beta\Lambda}{\mu}[c(1-c)p+(1-p)]\right] - \beta(1-p)\Lambda\right] = 0$$

Now, all the roots of equation (11.14) are negative if:

$$\beta(1-p)\Lambda < 2\mu^2 + (\eta_1+\eta_2)$$

and

$$\beta(1-p)\Lambda < \mu^2 + (\eta_1+\eta_2)\mu + \eta_1\left[\eta_2 - \frac{\beta\Lambda}{\mu}[c(1-c)p+(1-p)]\right]$$

The following lemma can be inferred from the discussion above.

Using the Routh-Hurtwitz stability criterion, the rumor-free state will be stable if the following conditions hold:

i. $\beta(1-p)\Lambda < 2\mu^2 + (\eta_1+\eta_2)\mu$

ii. $\beta(1-p)\Lambda < \mu^2 + (\eta_1+\eta_2)\mu + \eta_1\left[\eta_2 - \frac{\beta\Lambda}{\mu}[c(1-c)p+(1-p)]\right].$

Also, the characteristic polynomial around the prevailing state is given by:

$$\left(\lambda+\mu+\beta c(1-c)\hat{S}\right)\left(\lambda+\mu-\beta c\hat{S}\right)\eta_1\eta_2$$
$$-\phi^2\hat{A}\hat{I}_U\left[\left(\lambda+\mu-\beta c\hat{S}\right)\left(\lambda+\mu+\eta_2-\beta c\hat{I}_U\right)+\beta^2c^2\hat{S}\hat{I}_U\right]$$
$$+[(1-c)c\beta(\lambda+\mu)\left(\lambda+\mu+\beta c\hat{S}\right)\eta_1\hat{I}_E$$
$$-\left(\lambda+\mu+\beta c(1-c)\hat{S}\right)\left(\lambda+\mu+\beta c\hat{S}\right)\left(\lambda+\mu+\phi_1\hat{A}+\eta_1\right)\left(\lambda+\mu+\eta_2-\beta c\hat{I}_U\right)$$
$$-\beta^2c^2\hat{S}\hat{I}_U]\left(\lambda+\mu-\phi_1\hat{L}\right) \tag{11.15}$$

By applying the Routh-Hurwitz stability criterion on (11.15), we can derive the condition for the local stability of prevailing state.

Theorem 11.3.1

If $R_0 > 1$, the endemic equilibrium of the model (11.1) is globally asymptotically stable in the solution space $\Gamma = \{(I_U, I_E, L, S, A) \in R^5$: with $I_U(t) \geq 0, I_E(t) \geq 0,$ $L(t) \geq 0, S(t) \geq 0, A(t) \geq 0$ and sup $N(t) \leq \frac{\Lambda}{\mu}\ \forall t\}$ if a_i's > 0 in (10).

Proof: Let us consider the Lyapunov function $W(t) = \left[\left(I_U - \hat{I}_U\right) + \left(I_E - \hat{I}_E\right) + \left(L - \hat{L}\right) + \left(S - \hat{S}\right) + \left(A - \hat{A}\right)\right]^2$. Then, $W(t) \geq 0$ in the solution space $\Gamma = \{(I_U, I_E, L, S, A) \in R^5 : I_U(t) \geq 0, I_E(t) \geq 0, L(t) \geq 0, S(t) \geq 0, A(t) \geq 0\}$. Clearly

$$W(t)\,|_{\hat{I}_U,\,\hat{I}_E,\,\hat{L},\,\hat{S},\,\hat{A}} = 0$$

and

$$W(t) > 0$$

holds at any point in Γ except $\left\{I_U = \hat{I}_U, I_E = \hat{I}_E, L = \hat{L}, S = \hat{S}, A = \hat{A}\right\}$. Then,

$$\dot{W}(t) = 2\left[\left(I_U - \hat{I}_U\right) + \left(I_E - \hat{I}_E\right) + \left(L - \hat{L}\right) + \left(S - \hat{S}\right) + \left(A - \hat{A}\right)\right]\left(\dot{I}_U + \dot{I}_E + \dot{L} + \dot{S} + \dot{A}\right)$$

we have,

$$\left(\hat{I}_U + \hat{I}_E + \hat{L} + \hat{S} + \hat{A}\right) = \frac{\Lambda}{\mu} \tag{11.16}$$

Then, we get the following form of $\dot{V}$.

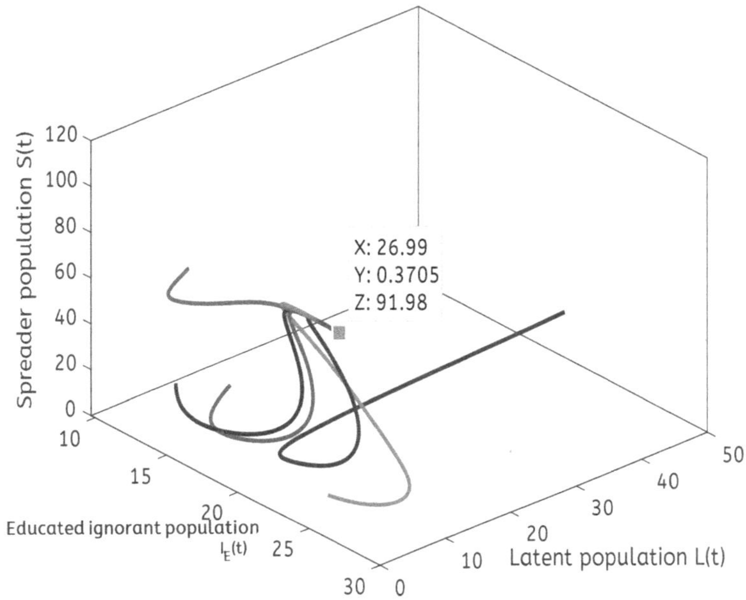

FIGURE 11.2 Solutions converge to a same point with different initial values.

$$\dot{W}(t) = 2\left[N(t) - \left(\hat{I}_U + \hat{I}_E + \hat{L} + \hat{S} + \hat{A}\right)\right]\left[\Lambda - \mu N(t)\right] \tag{11.17}$$

$$\dot{W}(t) = 2\left[N(t) - \frac{\Lambda}{\mu}\right]\left[\Lambda - \mu N(t)\right] \tag{11.18}$$

$$\dot{W}(t) = \frac{2}{\mu}\left[\mu N(t) - \Lambda\right]\left[\Lambda - \mu N(t)\right] \tag{11.19}$$

$$= -\frac{2}{\mu}\left[\Lambda - \mu N(t)\right]^2 \tag{11.20}$$

From inequality (11.20), we obtain $\dot{W}(t) \leq 0$ whenever $N(t) \leq \frac{\Lambda}{\mu}$ in Γ. Therefore, the system (11.1) is globally stable around the RPE by Lyapunov's global stability theorem. Hence, the theorem is proved (Figure 11.2).

11.4 NUMERICAL SIMULATION

This section is devoted to the numerical study of the analytical findings of model (11.1). All simulations are done using MATLAB2018.

Threshold value of influence provides the conditions under which the rumor prevails among users. Here, we examine the influence of the transmission rate of the rumor and the credibility factor on R_0, which triggers the stability switch.

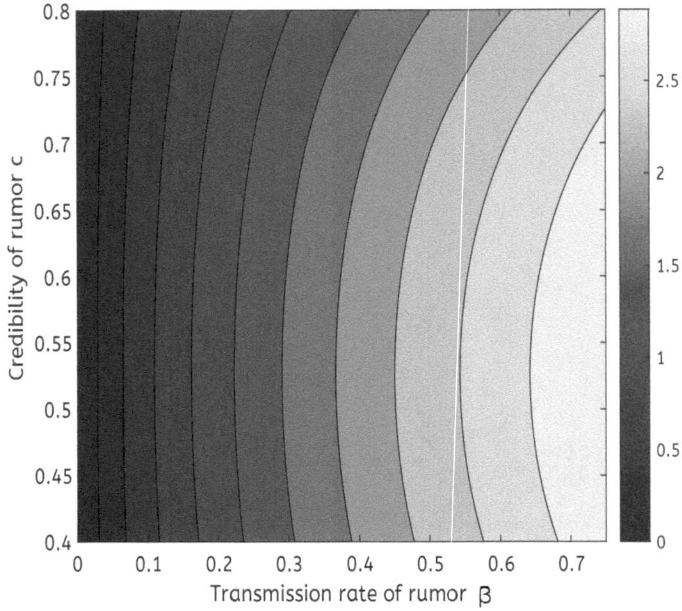

FIGURE 11.3 Contour plot of R_0 with c, β changing and keeping others fixed.

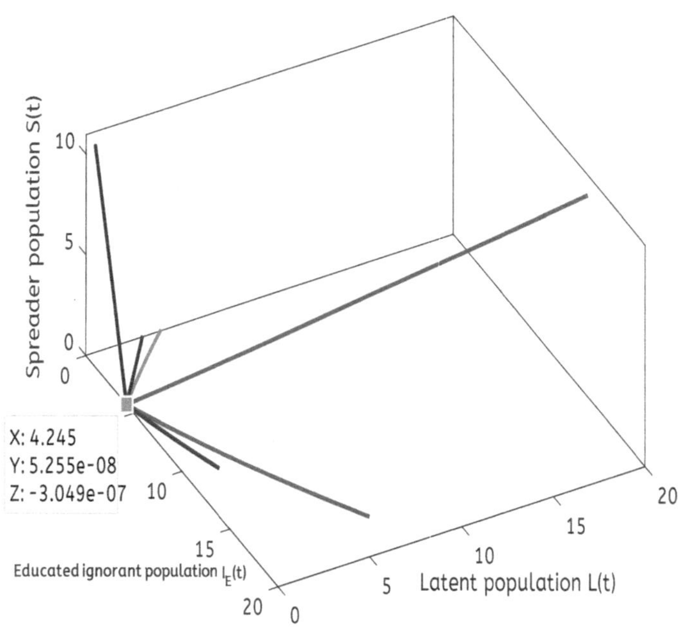

FIGURE 11.4 $R_0 = 0.98$ with parameter values $p = 0.25$, $\Lambda = 3$, $\beta = 0.03$, $c_1 = 0.25, \eta_1 = 0.01$, $\eta_2 = 0.05$, $\phi_1 = 0.012$, $\mu = 0.53$.

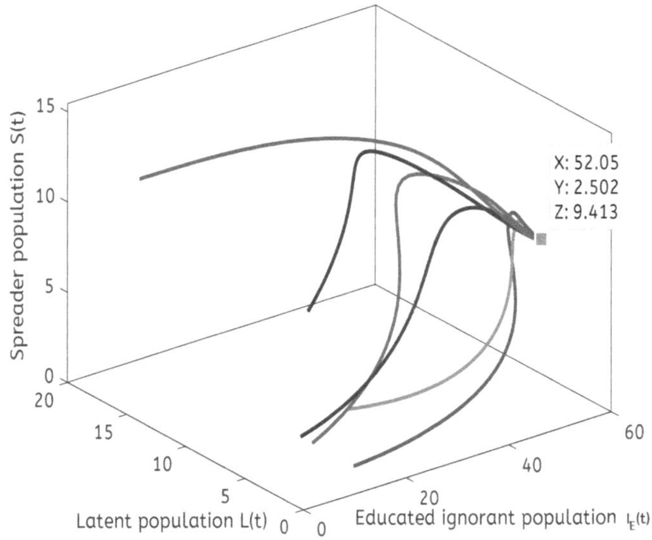

FIGURE 11.5 $R_0 = 2.355$, with parameter values $p = 0.75$, $\Lambda = 3$, $\beta = 0.6$, $c_1 = 0.5$, $\eta_1 = 0.01$, $\eta_2 = 0.01$, $\phi_1 = 0.008$, $\mu = 0.03$.

Figure 11.3 presents the effect of the transmission rate c and the credibility factor of the rumor c on R_0. β helps to increase R_0 and c has a non-linear effect on R_0. The phase portraits in Figures 11.4 and 11.5 illustrate the dynamics.

These figures show that the trajectories with different initial conditions meet at a single point on the I_U axis ($L(t) = S(t) = 0$) in Figure 11.4, and in Figure 11.5, the trajectories converge to a point $L(t) \neq 0$ and $S(t) \neq 0$. These figures present the stability switch from the rumor-free state to the prevailing state as R_0 crosses 1.

Figure 11.6 shows how the solutions of system (11.1) evolve over time. We have investigated the effect of delay in the normalized system. Here, we have incorporated the fact that the interaction term between spreaders' and immature ignorant people's individuals is delayed by a $\beta c S(t-\tau) I_U(t-\tau)$. We introduced this delay to explain that even immature ignorant individuals take time to think before spreading, and mature individuals join the hesitating or latent class. Figure 11.7 depicts how the normalized system starts to lose its stability when $\tau = 1.5$, and the time series becomes periodic instead of a straight line.

11.5 CONCLUSION

In this chapter, we study the rumor propagation process by considering user-maturity and the credibility factor of the rumor. We calculated the steady states of the system and identified the conditions for the existence of the prevailing state. Local stability analyses were performed with the help of the threshold value of influence, which is the most significant parametric expression that triggers the stability switch from the rumor free equilibrium to the prevailing equilibrium. The stability of the rumor-free

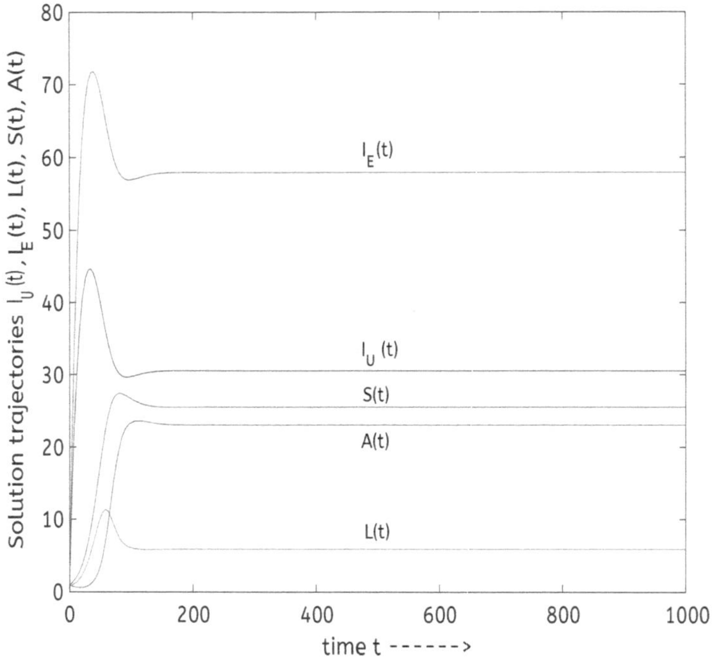

FIGURE 11.6 Evolution of populations with time with $N \neq 1$.

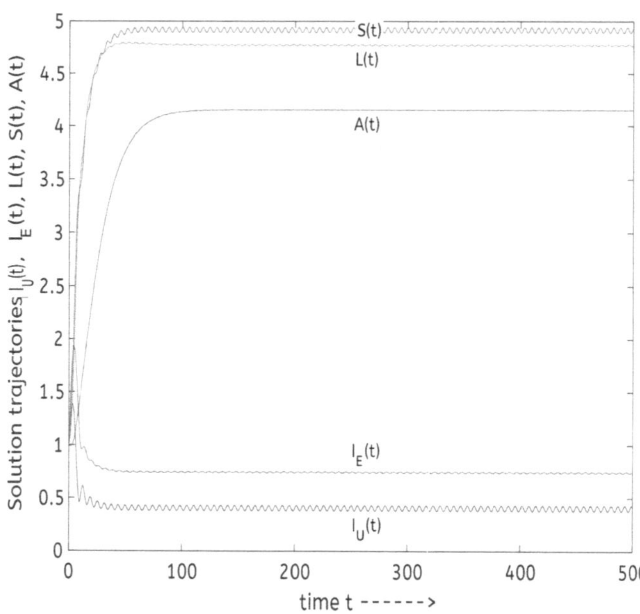

FIGURE 11.7 Evolution of populations with time with $N = 1$.

state signifies that the rumor will eventually wipe out. On the contrary, the stability of the prevailing state suggests that the rumor will persist. Moreover, we have obtained the necessary conditions for the system to be globally stable at the prevailing state, which implies that the rumor will persist regardless of the initial number of spreaders.

Here we analyzed only the spreading mechanism of rumor propagation through social networking sites. This work can be improved by applying control to rumor spreaders in emergency situations. Nowadays, reputed news agency, such as AFP has launched a website called 'AFP Covid-19 verification hub' [17] and has set up a partnership with Facebook in France to create an instructive video campaign about rumors related to COVID-19 [18]. Therefore, in future studies, we can extend our current model by incorporating optimal control techniques to regain the stability of the system.

REFERENCES

[1] Zubiaga, A., Liakata, M., Procter, R. et al. (2015). Towards detecting rumours in social media. In *Association for the Advancement of Artificial Intelligence Workshop 04.*

[2] U.S. Securities and Exchange Commission (2015). Charges: False Tweets Sent Two Stocks Reeling in Market Manipulation. SEC Press Release. https://www.sec.gov/newsroom/press-releases/2015-254

[3] BBC (2016). Twitter Account Hacked in Fake 'White House Blasts' Post. BBC World News. https://www.bbc.com/news/world-us-canada-21508660

[4] Garrett, R.K. (2019). Social media's contribution to political misperceptions in U.S. Presidential elections. *PLoS ONE*, 7(2):306–318.

[5] Daley, D. J., and Kendall, D. G. (1964). Epidemics and rumours. *Nature*, 204(4963):1118.

[6] Maki, D. P., and Thompson, M. (1973). Mathematical models and applications: with emphasis on the social life, and management sciences. Upper Saddle River: Prentice Hall.

[7] Hu, Y., Pan, Q., Hou, W., and He, M., (2018). Rumor spreading model with the different attitudes towards rumors. *Physica A: Statistical Mechanics and its Applications*, 502:331–344.

[8] Wang, J., Zhao, L., and Huang, R. (2014). 2SI2R rumor spreading model in homogeneous networks. *Physica A: Statistical Mechanics and its Applications*, 413:153–161.

[9] Xia, L.L., Jiang, G.P., Song, B., and Song, Y.R. (2015). Rumor spreading model considering hesitating mechanism in complex social networks. *Physica A: Statistical Mechanics and its Applications*, 437:295–303.

[10] Dhar, J., Jain, A., and Gupta, V. (2016). A mathematical model of news propagation on online social network and a control strategy for rumor spreading. *Social Network Analysis and Mining*, 6:57.

[11] Ghosh, M., Das, S., and Das, P. (2021). Dynamics and control of delayed rumor propagation through social networks. *Journal of Applied Mathematics and Computing*, 68:3011–3040.

[12] Huo, L., Lin, T., Fan, C., Liu, C., and Zhao, J. (2015). Optimal control of a rumor propagation model with latent period in emergency event. *Advances in Difference Equations*, 2015(1):54.

[13] Moreno, Y., Nekovee, M., and Pacheco, A. F. (2004). Dynamics of rumor spreading in complex networks. *Physical Review E*, 69:066130.

[14] Kawachi, K. (2008). Deterministic models for rumor transmission. *Nonlinear Analysis: Real World Applications*, 9(5):1989–2028.
[15] Ma, J. and Zhu, H. (2018). Rumor diffusion in heterogeneous networks by considering the individuals' subjective judgment and diverse characteristics. *Physica A: Statistical Mechanics and its Applications*, 499:276–287.
[16] Chen, X. and Wang, N. (2020). Rumor spreading model considering rumor credibility, correlation and crowd classification based on personality. *Scientific Reports*, 10(1):5887.
[17] AFP (Published on Thursday 07 May 2020a). AFP Covid-19 verification hub. https://factcheck.afp.com/afp-covid-19-verification-hub.
[18] AFP (16 APR 2020b). COVID-19: AFP launches media literacy initiative in France with Facebook. https://www.afp.com/en/agency/inside-afp/press-release/covid-19-afp-launches-media-literacy-initiative-france-facebook
[19] Driessche, P., and Watmough, J. (2002). Reproduction numbers and sub-threshold endemic equilibria for compartmental models of disease transmission. *Mathematical Biosciences*, 180:29–48.

12 Heat Transfer Analysis of Buoyancy-Driven MHD Flow with Heat Generation/Absorption in a Nanofluid-Filled Inclined Enclosure

Rujda Parveen and Tapas Ray Mahapatra

12.1 INTRODUCTION

The study of convective ferrofluid inside closed containers is of particular importance due to its wide range of engineering applications including, the food industry, heat exchangers, cooling of electronic equipment, etc. Davis [1] was the first to numerically investigate natural convection inside a square enclosure filled with air and subjected to various thermal boundary conditions at the walls. Various classical fluids (such as air, oil, water, and so on) are used to conduct heat transfer inside an enclosure. It has been stated that the low thermal conductivity of these fluids results in reduced heat transmission. To enhance thermal performance, nanoscale particles are suspended in the base fluid, creating a nanofluid, which improves the thermal conductivity of traditional fluids. Many industrial and energy applications have highlighted the significance of nanofluids such as nuclear reactor coolant, solar cells, chemical production, building heating, etc., to meet the rising demand for modern technology. References [2,3] provide a comprehensive review of the work on nanofluids and their effects on heat transfer enhancement applications in detail. Buoyancy-driven flow in closed cavities using nanofluid as the medium is the topic of interest for many researchers nowadays. Numerous articles in the literature consider classical cavities filled with nanofluid [4–6].

Natural convection in non-square cavities is more complicated than in square cavities. Heat transfer in complex and irregularly shaped enclosures is one of the contemporary techniques used to enhance the heat transfer rate. Such enclosures have many practical applications in engineering, such as in electronic devices, solar collectors, robots, fuel cells, etc. Various numerical studies have been conducted to examine the flow behavior in irregular cavities filled with nanofluid [7–9]. The investigation of Al_2O_3-water nanosuspension inside an H-shaped enclosure with a V-shaped baffle

DOI: 10.1201/9781003638414-12

was numerically performed by Keramat et al. [10] to analyze the heat transmission and fluid characteristics. They concluded that the heat transfer rate improves with the enhancement of nanoparticle volume fraction. Free convection in a wavy cavity filled with a hybrid nanofluid and embedded with a heat-generating cylindrical blockage at the center of the cavity was numerically studied by Ali et al. [11]. They used the finite element method to solve their governing equations. Their study showed that the Rayleigh number and volume fraction of the hybrid nanofluid have a direct relationship with the nanoparticle heat transfer rate, while they have an inverse relationship with the Hartmann number. Using a non-homogeneous dynamic model, Uddina et al. [12] reported a numerical analysis of free convection inside a square-shaped enclosure filled with copper oxide nanofluid and featuring a wavy upper wall, subjected to a uniform magnetic field. Their findings demonstrated that adding nanoparticles to the base fluid and increasing the number of undulations enhances the density of the nanoparticles and speeds up the heat transfer performance. Aghakhani et al. [13] studied the impact of magnetic fields and radiation on entropy production and heat transfer of alumina/water nanofluid in a tilted enclosure with a curved fin at the lower part. They found that increasing the Hartmann number causes a reduction in the heat transfer rate and entropy generation.

Ferrofluids, also called magnetic nanofluids, are composed of magnetic nanoparticles (Fe, Co, Fe_2O_3, etc.) suspended in a base fluid. Compared to non-magnetic nanofluids, ferrofluids exhibit distinct behavior under the influence of a magnetic field (Cu/H_2O, Al_2O_3/H_2O, etc.). Water-based magnetic fluids have significant applications in medical science and drug delivery. Lee and Kim [14] conducted a heat transfer enhancement study using ferrofluid inside a square enclosure subjected to various obstacles (square, rhombus, circle). Their findings demonstrated how the enclosure's flow phenomena and heat transmission are impacted by modifications to the obstacle's internal shape. Sadeghi et al. [15] analyzed buoyancy-driven flow within a ferrofluid-filled enclosure of wavy shape, equipped with two circular cylinders, under the influence of a constant magnetic field and thermal radiation. The results showed that applying a magnetic field and thermal radiation can improve the convective flow inside the enclosure. Under the influence of a uniformly inclined magnetic field, Kushawaha et al. [16] performed double-diffusive natural convection using magnetic and non-magnetic nanofluids inside a square cavity. They reported that the effects of magnetic fields on mass and heat transport are more significant at larger and lower Rayleigh numbers, respectively.

Numerous researchers have considered natural convective heat transfer with the heat generation/absorption effect an interesting area of research because of its significance in numerous engineering applications, including geophysics and energy-related domains. In the presence of an internal heat source and a magnetic field, Selimefendigil and Oztop [17] examined the formation of entropy and natural convection inside a cavity filled with nanofluid with variously shaped walls. Using the finite difference approach, Chamkha and Naser [18] investigated double-diffusive hydromagnetic convective flow numerically in a rectangular cavity. They discovered that the mass and heat transfer processes, along with the flow characteristics within the enclosure, are significantly influenced by the intensity of the magnetic field and heat generation/absorption. In an inclined chamber with an adiabatic obstacle and heat generation/absorption effect, Sam and Elshehabey [19] investigated the buoyancy-driven flow of nanofluids. Both square and circular obstacles have been taken into consideration. They discovered that the heat transmission rate of the square obstacle is higher than that of the circular cylinder. Reference [20] analyzed entropy generation in a nanofluid-filled cavity having heat generation or absorption under the influence of the magnetic field. They observed that the addition of nanoparticles tends to reduce entropy generation. The influence of the magnetic field and heat generation/absorption due to free convection in a square cavity filled with nanofluid was examined numerically by Teamah and El-Maghlany [21]. Cu-water nanofluid in a wavy enclosure built with a circular hot cylinder was studied by Abdulkadhim et al. [22] to see how heat generation and absorption affected the nanofluid's natural convection behavior. They stated that heat generation caused a remarkable decrement in heat transfer performance, whereas heat absorption improved thermal performance. From the above literature review, it is worth noting that limited research has been done considering the magnetic field, ferrofluid (Fe_3O_4/H_2O), and heat generation/absorption in a wavy enclosure. Thus, in our view, this research makes an original contribution to this significant field. Therefore, the impacts of heat generation/absorption, magnetic field influence, and enclosure inclination angle on heat transfer are explored in the present chapter using a finite difference approach followed by the Biconjugate Gradient Stabilized method (BICGStab) [23]. The considered model has a wide range of practical applications in heat exchangers and solar systems.

12.2 PHYSICAL MODEL

Figure 12.1 shows the configuration of a two-dimensional inclined enclosure along with the related boundary conditions. The left wall of the enclosure is sinusoidally heated ($T_c + (T_h - T_c)\sin\pi y\ L$), and the right wall is considered to be at a constant cold

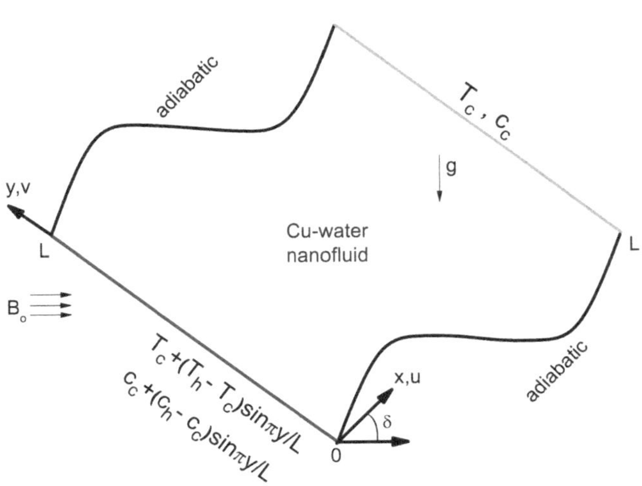

FIGURE 12.1 Schematic representation of fluid model and boundary conditions in dimensional form.

temperature (T_c). The wavy walls at the top and bottom are adiabatic. Horizontally, a uniform magnetic field (B_0) is applied. The acceleration due to gravity is taken in the vertically downward direction. The enclosure has an angle of δ with respect to the horizontal axis. The working fluid is a Newtonian ferrofluid (Fe_3O_4/ H_2O), and the enclosure is undergoing internal heat generation/absorption. The flow is described as laminar, incompressible, and steady. Additionally, the effects of radiation and viscous dissipation are assumed to be negligible. Furthermore, the amplitude (a) and the undulation number are kept constant. The density variation is described by:

$$\rho = \rho_0\left[1-\beta_T(T - T_c)\right]$$

Here, β_T is the coefficient of thermal expansion.

12.3 MATHEMATICAL FORMULATIONS

The properties of the nanofluid are presented in Table 12.1, where ϕ denotes the nanoparticles' volume fraction, and the subscripts *bf*, *s*, and *nf* denote the base fluid, solid particle, and nanofluid, respectively. Table 12.2 shows the thermophysical properties of the base fluid and nanoparticles.

The governing dimensionless equations for buoyancy-driven ferrofluid flow [22] are as follows:

$$\frac{\partial U}{\partial X}+\frac{\partial V}{\partial Y}=0, \tag{12.1}$$

$$U\frac{\partial U}{\partial X}+V\frac{\partial U}{\partial Y}=-\frac{\partial P}{\partial X}+\frac{\mu_{nf}}{\rho_{nf}\alpha_{bf}}\left(\frac{\partial^2 U}{\partial X^2}+\frac{\partial^2 U}{\partial Y^2}\right)+\frac{(\rho\beta_T)_{nf}}{\rho_{nf}\beta_{bf}}Ra\ Pr\theta\ \sin\delta \tag{12.2}$$

TABLE 12.1
The Effective Properties of Nanofluid [4]

Properties	Relation
Effective density	$\rho_{nf}=(1-\phi)\rho_f+\phi\rho_s$
Specific heat capacitance	$(\rho C_p)_{nf}=(1-\phi)(\rho C_p)_f+\phi(\rho C_p)_s$
Thermal expansion coefficient	$(\rho\beta_T)_{nf}=(1-\phi)(\rho\beta_T)_f+\phi(\rho\beta_T)_s$
Thermal diffusivity	$\alpha_{nf}=\frac{k_{nf}}{(\rho C_p)_{nf}}$
Effective dynamic viscosity [24]	$\mu_{nf}=\mu_f(1-\phi)^{-2.5}$
Effective thermal conductivity [25]	$k_{nf}=k_f\left(\frac{k_s+2k_f-2\phi(k_f-k_s)}{k_s+2k_f+\phi(k_f-k_s}\right)$
Electrical conductivity	$\sigma_{nf}=\sigma_f\left(\frac{\sigma_s+2\sigma_f-2\phi(\sigma_f-\sigma_s)}{\sigma_s+2\sigma_f+\phi(\sigma_f-\sigma_s}\right)$

TABLE 12.2
Thermophysical Characteristics of the Base Fluid and Magnetic Nanoparticles

Physical Properties	Pure Water	Fe_3O_4
C_p	4179	670
ρ	997.1	5200
k	0.613	6
$\alpha \times 10^7$	1.47	1163.1
β	21×10^{-5}	1.3×10^{-5}
σ	0.05	2.5×10^4
μ	0.001003	–

$$U\frac{\partial V}{\partial X}+V\frac{\partial V}{\partial Y}=-\frac{\partial P}{\partial Y}+\frac{\mu_{nf}}{\rho_{nf}\alpha_{bf}}\left(\frac{\partial^2 V}{\partial X^2}+\frac{\partial^2 V}{\partial Y^2}\right)+\frac{(\rho\beta_T)_{nf}}{\rho_{nf}\beta_{bf}}Ra\,Pr\theta\cos\delta-\frac{\sigma_{nf}\rho_{bf}}{\sigma_{bf}\rho_{nf}}Ha^2PrV \tag{12.3}$$

$$U\frac{\partial \theta}{\partial X}+V\frac{\partial \theta}{\partial Y}=-\frac{\alpha_{nf}}{\alpha_{bf}}\left(\frac{\partial^2 \theta}{\partial X^2}+\frac{\partial^2 \theta}{\partial Y^2}\right)+Q_0\theta \tag{12.4}$$

where $Pr\left(=\frac{\nu_{bf}}{\alpha_{bf}}\right)$ is the Prandtl number, $Ra\left(=\frac{g\beta_{bf}(T_h-T_c)L^3}{\nu_{bf}\alpha_{bf}}\right)$ is the Rayleigh number, $Ha\left(=\sqrt{\frac{\sigma_{bf}}{\mu_{bf}}}B_0L\right)$ is the Hartmann number, and, $Q_0\left(=\frac{q_0L^2}{(\rho C_p)_{nf}\alpha_{bf}}\right)$ is the dimensionless heat generation/absorption parameter.

The following dimensionless variables are used in the above governing equation:

$$X=\frac{x}{L},\ Y=\frac{y}{L},\ U=\frac{uL}{\alpha_{bf}},\ V=\frac{vL}{\alpha_{bf}},\ P=\frac{pL^2}{\rho_{nf}},\ \theta=\frac{T-T_c}{\Delta T}$$

Along the sinusoidally heated left wall, the local and average Nusselt numbers are computed as follows:

$$Nu=-\frac{k_{nf}}{k_f}\frac{\partial\theta}{\partial X}\ \text{and}\ Nu_{\text{avg}}=\int_0^1 Nu\,dY \tag{12.5}$$

Simpson's third rule of integration is applied to assess the integrals. Nusselt number is associated with the heat transfer rate from the enclosure.

12.4 NUMERICAL TECHNIQUE

The relationship between stream function and velocity components represents the flow field and is given by:

$$U = \frac{\partial \psi}{\partial Y}, \; V = -\frac{\partial \psi}{\partial X} \tag{12.6}$$

The vorticity (ω) in non-dimensional form is given by

$$\omega = \frac{\partial V}{\partial X} - \frac{\partial U}{\partial Y} \tag{12.7}$$

which gives a single non-dimensional equation

$$\frac{\partial^2 \psi}{\partial X^2} + \frac{\partial^2 \psi}{\partial Y^2} = -\omega \tag{12.8}$$

We have considered the following coordinate transformation in this study:

$$\xi = X, \; \eta = \frac{Y - F_1(X)}{F_2(X) - F_1(X)} \tag{12.9}$$

where,

$$F_1(X) = a\sin(2\pi X), \text{ and } F_2(X) = 1 + a\sin(2\pi X).$$

The dimensionless amplitude of the sinusoidal walls is taken as $a = A/L$ (=0.2). The top and bottom sinusoidal boundaries are transformed into straight lines using the above transformation.

The relationship below can be used to evaluate the governing dimensionless equations in the ξ–η domain:

$$\begin{pmatrix} \frac{\partial \xi}{\partial X} & \frac{\partial \xi}{\partial Y} \\ \frac{\partial \eta}{\partial X} & \frac{\partial \eta}{\partial Y} \end{pmatrix} = \frac{1}{J} \begin{pmatrix} \frac{\partial Y}{\partial \eta} & -\frac{\partial X}{\partial \eta} \\ -\frac{\partial Y}{\partial \xi} & \frac{\partial X}{\partial \xi} \end{pmatrix}$$

where,

$$J = \frac{\partial(X,Y)}{\partial(\xi,\eta)} = \begin{vmatrix} \frac{\partial Y}{\partial \eta} & -\frac{\partial X}{\partial \eta} \\ -\frac{\partial Y}{\partial \xi} & \frac{\partial X}{\partial \xi} \end{vmatrix}$$

is the Jacobian of the transformation.

It is difficult to solve the problem analytically due to the existence of non-linear terms in the governing equations (12.1–12.4). Using the aforesaid transformation, the equations are changed into a bi-harmonic equation in the stream function-velocity formulation. The transformed equations are discretized using a finite difference scheme. An outer-inner iteration procedure, biconjugate gradient stabilized method (BiCGStab) is used to solve the discretized stream function ψ equation. The tridiagonal system

$$\left(\frac{\partial\psi}{\partial X}\right)_{i+1,j}+4\left(\frac{\partial\psi}{\partial X}\right)_{i,j}+\left(\frac{\partial\psi}{\partial X}\right)_{i-1,j}=\frac{3}{d}\left(\psi_{i+1,j}-\psi_{i-1,j}\right),$$

$$\left(\frac{\partial\psi}{\partial \eta}\right)_{i,j+1}+4\left(\frac{\partial\psi}{\partial \eta}\right)_{i,j}+\left(\frac{\partial\psi}{\partial \eta}\right)_{i,j-1}=\frac{3}{d}\left(\psi_{i,j+1}-\psi_{i,j-1}\right).$$

Here d is taken as the step length on a uniform rectangular mesh, it is solved by using the Thomas algorithm to get ψ_ξ and ψ_η. The computed ψ_ξ and ψ_η values are then used to derive velocity values, u^* and v^*. More details of the used numerical scheme are presented in [8]. The numerical method was implemented in FORTRAN software. The cycle of numerical iteration continues until the convergence criterion (0.5×10^{-6}) is satisfied. The graphical presentation has been performed by using GNUPLOT.

12.5 RESULTS AND DISCUSSIONS

MHD free convection inside a tilted curved enclosure filled with Fe_3O_4/H_2O nanofluid and undergoing the heat generation/absorption effect is analyzed by considering the controlling parameters Rayleigh numbers, heat generation/absorption s, volume fraction of nanoparticles, and the inclination angle of the enclosure. The average Nusselt number, streamlines, and isotherms are used to illustrate the results. The Prandtl number ($Pr=6.2$) is kept fixed for all simulations.

12.5.1 Influence of Rayleigh Number and Hartmann Number

For various Hartmann and Rayleigh numbers, the flow field is depicted in Figure 12.2. It is observed that the enclosure is filled with primary clockwise circulation and secondary anticlockwise circulation at the top-left corner. For constant Ha, conduction is the predominant mode of heat transmission at low Ra. However, as Ra increases, the buoyancy forces overcome the viscous forces, implying that convection becomes the primary route of heat transfer. It can also be deduced from the streamline pattern that as Ra rises, the performance of natural convection improves, the increasing intensity of the flow circulation inside the enclosure. The value of the stream function, $|\psi_{max}|$ increases from 0.38 for $Ra=10^3$ to 11.37 for $Ra=10^5$ at $Ha=20$.

With an increase in Hartmann's number at constant Ra, the Lorentz force increases, which tends to resist the fluid flow. This results in a decrease in buoyancy force, and hence the fluid velocity decreases. As the velocity decreases, the flow strength inside the enclosure also decreases, and the value of the stream function for $Ra=10^5$ declines from $|\psi_{max}|=17.15$ for $Ha=0$ to $|\psi_{max}|=7.44$ for $Ha=40$.

The temperature field for various Hartmann and Rayleigh numbers is depicted in Figure 12.3. It is observed that as the Rayleigh number increases, the temperature contours change. The isotherm lines are regular at low Rayleigh numbers, as the heat

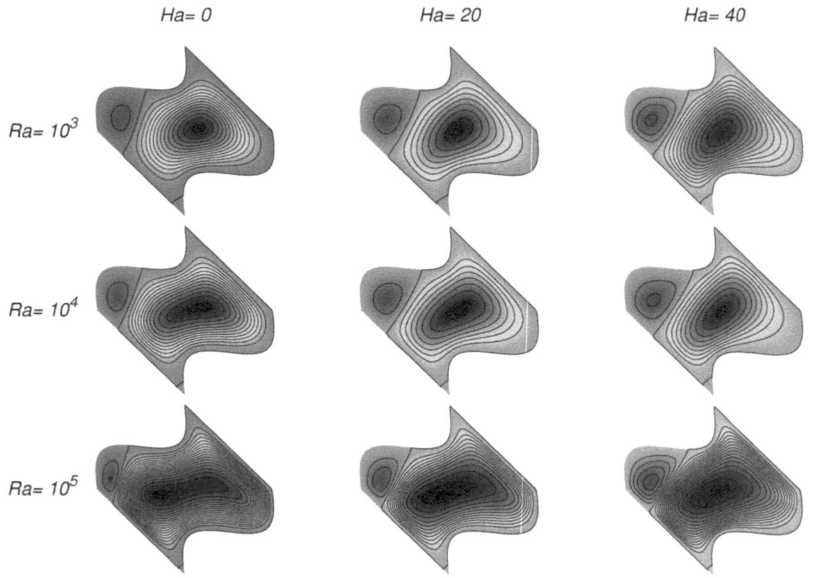

FIGURE 12.2 Streamlines for different Rayleigh numbers and Hartmann numbers at $\delta = 45^{\circ}$, $Q_o = 5$, and $\phi = 0.02$.

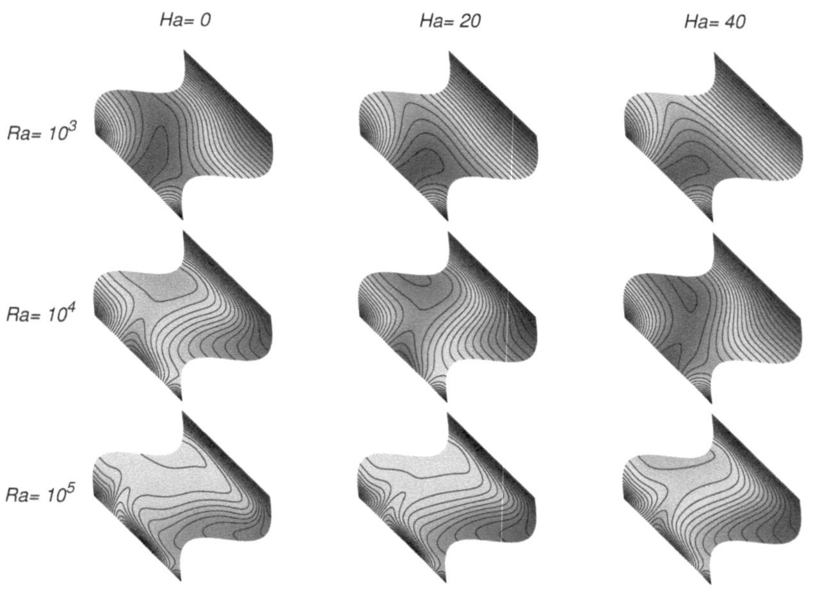

FIGURE 12.3 Isotherms for different Rayleigh numbers and Hartmann numbers at $\delta = 45^{\circ}$, $Q_o = 5$, and $\phi = 0.02$.

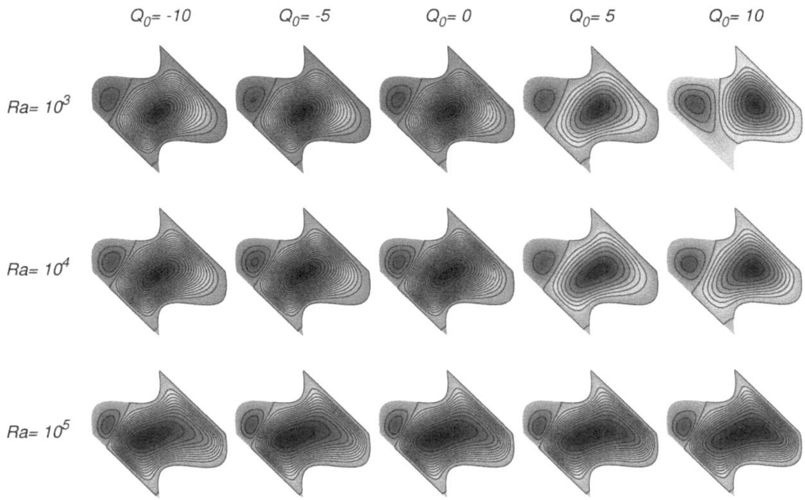

FIGURE 12.4 Streamlines for different Rayleigh numbers and heat generation/absorption coefficient at $\delta = 45^{\circ}$, $Ha = 20$, and $\phi = 0.02$.

transfer mechanism is due to conduction. However, with the rise in Rayleigh number, the contours become distorted and irregular, indicating a change in the heat transfer mode from conduction to convection. Dense isotherm lines are formed along the left and right walls due to an increased temperature gradient, resulting in the formation of a thermal boundary layer. As the magnetic strength increases, the isotherm lines become linear and parallel because of the decrease in the temperature gradient. Consequently, there is a decrease in the convection mode of heat transfer and a reduction in the thermal boundary layer along the side walls.

12.5.2 Influence of Heat Generation/Absorption Parameter

Figure 12.4 represents streamlines for different values of Rayleigh number and heat generation/absorption parameter. For the lower value of Ra (= 10^3), the conduction regime is dominant. It is observed that as the heat coefficient reduces from ($Q_o = 0$) to ($Q_o = -10$), i.e. in the heat absorption case, the flow velocity decreases, as seen from the value of the stream function, which drops from $|\psi_{max}| = 0.35$ to 0.32. On the other hand, the flow intensity gradually increases when the heat coefficient goes from ($Q_o = 0$) to ($Q_o = 10$), i.e. in the heat generation case. In addition, the value of $|\psi_{max}|$ rises from 0.35 to 1.80. Furthermore, when the Rayleigh number is increased to $Ra = 10^5$, the flow inside the enclosure shows distinct behavior. The value of the stream function increases from 11.56 to 11.82 when the heat coefficient goes from ($Q_o = 0$) to ($Q_o = -10$), and decreases from 11.56 to 11.31 when the heat coefficient goes from ($Q_o = 0$) to ($Q_o = 10$). This occurs because convection effectively dominates heat transmission at high Ra, and as a result, the heat coefficient becomes less significant as Ra grows.

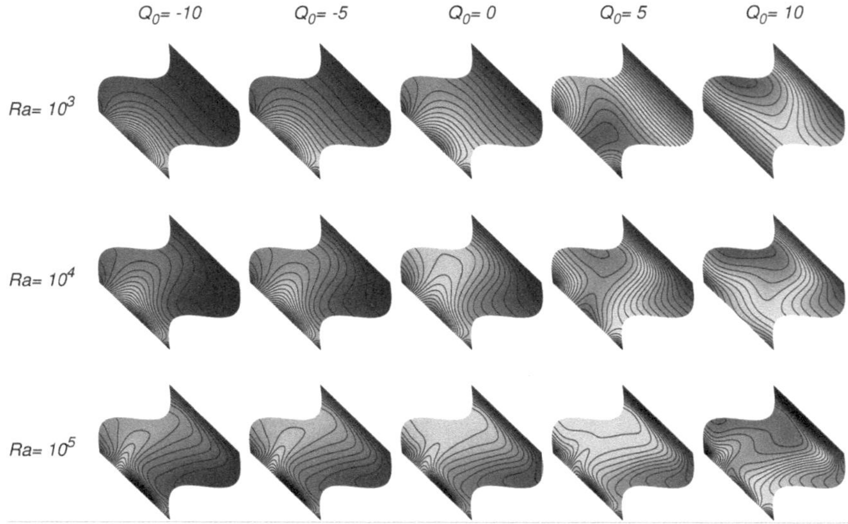

FIGURE 12.5 Isotherms for different Rayleigh numbers and heat generation/absorption coefficient at $\delta = 45^o$, $Ha = 20$, and $\phi = 0.02$.

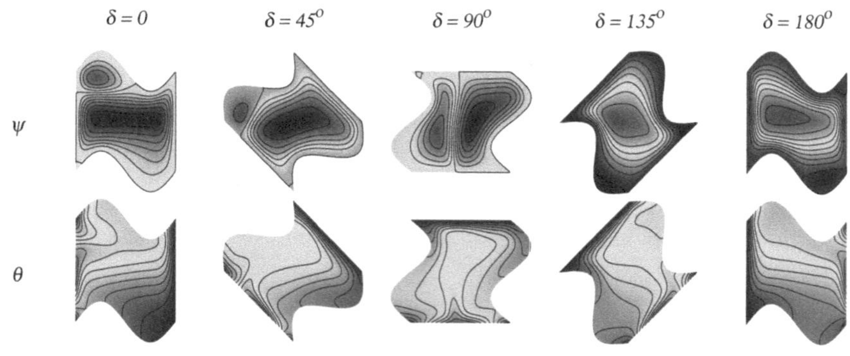

FIGURE 12.6 Streamlines and Isotherms for different inclination angles of the enclosure at $Ra = 10^5$, $Ha = 20$, $\phi = 0.02$, and $Q_o = 5$.

Figure 12.5 describes the isotherms for different values of Rayleigh number and heat generation/absorption parameter. For $Q_0 > 0$, the isothermal lines become parallel along the right wall due to a decrease in the temperature gradient. For $Q_0 < 0$, the thermal boundary layer is formed due to a strong temperature gradient. With the rise of Ra, the buoyancy forces increase, which causes convection to dominate in the enclosure.

12.5.3 Influence of Inclination Angle of the Enclosure

Figure 12.6 represents streamlines and isotherms for dissimilar inclination angles of the enclosure. From the figure, it is observed that the inclination angle plays a crucial role in improving the flow circulation within the wavy cavity. Unsymmetrical major clockwise and minor anticlockwise vortices are detected inside the cavity when $\delta = 0°$. The clockwise circulations change to anticlockwise circulations with the increase in δ. The secondary anticlockwise vortices begin to enlarge and fill half of the enclosure as the inclination angle increases to $\delta = 90°$, while the primary vortices begin to compress. This augments the flow velocity's strength, and the value of $|\psi_{max}|$ rises from 6.39 at $\delta = 0°$ to 11.58 at $\delta = 90°$. Again, with the rise in inclination angle from $\delta = 90°$ to $\delta = 135°$, the enclosure is filled with primary anticlockwise circulation, while the secondary clockwise circulation begins to vanish. The flow circulation strength increases for $\delta = 135°$, which then decreases for $\delta = 180°$. A clear change in the stream function value occurs, $|\psi_{max}| = 13.69$ for $\delta = 135°$ and $|\psi_{max}| = 8.30$ for $\delta = 180°$. It is observed that, in the case of an inclined enclosure, the fluid velocity is higher, resulting in a greater convection effect. The temperature and concentration fields begin to distort with the rise in inclination angle from $\delta = 0°$ to $\delta = 45°$ due to an increase in the temperature gradient, hence enhancing the convection effect. However, with a further increase in inclination angle, the contours tend to become linear as the direction of the heated wall moves upward, which reduces the buoyancy effect, in turn reducing the convection mode.

12.5.4 Nusselt Number

Figure 12.7 shows the effect of the Rayleigh number and Hartmann number on the average Nusselt number. For all *Ha*, it appears that the rate of heat transmission increases as the Rayleigh number grows. This occurs due to an increase in buoyancy force, which raises fluid velocity and switches the heat transfer method from conduction to convection. On the contrary, the average Nusselt number decreases with increasing Hartmann number. This is because the convection effect is diminished as a result of the Lorentz force, which lowers the flow velocity.

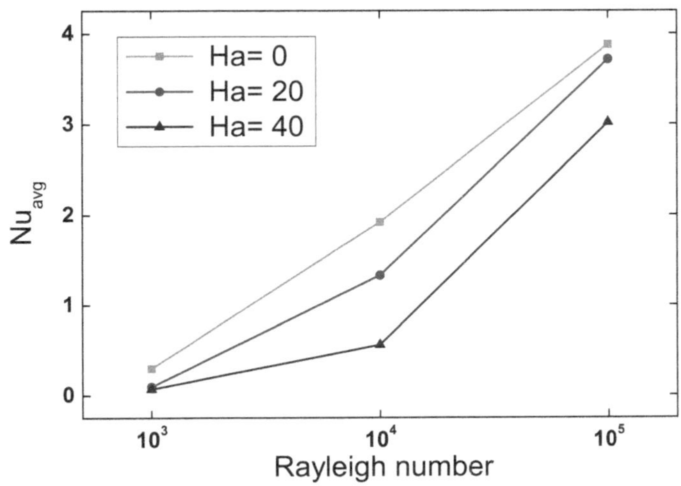

FIGURE 12.7 Variation of average Nusselt number for different values of Rayleigh number and Hartmann number at $\delta = 45°$, $Q_o = 5$, and $\phi = 0.02$.

The impact of the solid volume percentage, Rayleigh number, and Hartmann number of nanoparticles on the average Nusselt number on the left hot wall is shown in Figure 12.8. The addition of nanoparticles' volume fraction causes the heat transfer rate to de-escalate for all values of Rayleigh numbers and Hartmann numbers. This is because the thermal conductivity of the nanofluid increases with the suspension of nanoparticles, which subsequently dominates the conduction effect. It is found that at a high Rayleigh number, $Ra=10^5$, and at a low Hartmann number $Ha=0$, higher heat transfer is obtained.

The average Nusselt number fluctuation for different inclination angles and volume fraction of nanoparticles is shown in Figure 12.9. It is observed that as the nanoparticle volume fraction increases, the heat transfer rate decreases for all inclination angles. The heat transfer rate increases with the increase in inclination angle from

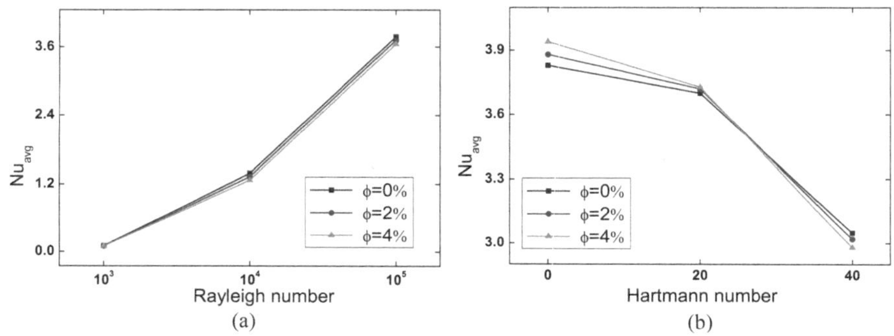

FIGURE 12.8 Variation of average Nusselt number for different Rayleigh numbers, Hartmann numbers, and ϕ at $Ha=20$, $Q_o=5$, and $\delta=45°$.

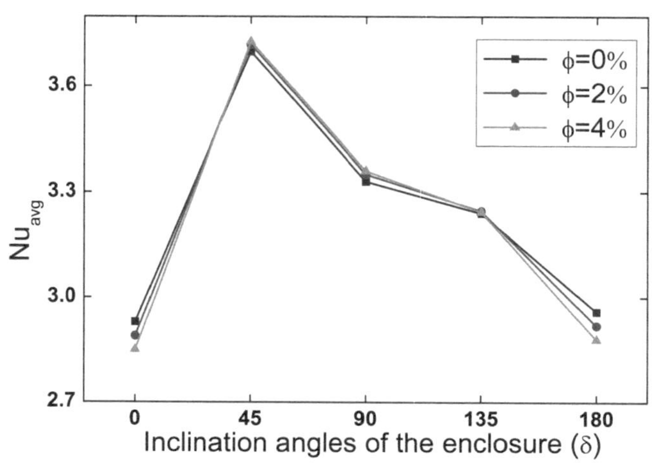

FIGURE 12.9 Variation of average Nusselt number for different inclination angles of the enclosure and ϕ at $Ra=10^5$, $Ha=20$, and $Q_o=5$.

$\delta = 0°$ to $\delta = 45°$ due to the increase in buoyancy force. The trend reverses for further increase in the inclination angle due to the reverse buoyancy effect for all values of ϕ.

12.6 CONCLUSIONS

This chapter describes the use of numerical simulation to examine the effects on the temperature field and flow pattern in a wavy cavity caused by the heat generation/absorption parameter, Rayleigh number, nanoparticle volume fraction, Hartmann number, and enclosure inclination angle. It is found that the rate of heat transmission improves significantly as the Rayleigh number increases. Increasing the nanoparticle volume fraction and Hartmann number results in a considerable reduction in heat transfer. The mass transfer rate, heat transfer rate, and fluid flow pattern are all influenced by heat generation. By increasing the heat generation parameter value, the flow strength decreases, but when the heat absorption parameter value decreases, the flow strength increases. By changing the locations of the hot and cold walls, the geometry inclination angle greatly enhances convective heat transfer rates. The highest transfer rate is attained in the case of the inclined enclosure, at an angle of 45°.

REFERENCES

[1] Davis, G. (1983), Natural convection of air in a square cavity: A benchmark numerical solutions. *Int. J. Numer. Methods Fluids*, 3, 249–264.

[2] Murshed, S., Leong, K., & Yang, C. (2008). Thermophysical and electrokinetic properties of nanofluids- A critical review. *App. Therm. Eng.*, 28, 2109–2125.

[3] Narankhishig, Z., Jeonggyun, H., Lee, H., & Cho, H. (2021). Convective heat transfer characteristics of nanofluids including the magnetic effect on heat transfer enhancement-a review. *App. Therm. Eng.*, 193, 116987.

[4] Thangavelu, M., Nithyadevi, N., Oztop, H. F., & Abu-Hamdeh, N. H. (2018). Natural convective heat transfer of Ag-water nanofluid flow inside enclosure with center heater and bottom heat source. *Chin. J. Phy.*, 56, 1497–1507.

[5] Teamah, M. A., Elsafty, A. F., Elfeky, M. Z., & El-Gazzar, E. Z (2011). Numerical simulation of double-diffusive natural convective flow in an inclined rectangular enclosure in the presence of magnetic field and heat source, part A: Effect of Rayleigh number and inclination angle. *Alexandria Eng. J.*, 50, 269–282.

[6] Sheikhzadeh, G. A., Arefmanesh, A., Kheirkhah, M. H., & Abdollahi, R. (2011). Natural convection of Cu-water nanofluid in a cavity with partially active side walls. *Eur. J. Mech. B Fluids*, 30, 166–176.

[7] Nia, S. N., Rabiei, F., Rashidi, M. M. & Kwang, T. M. (2020). Lattice Boltzmann simulation of natural convection heat transfer of a nanofluid in a L-shape enclosure with a baffle. *Results Phys.*, 19, 103413.

[8] Parveen, R., & Mahapatra, T. R. (2019). Numerical simulation of MHD double diffusive natural convection and entropy generation in a wavy enclosure filled with nanofluid with discrete heating. *Heliyon*, 5, e02496.

[9] Raizah, Z. A., Ahmed, S. E., & Aly, A. M. (2020). ISPH simulations of natural convection flow in E-enclosure filled with a nanofluid including homogeneous/heterogeneous porous media and solid particles. *Int. J. Heat Mass Trans.*, 160, 120153.

[10] Keramat, F., Dehghan, P., Mofarahi, M., & Lee, C. (2020). Numerical analysis of natural convection of alumina–water nanofluid in an H-shaped enclosure with a V-shaped baffle. *J. Taiwan Ins. Chem. Eng.*, 111, 63–72.

[11] Ali, M. M., Akhter, R., & Alim, M. A. (2021). Hydromagnetic natural convection in a wavy walled enclosure equipped with hybrid nanofluid and heat generating cylinder. *Alex. Eng. J.*, 60, 5245–5265.

[12] Uddina, M. J., Rasela, S. K., Rahman, M. M., & Vajravelu, K. (2020). Natural convective heat transfer in a nanofluid-filled square vessel having a wavy upper surface in the presence of a magnetic field. *Ther. Sci. Eng. Prog.*, 19, 100660.

[13] Aghakhani, S., Pordanjani, A. H., Afrand, M., Sharifpur, M., & Meyer, J.P. (2020). Natural convective heat transfer and entropy generation of alumina/water nanofluid in a tilted enclosure with an elliptic constant temperature: Applying magnetic field and radiation effects. *Int. J. Mech. Sci.*, 174, 105470.

[14] Lee, M., & Kim, Y. (2020). Effect of non-uniform magnetic fields on the characteristics of ferrofluid flow in a square enclosure. *J. Mag. Mag. Mat.*, 506, 166697.

[15] Sadeghi, M. S., Tayebi, T., Dogonchi, A. S., Nayak, M. K., & Waqas, M. (2021). Analysis of thermal behavior of magnetic buoyancy-driven flow in ferrofluid–filled wavy enclosure furnished with two circular cylinders. *Int. Comm. Heat Mass Trans.*, 120, 104951.

[16] Kushawaha, D., Yadav, S., & Singh, D. K. (2021). Magnetic field effect on double-diffusion with magnetic and non-magnetic nanofluids. *Int. J. Mech. Sci.*, 191, 106085.

[17] Selimefendigil, F., & Oztop, H. F. (2015). Natural convection and entropy generation of nanofluid filled cavity having different shaped obstacles under the influence of magnetic field and internal heat generation. *J. Taiwan Ins. Chem. Eng.*, 56, 42–56.

[18] Chamkha, A. J., & Al-Naser, H. (2002). Hydromagnetic double-diffusive convection in a rectangular enclosure with opposing temperature and concentration gradients. *Int. J. Heat Mass Trans.*, 45, 2465–2483.

[19] Ahmed, S. E., & Elshehabey, H. (2018). Buoyancy-driven flow of nanofluids in an inclined enclosure containing an adiabatic obstacle with heat generation/absorption: Effects of periodic thermal conditions. *Int. J. Heat Mass Trans.*, 124, 58–73.

[20] Mahmoudi, A., Mejri, I., Abbassi, M. A., & Omri, A. (2014). Analysis of the entropy generation in a nanofluid-filled cavity in the presence of magnetic field and uniform heat generation/absorption. *J. Mol. Liq.*, 198, 63–77.

[21] Teamah, M. A., & El-Maghlany, W. M. (2021). Augmentation of natural convective heat transfer in square cavity by utilizing nanofluids in the presence of magnetic field and uniform heat generation/absorption. *Int. J. Therm. Sci.*, 58, 130–142.

[22] Abdulkadhim, A., Hamzah, H. K., Ali, H., Yıldız, C., Abed, A. M., Abed, I. M., & Arıcı, M. (2021). Effect of heat generation and heat absorption on natural convection of Cu water nanofluid in a wavy enclosure under magnetic field. *Int. Comm. Heat Mass Trans.*, 120, 105024.

[23] Gupta, M. M., & Kalita, J. C. (2005). A new paradigm for solving Navier-Stokes equation: Stream function-vorticity formulation. *J. Comp. Phys.*, 207, 52–68.

[24] Brinkman, H. C. (1952). The viscosity of concentrated suspensions and solutions. *J. Chem. Phys.*, 20, 571–581.

[25] Maxwell, J. C. (1873). *A Treatise on Electricity and Magnetism, II*, Oxford University Press, Cambridge, UK.

13 Application of a New Multi-Domain Local Linearization Method in Investigating the Soret and Dufour Effects on Natural Convection over the Vertical Frustum of a Cone in a Nanofluid

Ibukun Sarah Oyelakin, Shina Daniel Oloniiju, Yusuf Olatunji Tijani, and Olumuyiwa Otegbeye

13.1 INTRODUCTION

Differential models that describe the dynamics of physical phenomena mostly appear in nonlinear forms. Due to the nonlinearity and complexity of these differential models, numerical solutions are often sought instead of closed-form solutions. A type of physical process that involves the analysis of nonlinear, multidimensional dynamical systems is heat and mass transfer processes. These processes are primarily modelled by:

i. The Navier-Stokes equations describe the flow of liquid substances and are mathematically expressed as the mass and linear momentum conservation equations. The assumption that mass and momentum are conserved allows for the development of equations that describe the velocity and density of the fluid in terms of properties such as shear stress, pressure, and other body forces acting on the fluid [1].

ii. The energy conservation equation, based on the law of thermodynamics, assumes that no energy is created, destroyed, or lost during a heat transfer process. This conservation allows the quantification of the fluid temperature in terms of its dependencies on specific thermophysical properties [2].

DOI: 10.1201/9781003638414-13

iii. The concentration conservation equation, similar to the energy balance equation, assumes that the fluid concentration is preserved during the heat and mass transfer process. It quantifies the fluid concentration in terms of properties such as mass diffusivity [3].

These equations, whose complexity depends on the assumptions made for the process or fluid properties, coupled with specified initial and boundary conditions, are analyzed as a system. In the case of a system with nonlinearity and multidimensionality, numerical solutions are often sought. Such nonlinear, multidimensional systems are ubiquitous in analyzing heat and mass transfer processes. This chapter aims to introduce readers to the multi-domain, local linearization spectral-based numerical method to solve such systems of differential equations. To demonstrate the applicability of this method, a heat and mass transfer model that describes the effects of Soret-Dufour type diffusivity on natural convection in nanofluid streaming over a cone's vertical frustum is considered. Following Motsa et al. [4], the non-dimensionalized system of nonlinear partial differential equations are given by

$$\left.\begin{aligned}
&f''' + \left(R + \frac{3}{4}\right) ff'' - \frac{1}{2} f'^2 + \theta + NcS - Nr\gamma = \xi\left(f'\frac{\partial f'}{\partial \xi} - f''\frac{\partial f}{\partial \xi}\right),\\
&\frac{1}{Pr}(1 + Rd)\theta'' + \left(R + \frac{3}{4}\right) f\theta' + Nb\gamma'\theta' + Nt\theta'^2 + D_f S'' = \xi\left(f'\frac{\partial \theta}{\partial \xi} - \theta'\frac{\partial f}{\partial \xi}\right),\\
&\frac{1}{Le}\gamma'' + \left(R + \frac{3}{4}\right) f\gamma' + \frac{1}{Le}\frac{Nt}{Nb}\theta'' = \xi\left(f'\frac{\partial \gamma}{\partial \xi} - \gamma'\frac{\partial f}{\partial \xi}\right),\\
&\frac{1}{Sc}S'' + \left(R + \frac{3}{4}\right) fS' + S_T\theta'' = \xi\left(f'\frac{\partial S}{\partial \xi} - S'\frac{\partial f}{\partial \xi}\right),
\end{aligned}\right\} \tag{13.1}$$

solved subject to the following boundary conditions:

$$\left.\begin{aligned}
&\text{at } \eta = 0 : f = 0, \quad f' = 0, \quad \theta = \gamma = S = 1,\\
&\text{as } \eta \to \infty : f' \to 0, \quad \theta \to 0, \quad \gamma \to 0, \quad S \to 0,
\end{aligned}\right\} \tag{13.2}$$

where the dimensionless parameters are defined as: D_f is the Dufour parameter, Le is the Lewis number, Nb is the Brownian motion parameter, Nc is the regular buoyancy parameter, Nr is the nanofluid buoyancy ratio, Nt is the thermophoresis parameter, Sc is the Schmidt number, and S_T is the Soret parameter, see reference [4].

As mentioned earlier, solutions to equations such as those in equation (13.1) may not exist in closed form. If they do exist, they are, perhaps, in terms of some special functions, special integrals, or infinite series. Owing to this, applied mathematicians more often approach such problems numerically. The literature contains several, and an exhaustive list of, numerical techniques for solving systems of equations that describe heat and mass transfer processes. A case in point is Lewis et al. [5], who provided an excellent introduction to heat transfer analysis through the finite element method. The finite difference analysis of a model that describes heat and mass flux in liquid substances with transpiration and internal heat generation was studied by Chau and Gaffney [6]. The studies by Chai and Pantankar [7] and Batool et al. [8] used the finite volume method to analyze heat and mass transfer processes numerically.

The numerical methods that form the premise on which the technique described in this chapter is developed are the spectral and collocation methods. On the one hand, spectral methods are an exponentially convergent class of numerical techniques used in discretizing differential equations. An essential component of spectral methods is the trial function, which is typically global, smooth, and almost orthogonal. This method assumes that the solution of a differential equation can be represented in terms of a trial function, as seen in Canuto et al. [9]. This assumption makes spectral methods suitable for finding the solutions of smooth partial differential equations, such as those that model heat and mass transfer processes, thus providing highly accurate numerical solutions that minimize computer memory and computational time. A significant limitation of spectral methods is their low accuracy for non-smooth partial differential equations and problems defined on irregular and complex geometries, which are model equations that are more easily solvable by methods such as finite elements. Spectral methods have been employed to solve problems in other fields of applied mathematics, such as meteorology [10] and quantum field theory [11]. On the other hand, collocation involves approximating the solution of differential equations in terms of a predefined function at some predefined points, called collocation points, in a defined domain. The solution of the differential equation is sought to satisfy the differential equation and its boundary or initial conditions at these distinct collocation points [12]. Over the last few years, the spectral and collocation methods, along with specific linearization techniques, have been combined to solve differential equations that model heat and mass transfer. For example, see studies such as (Soomro et al. [13], Hernandez-Calderon et al. [14], Li et al. [15], Lekoko et al. [16], and Oloniiju et al. [17]).

This chapter uses a spectral collocation method on a piecewise-defined domain with local linearization techniques to solve the system in equation (13.1). We remark here that our focus is on the numerical method and not on the physics of the problem. The local linearization technique partly decouples the system based on the generalized Newton-Raphson method in Bellman and Kalaba [18]. The spectral collocation-based local linearization technique has been applied to solve nonlinear ordinary and partial differential equations (Motsa [19], Motsa, RamReddy, and Rao [4], Otegbeye et al. [20]). As we shall later see in this chapter, the method is accurate and computationally efficient.

13.2 DESCRIPTION OF THE MULTI-DOMAIN SPECTRAL LOCAL LINEARIZATION METHOD (MD-SLLM)

This section presents the description of the MD-SLLM for a general nonlinear system of coupled partial differential equations (PDEs) given as follows:

$$\mathbf{G}_l[\mathbf{v}_1(\eta,\xi),\mathbf{v}_2(\eta,\xi),\ldots,\mathbf{v}_N(\eta,\xi)]=\mathbf{0}, \qquad l=1,2,\ldots,N. \tag{13.3}$$

It is important to start from the development of the spectral local linearization method (SLLM) introduced in (Motsa [19]) to understand the implementation of the MD-SLLM. The SLLM for a system of ordinary differential equations involves linearizing the l-th set of $(\mathbf{v}_1, \mathbf{v}_2, \cdots, \mathbf{v}_N)$ in the l-th equation of the system (13.3) using a first-order univariate Taylor series expansion, which decouples the system. The decoupling is done by expanding the nonlinear terms of the v_l-th variable and its derivatives in the corresponding $\mathbf{G}_l$-th equation through a first-order Taylor expansion. The approximations $(\mathbf{v}_{1,r}, \mathbf{v}_{2,r}, \cdots, \mathbf{v}_N)$ are selected so that

$$\left.\begin{aligned}
&\mathbf{G}_1[(\mathbf{v}_{1,r},\ldots,\mathbf{v}_{N,r})+\Delta(\mathbf{v}_1,\ldots,\mathbf{v}_N)] \approx \mathbf{G}_1[\mathbf{v}_{1,r},\ldots,\mathbf{v}_{N,r}] \\
&\qquad + \sum_{p=0}^{m_{1,1}} \frac{\partial^{(p)}\mathbf{G}_1[\mathbf{v}_{1,r},\ldots,\mathbf{v}_{N,r}]}{\partial \mathbf{v}_1^{(p)}}\left(\Delta \mathbf{v}_1^{(p)}\right), \\
&\mathbf{G}_2[(\mathbf{v}_{1,r+1},\mathbf{v}_{2,r},\ldots,\mathbf{v}_{N,r})+\Delta(\mathbf{v}_1,\ldots,\mathbf{v}_N)] \approx \mathbf{G}_2[\mathbf{v}_{1,r+1},\mathbf{v}_{2,r},\ldots,\mathbf{v}_{N,r}] \\
&\qquad + \sum_{p=0}^{m_{2,2}} \frac{\partial^{(p)}\mathbf{G}_2[\mathbf{v}_{1,r+1},\mathbf{v}_{2,r},\ldots,\mathbf{v}_{N,r}]}{\partial \mathbf{v}_2^{(p)}}\left(\Delta \mathbf{v}_2^{(p)}\right), \\
&\quad\vdots \qquad\qquad \vdots \\
&\mathbf{G}_N[(\mathbf{v}_{1,r+1},\ldots,\mathbf{v}_{N-1,r+1},\mathbf{v}_{N,r})+\Delta(\mathbf{v}_1,\ldots,\mathbf{v}_N)] \approx \mathbf{G}_N[\mathbf{v}_{1,r+1},\ldots,\mathbf{v}_{N-1,r+1},\mathbf{v}_{N,r}] \\
&\qquad + \sum_{p=0}^{m_{N,N}} \frac{\partial^{(p)}\mathbf{G}_N[\mathbf{v}_{1,r+1},\ldots,\mathbf{v}_{N-1,r+1},\mathbf{v}_{N,r}]}{\partial \mathbf{v}_N^{(p)}}\left(\Delta \mathbf{v}_N^{(p)}\right),
\end{aligned}\right\} \quad (13.4)$$

where $\Delta \mathbf{v}_l^{(p)} = \mathbf{v}_{l,r+1}^{(p)} - \mathbf{v}_{l,r}^{(p)}$ (with subscript r denoting the current iteration level and $r+1$ –subscript, the next iteration level). The system (13.4) expands to

$$\left.\begin{aligned}
&\mathbf{G}_1[(\mathbf{v}_{1,r},\ldots,\mathbf{v}_{N,r})+\Delta(\mathbf{v}_1,\ldots,\mathbf{v}_N)] \approx \mathbf{G}_1[\mathbf{v}_{1,r},\ldots,\mathbf{v}_{N,r}] \\
&\qquad + \sum_{p=0}^{m_{1,1}} \frac{\partial^{(p)}\mathbf{G}_1[\mathbf{v}_{1,r},\ldots,\mathbf{v}_{N,r}]}{\partial \mathbf{v}_1^{(p)}}\mathbf{v}_{1,r+1}^{(p)} - \sum_{p=0}^{m_{1,1}} \frac{\partial^{(p)}\mathbf{G}_1[\mathbf{v}_{1,r},\ldots,\mathbf{v}_{N,r}]}{\partial \mathbf{v}_1^{(p)}}\mathbf{v}_{1,r}^{(p)}, \\
&\mathbf{G}_2[(\mathbf{v}_{1,r+1},\mathbf{v}_{2,r},\ldots,\mathbf{v}_{N,r})+\Delta(\mathbf{v}_1,\ldots,\mathbf{v}_N)] \approx \mathbf{G}_2[\mathbf{v}_{1,r+1},\mathbf{v}_{2,r},\cdots,\mathbf{v}_{N,r}] \\
&\qquad + \sum_{p=0}^{m_{2,2}} \frac{\partial^{(p)}\mathbf{G}_2[\mathbf{v}_{1,r+1},\cdots,\mathbf{v}_{N,r}]}{\partial \mathbf{v}_2^{(p)}}\mathbf{v}_{2,r+1}^{(p)} - \sum_{p=0}^{m_{2,2}} \frac{\partial^{(p)}\mathbf{G}_2[\mathbf{v}_{1,r+1},\cdots,\mathbf{v}_{N,r}]}{\partial \mathbf{v}_2^{(p)}}\mathbf{v}_{2,r}^{(p)}, \\
&\quad\vdots \qquad\qquad \vdots \\
&\mathbf{G}_N[(\mathbf{v}_{1,r+1},\cdots,\mathbf{v}_{N-1,r+1},\mathbf{v}_{N,r})+\Delta(\mathbf{v}_1,\cdots,\mathbf{v}_N)] \approx \mathbf{G}_N[\mathbf{v}_{1,r+1},\cdots,\mathbf{v}_{N-1,r+1},\mathbf{v}_{N,r}] \\
&\qquad + \sum_{p=0}^{m_{N,N}} \frac{\partial^{(p)}\mathbf{G}_N[\mathbf{v}_{1,r+1},\cdots,\mathbf{v}_{N,r}]}{\partial \mathbf{v}_N^{(p)}}v_{N,r+1}^{(p)} - \sum_{p=0}^{m_{N,N}} \frac{\partial^{(p)}G_N[\mathbf{v}_{1,r+1},\cdots,\mathbf{v}_{N,r}]}{\partial \mathbf{v}_N^{(p)}}\mathbf{v}_{N,r}^{(p)}.
\end{aligned}\right\}$$

(13.5)

The coupled system (13.3) is now rewritten using the expansion given in equation (13.5) so that the following decoupled linear system is obtained

$$\left.\begin{aligned}
&\sum_{p=0}^{m_{1,1}} \frac{\partial^{(p)}\mathbf{G}_1[\mathbf{v}_{1,r},\cdots,\mathbf{v}_{N,r}]}{\partial \mathbf{v}_1^{(p)}}\mathbf{v}_{1,r+1}^{(p)} = R_1,\\
&\sum_{p=0}^{m_{2,2}} \frac{\partial^{(p)}\mathbf{G}_2[\mathbf{v}_{1,r+1},\mathbf{v}_{2,r},\cdots,\mathbf{v}_{N,r}]}{\partial \mathbf{v}_2^{(p)}}\mathbf{v}_{2,r+1}^{(p)} = R_2,\\
&\vdots\\
&\sum_{p=0}^{m_{N,N}} \frac{\partial^{(p)}\mathbf{G}_N[\mathbf{v}_{1,r+1},\cdots,\mathbf{v}_{N-1,r+1},\mathbf{v}_{N,r}]}{\partial \mathbf{v}_N^{(p)}}\mathbf{v}_{N,r+1}^{(p)} = R_N,
\end{aligned}\right\} \tag{13.6}$$

where

$$R_1 = \sum_{p=0}^{m_{1,1}} \frac{\partial^{(p)}\mathbf{G}_1[\mathbf{v}_{1,r},\cdots,\mathbf{v}_{N,r}]}{\partial \mathbf{v}_1^{(p)}}\mathbf{v}_{1,r}^{(p)} - \mathbf{G}_1[\mathbf{v}_{1,r},\cdots,\mathbf{v}_{N,r}],$$

$$R_2 = \sum_{p=0}^{m_{2,2}} \frac{\partial^{(p)}\mathbf{G}_2[\mathbf{v}_{1,r+1},\mathbf{v}_{2,r},\cdots,\mathbf{v}_{N,r}]}{\partial \mathbf{v}_2^{(p)}}\mathbf{v}_{2,r}^{(p)} - \mathbf{G}_2[\mathbf{v}_{1,r+1},\mathbf{v}_{2,r},\cdots,\mathbf{v}_{N,r}],$$

$$R_N = \sum_{p=0}^{m_{N,N}} \frac{\partial^{(p)}\mathbf{G}_N[\mathbf{v}_{1,r+1},\cdots,\mathbf{v}_{N-1,r+1},\mathbf{v}_{N,r}]}{\partial \mathbf{v}_N^{(p)}}\mathbf{v}_{N,r}^{(p)} - \mathbf{G}_N[\mathbf{v}_{1,r+1},\cdots,\mathbf{v}_{N-1,r+1},\mathbf{v}_{N,r}].$$

The multi-domain approach involves decomposing the temporal domain of the system (13.6) into smaller non-overlapping sub-intervals $\rho_e = [\xi_{e-1},\xi_e], e = 1,2,\ldots,T$, where $\xi \in \rho$, $\rho = [0,T]$. A detailed description of this approach was given in (Magagula et al. [21], Agbaje et al. [22], Goqo et al. [23]), among other studies. Figure 13.1 further illustrates the decomposition of the temporal domain into multiple domains.

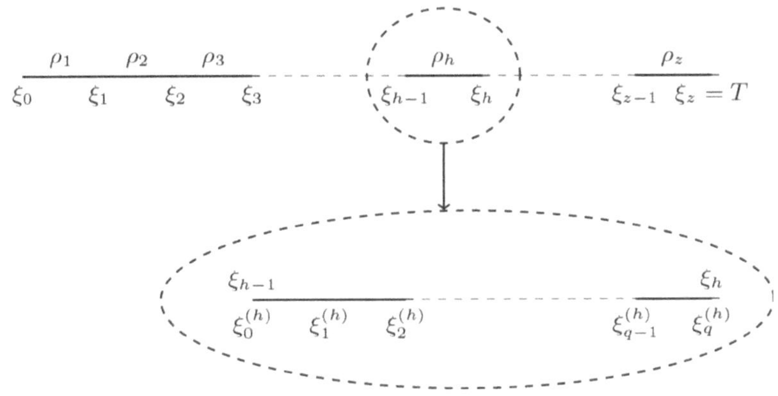

FIGURE 13.1 Non-overlapping multi-domain grid.

Figure 13.1 shows that the time domain $[0,T]$ is decomposed into equally spaced sub-intervals. The solution at the terminal of each sub-interval is taken as the initial solution of the next sub-interval. This process ensures that no overlap occurs. The system (13.6) is solved in each interval $\left[\xi_{e-1},\xi_e\right]$ in the form

$$\left.\begin{aligned}
&\sum_{p=0}^{m_{1,1}} \frac{\partial^{(p)}\mathbf{G}_1[\mathbf{v}_{1,r}^e,\ldots,\mathbf{v}_{N,r}^e]}{\partial \mathbf{v}_1^{(p)}}\mathbf{v}_{1,r+1}^{e(p)} = R_1,\\
&\sum_{p=0}^{m_{2,2}} \frac{\partial^{(p)}\mathbf{G}_2[\mathbf{v}_{1,r+1}^e,\mathbf{v}_{2,r}^e,\ldots,\mathbf{v}_{N,r}^e]}{\partial \mathbf{v}_2^{(p)}}\mathbf{v}_{2,r+1}^{e(p)} = R_2,\\
&\quad\vdots\\
&\sum_{p=0}^{m_{N,N}} \frac{\partial^{(p)}\mathbf{G}_N[\mathbf{v}_{1,r+1}^e,\ldots,\mathbf{v}_{N-1,r+1}^e,\mathbf{v}_{N,r}^e]}{\partial \mathbf{v}_N^{(p)}}\mathbf{v}_{N,r+1}^{e(p)} = R_N,
\end{aligned}\right\} \tag{13.7}$$

where $\mathbf{v}_{l,r+1}^e$ and $\mathbf{v}_{l,r}^e$, $l=1,2,\ldots,N$ denote the solution to the system (13.7) at each sub-interval ξ_e. The multi-domain process involves starting with an initial condition in the first sub-interval and using the solution at the terminal point of the first sub-interval as the initial condition of the next sub-interval, and subsequently following a similar sequence until the last sub-interval. The continuity condition, which implements the initial condition in each sub-interval is defined as in Agbaje et al. [22]

$$\mathbf{v}^{(e)}\left(\eta,\xi_{e-1}\right)=\mathbf{v}^{(e-1)}\left(\eta,\xi_e\right),$$

and implemented over the interval $\left[\xi_{e-1},\xi_e\right]$. The decoupled linear system (13.7) is solved by applying the Chebyshev spectral collocation method. The spectral collocation method is a well-known global method that uses orthogonal functions as trial functions and requires few grid points to produce highly accurate and convergent solutions (Canuto et al [24], Trefethen [25]). A detailed description of the development of the method and the derivation of the differentiation matrix can be found in Trefethen [24]. To apply the Chebyshev spectral collocation method, the physical domains $\eta\in[a,b]$ and $\xi\in[0,T]$ are mapped onto the computational domains $x\in[-1,1]$ and $y\in[-1,1]$, respectively. The numerical solutions $v_{1,r}(x,y),\ldots,v_{N,r}(x,y)$ of the system of linear equations (13.7) are obtained by writing the solutions in terms of the bivariate Lagrange interpolation polynomials, which is of the form

$$\left.\begin{aligned}
v_1(x,y) &\approx \sum_{i=0}^{M_x}\sum_{j=0}^{M_y} v_1\left(x_i,y_j\right)\mathbf{L}_i(x)\mathbf{L}_j(y),\\
&\quad\vdots\\
v_N(x,y) &\approx \sum_{i=0}^{M_x}\sum_{j=0}^{M_y} v_N\left(x_i,y_j\right)\mathbf{L}_i(x)\mathbf{L}_j(y),
\end{aligned}\right\} \tag{13.8}$$

where M_x and M_y are the numbers of grid-points in space and time, respectively, $\mathbf{L}_i$ and $\mathbf{L}_j$ are the Lagrange cardinal functions defined as

$$\left.\begin{aligned} \mathbf{L}_i(x) &= \prod_{i=0,i\neq k}^{M_x} \frac{x-x_k}{x_i-x_k}, \\ \mathbf{L}_j(y) &= \prod_{j=0,j\neq k}^{M_y} \frac{y-y_k}{y_j-y_k}, \end{aligned}\right\} \tag{13.9}$$

and

$$\left.\begin{aligned} \mathbf{L}_i\left(x_k\right) &= \delta_{ik} = \begin{cases} 0 \text{ if } i\neq k \\ 1 \text{ if } i=k \end{cases}, \\ \mathbf{L}_j\left(y_k\right) &= \delta_{jk} = \begin{cases} 0 \text{ if } j\neq k \\ 1 \text{ if } j=k \end{cases}. \end{aligned}\right\} \tag{13.10}$$

The Chebyshev–Gauss–Lobatto points are chosen as grid points x_i and y_j as they are known to be suitable in the smooth conversion of continuous time and spatial derivatives into discrete derivatives, and these points are defined thus;

$$\left.\begin{aligned} x_i &= \cos\frac{\pi i}{M_x}, \quad i=0,1,\ldots,M_x, \\ y_j &= \cos\frac{\pi j}{M_y}, \quad i=0,1,\ldots,M_y. \end{aligned}\right\} \tag{13.11}$$

Following Trefethen [25], differentiation matrices (denoted D and d) are used to collocate, and they are applied as follows:

$$\left.\begin{aligned} \left.\frac{\partial^{(p)} v_1}{\partial x^{(p)}}\right|_{x_i,y_j} &= \mathbf{D}^{(p)}V_{1,i}, & \left.\frac{\partial v_1}{\partial y}\right|_{x_i,y_j} &= \sum_{j=0}^{M_y}\mathbf{d}_{ij}V_{1,j}, \\ \left.\frac{\partial^{(p)} v_2}{\partial x^{(p)}}\right|_{x_i,y_j} &= \mathbf{D}^{(p)}V_{2,i}, & \left.\frac{\partial v_2}{\partial y}\right|_{x_i,y_j} &= \sum_{j=0}^{M_y}\mathbf{d}_{ij}V_{2,j}, \\ &\vdots \\ \left.\frac{\partial^{(p)} v_N}{\partial x^{(p)}}\right|_{x_i,y_j} &= \mathbf{D}^{(p)}V_{N,i}, & \left.\frac{\partial v_N}{\partial y}\right|_{x_i,y_j} &= \sum_{j=0}^{M_y}\mathbf{d}_{ij}V_{N,j}, \end{aligned}\right\} \tag{13.12}$$

where $V_1, V_2, \ldots, V_N$ are vectors of the form

$$V_1 = \left[v_1\left(x_0, y_j\right), v_1\left(x_1, y_j\right), \ldots, v_1\left(x_{N_x}, y_j\right)\right]^T,$$

$$V_2 = \left[v_2\left(x_0, y_j\right), v_2\left(x_1, y_j\right), \ldots, v_2\left(x_{N_x}, y_j\right)\right]^T,$$

$$\vdots$$

$$V_N = \left[v_N\left(x_0, y_j\right), v_N\left(x_1, y_j\right), \ldots, v_N\left(x_{N_x}, y_j\right)\right]^T.$$

By applying equation (13.12) to equation (13.6), we obtain the following

$$\left.\begin{array}{l}
\sum_{p=0}^{m_{1,1}} \mathbf{diag}\left[\dfrac{\partial^{(p)}\mathbf{G}_1[\mathbf{v}_{1,j,r},\ldots,\mathbf{v}_{N,j,r}]}{\partial\left(v_{1,j}\right)^{(p)}}\right]\left(\left(\mathbf{V}_{1,r+1}\right)^{(p)}\right) = \mathbf{R}_{1,j}, \\
\sum_{p=0}^{m_{2,2}} \mathbf{diag}\left[\dfrac{\partial^{(p)}G_2[\mathbf{v}_{1,j,r+1},\mathbf{v}_{2,j,r},\ldots,\mathbf{v}_{N,j,r}]}{\partial\left(v_{2,j}\right)^{(p)}}\right]\left(\left(\mathbf{V}_{2,r+1}\right)^{(p)}\right) = \mathbf{R}_{2,j}, \\
\vdots \\
\sum_{p=0}^{m_{N,N}} \mathbf{diag}\left[\dfrac{\partial^{(p)}\mathbf{G}_N[\mathbf{v}_{1,j,r+1},\ldots,\mathbf{v}_{N-1,j,r+1},\mathbf{v}_{N,j,r}]}{\partial\left(v_{N,j}\right)^{(p)}}\right]\left(\left(\mathbf{V}_{N,r+1}\right)^{(p)}\right) = \mathbf{R}_{N,j},
\end{array}\right\}$$

which is defined in compact form as

$$\left.\begin{array}{l}
\mathbf{A}_1\mathbf{V}_{1,r+1} = \mathbf{R}_{1,j}, \\
\mathbf{A}_2\mathbf{V}_{2,r+1} = \mathbf{R}_{2,j}, \\
\vdots \\
\mathbf{A}_N\mathbf{V}_{N,r+1} = \mathbf{R}_{N,j},
\end{array}\right\}$$

where the entries of $\mathbf{A}_l = \sum_{p=0}^{m_{l,l}} \mathbf{diag}\left[\dfrac{\partial^{(p)}\mathbf{G}_l[\mathbf{v}_{1,j,r},\ldots,\mathbf{v}_{N,j,r}]}{\partial\left(v_{l,j}\right)^{(p)}}\right]\mathbf{D}^{(p)}$ are matrices of size $M_y(M_x+1)\times M_y(M_x+1)$ and each entry of $\mathbf{V}_l$ and $\mathbf{R}_l$ has $M_y(M_x+1)$ dimension. To illustrate, consider the first equation of the system (13.13)

$$\mathbf{A}_1\mathbf{V}_{1,r+1} = \mathbf{R}_{1,j}, \tag{13.13}$$

Equation (13.13) can be written in matrix-vector form as

$$\begin{bmatrix} \mathbf{A}_{0,0} & \mathbf{A}_{0,1} & \cdots & \mathbf{A}_{0,Ny} \\ \mathbf{A}_{1,0} & \mathbf{A}_{1,1} & \cdots & \mathbf{A}_{1,Ny} \\ \vdots & \vdots & \ddots & \vdots \\ \mathbf{A}_{Ny,0} & \mathbf{A}_{Ny,1} & \cdots & \mathbf{A}_{Ny,Ny} \end{bmatrix} \begin{bmatrix} \mathbf{V}_0 \\ \mathbf{V}_1 \\ \vdots \\ \mathbf{V}_{Ny} \end{bmatrix} = \begin{bmatrix} \mathbf{R}_0 \\ \mathbf{R}_1 \\ \vdots \\ \mathbf{R}_{N_y} \end{bmatrix}, \tag{13.14}$$

where

$$\left.\begin{aligned} \mathbf{A}_{j,j} &= \sum_{p=0}^{m_{1,1}} \mathbf{diag}\left[\frac{\partial^{(p)}\mathbf{G}_1[\mathbf{v}_{1,j,r}]}{\partial u_j^{(p)}}\right]\mathbf{D}^{(p)} - d_{j,j}\mathbf{I}, \text{ when } j = k \\ \mathbf{A}_{j,k} &= -d_{j,k}\mathbf{I}, \quad \text{when } j \neq k, \end{aligned}\right\} \tag{13.15}$$

and I is an identity matrix of size $(M_x + 1)$.

13.3 NUMERICAL SOLUTION OF EQUATION (13.1)

Following the general description of the MD-SLLM in desc, the method is now applied to obtain the numerical solution to the system of equations (13.1). To implement the multi-domain spectral local linearization method on the coupled system of equations (13.1) and the corresponding boundary conditions (13.2), the nonlinear terms with respect to the function f in the first equation of the system are identified, and a first-order Taylor series expansion is applied to linearize the nonlinear terms while considering that the functions θ, S and γ are known from an initial guess of the approximation. This linearization of the first equation of the system results in equation (13.16) below

$$f'''_{r+1} + [a_{1,r}]f''_{r+1} + [a_{2,r}]f'_{r+1} + [a_{3,r}]f_{r+1} = [a_{4,r}]\frac{\partial f'_{r+1}}{\partial \xi} - [a_{5,r}]\frac{\partial f_{r+1}}{\partial \xi} + R_1, \tag{13.16}$$

where

$$a_{1,r} = \left(R + \frac{3}{4}\right)f_r + \xi\frac{\partial f_r}{\partial \xi}, \quad a_{2,r} = -f'_r - \xi\frac{\partial f'_r}{\partial \xi},$$

$$a_{3,r} = \left(R + \frac{3}{4}\right)f''_r, \; a_{4,r} = \xi f'_r, \quad a_{5,r} = \xi f''_r,$$

$$R_1 = \left(R + \frac{3}{4}\right)f_r f''_r - \frac{1}{2}f'^2_r - \xi f'_r\frac{\partial f'_r}{\partial \xi} + \xi f''_r\frac{\partial f_r}{\partial \xi} - \theta_r - NcS_r + Nr\gamma_r.$$

Thus, equation (13.16) is a linear equation in f and its derivatives. A similar linearization is, then, performed on the second to fourth equations of the system (13.1), so that we have:

$$\left(b_{1,r}\right)\theta''_{r+1}+\left[b_{2,r}\right]\theta'_{r+1}=\left[b_{3,r}\right]\frac{\partial\theta_{r+1}}{\partial\xi}+R_2, \tag{13.17}$$

$$\frac{1}{Le}\gamma''_{r+1}+\left[c_{1,r}\right]\gamma'_{r+1}=\left[c_{2,r}\right]\frac{\partial\gamma_{r+1}}{\partial\xi}+R_3, \tag{13.18}$$

$$\frac{1}{Sc}S''_{r+1}+\left[g_{1,r}\right]S'_{r+1}=\left[g_{2,r}\right]\frac{\partial S_{r+1}}{\partial\xi}+R_4, \tag{13.19}$$

where

$$b_{1,r}=\frac{1}{Pr}(1+Rd),$$

$$b_{2,r}=\left(R+\frac{3}{4}\right)f_{r+1}+Nb\gamma''_r+2Nt\theta'_r+\xi\frac{\partial f_{r+1}}{\partial\xi},$$

$$b_{3,r}=\xi f'_{r+1},\ R_2=Nt\theta_r'^2-D_fS''_r,$$

$$c_{1,r}=\left(R+\frac{3}{4}\right)f_{r+1}+\xi\frac{\partial f_{r+1}}{\partial\xi},\ c_{2,r}=\xi f'_{r+1},\ R_3=-\frac{1}{Le}\frac{Nt}{Nb}\theta''_{r+1},$$

$$g_{1,r}=\left(R+\frac{3}{4}\right)f_{r+1}+\xi\frac{\partial f_{r+1}}{\partial\xi},\ g_{2,r}=\xi f'_{r+1},\ R_4=-S_T\theta''_{r+1}.$$

The system equations (13.16) to (13.19) represents a decoupled system of linear equations, which is solved in each sub-interval $\left[\xi_{e-1},\xi_e\right]$ as

$$f'''_{r+1}+\left[a_{1,r}^{(e)}\right]f''_{r+1}+\left[a_{2,r}^{(e)}\right]f'_{r+1}+\left[a_{3,r}^{(e)}\right]f_{r+1}=\left[a_{4,r}^{(e)}\right]\frac{\partial f'_{r+1}}{\partial\xi}-\left[a_{5,r}^{(e)}\right]\frac{\partial f_{r+1}}{\partial\xi}+R_1^{(e)}, \tag{13.20}$$

$$\left(b_{1,r}^{(e)}\right)\theta''_{r+1}+\left[b_{2,r}^{(e)}\right]\theta'_{r+1}=\left[b_{3,r}^{(e)}\right]\frac{\partial\theta_{r+1}}{\partial\xi}+R_2^{(e)}, \tag{13.21}$$

$$\frac{1}{Le}\gamma''_{r+1}+\left[c_{1,r}^{(e)}\right]\gamma'_{r+1}=\left[c_{2,r}^{(e)}\right]\frac{\partial\gamma_{r+1}}{\partial\xi}+R_3^{(e)}, \tag{13.22}$$

$$\frac{1}{Sc}S''_{r+1}+\left[g_{1,r}^{(e)}\right]S'_{r+1}=\left[g_{2,r}^{(e)}\right]\frac{\partial S_{r+1}}{\partial\xi}+R_4^{(e)},\qquad e=1,2,\ldots,T. \tag{13.23}$$

The application of the Chebyshev spectral collocation method to equations (13.20) to (13.23) results in the following linear discrete system

$$\mathbf{A}_{1,r,j}^{(e)}\mathbf{F}_{r+1,j}=\mathbf{R}_{1,j}^{(e)}, \tag{13.24}$$

$$\mathbf{A}_{2,r,j}^{(e)}\mathbf{\Theta}_{r+1,j}=\mathbf{R}_{2,j}^{(e)}, \tag{13.25}$$

$$\mathbf{A}_{3,r,j}^{(e)}\mathbf{\Gamma}_{r+1,j} = \mathbf{R}_{3,j}^{(e)}, \tag{13.26}$$

$$\mathbf{A}_{4,r,j}^{(e)}\mathbf{S}_{r+1,j} = \mathbf{R}_{4,j}^{(e)}, \tag{13.27}$$

where

$$\mathbf{A}_{1,r,j}^{(e)} = \mathbf{D}^3 + \left[\mathbf{a}_{1,r,j}^{(e)}\right]\mathbf{D}^2 + \left[\mathbf{a}_{2,r,j}^{(e)}\right]\mathbf{D} + \left[\mathbf{a}_{3,r,j}^{(e)}\right] - \left[\mathbf{a}_{4,r,j}^{(e)}\right]\mathbf{D}\sum_{i=1}^{N_t}\mathbf{d}_{j,i} + \left[\mathbf{a}_{5,r,j}^{(e)}\right]\sum_{i=1}^{N_t}\mathbf{d}_{j,i},$$

$$\mathbf{A}_{2,r,j}^{(e)} = \left(\mathbf{b}_{1,r}^{(e)}\right)\mathbf{D}^2 + \left[\mathbf{b}_{2,r,j}^{(e)}\right]\mathbf{D} - \left[\mathbf{b}_{3,r,j}^{(e)}\right]\mathbf{d}_{j,Nt+1},$$

$$\mathbf{A}_{3,r,j}^{(e)} = \frac{1}{Le}\mathbf{D}^2 + \left[\mathbf{c}_{1,r,j}^{(e)}\right]\mathbf{D} - \left[\mathbf{c}_{2,r,j}^{(e)}\right]\mathbf{d}_{j,Nt+1},$$

$$\mathbf{A}_{4,r,j}^{(e)} = \frac{1}{Sc}\mathbf{D}^2 + \left[\mathbf{g}_{1,r,j}^{(e)}\right]\mathbf{D} - \left[\mathbf{g}_{2,r,j}^{(e)}\right]\mathbf{d}_{j,Nt+1},$$

$$\mathbf{R}_{1,j}^{(e)} = \left[\mathbf{a}_{4,r,j}^{(e)}\right]\mathbf{d}_{j,Nt+1}\mathbf{F}_{r,Nt+1} - \left[\mathbf{a}_{5,r,j}^{(e)}\right]\mathbf{d}_{j,Nt+1}\mathbf{F}_{r,Nt+1} + R_1,$$

$$\mathbf{R}_{2,j}^{(e)} = \left[\mathbf{b}_{3,r,j}^{(e)}\right]\mathbf{d}_{j,Nt+1}\mathbf{\Theta}_{r,Nt+1} + R_2,$$

$$\mathbf{R}_{3,j}^{(e)} = \left[\mathbf{c}_{2,r,j}^{(e)}\right]\mathbf{d}_{j,Nt+1}\mathbf{\Gamma}_{r,Nt+1} + R_3,$$

$$\mathbf{R}_{4,j}^{(e)} = \left[\mathbf{g}_{2,r,j}^{(e)}\right]\mathbf{d}_{j,Nt+1}\mathbf{S}_{r,Nt+1} + R_4.$$

Equations (13.24) to (13.27) are solved with the corresponding discrete form of the boundary conditions.

13.4 RESULTS AND DISCUSSION

This section presents some numerical results obtained through the method proposed in this study. Error analyses of the solutions using the norm of the residual error and convergence error are also presented. These errors indicate the convergence and accuracy of the numerical solutions obtained through the MD–SLLM. Note here that the collocation points in η and ξ are set to 20 and 10, respectively, and the semi-infinite domain $[0,\infty)$ is truncated to $[0, L_x = 8]$, terminal time is set $T = 20$, and the number of sub-intervals is chosen as 5. The parameter values are chosen to be $Nc = 0.5$, $Nr = 0.5$, $Pr = 1$, $Nb = 0.2$, $Nt = 0.3$, $Le = 10$, $Sc = 0.6$, and $St = 1$. The analysis begins by presenting the convergence of the local skin friction at different values of ξ in Table 13.1.

TABLE 13.1
Convergence of the Local Skin Friction Coefficient at Different Values of ξ

$\xi = 0$		$\xi = 6$		$\xi = 12$		$\xi = 18$	
iter	$f''(0)$	iter	$f''(0)$	iter	$f''(0)$	iter	$f''(0)$
1 1	1.917106500	1	1.101216400	1	1.038974100	1	1.018347500
2	1.372333800	2	1.010671400	2	1.012596100	2	1.015676400
	1.180109200	3	1.045141900	3	1.026741800	3	1.016925600
⋮	⋮	⋮	⋮	⋮	⋮	⋮	⋮
⋮	⋮	⋮	⋮	⋮	⋮	⋮	⋮
	1.142976300	21	1.035143700	14	1.021716500	10	1.06593900
	1.142976300	22	1.035143700	15	1.021716500	11	1.06593900

Table 13.1 shows that the skin friction converges to 9 decimal places at different iterations for the selected values of ξ. The system reduces to ordinary differential equations when $\xi = 0$, and as expected, few iterations are required to obtain a solution that converges for the skin friction coefficient. The table also shows that as ξ increases, the number of iterations needed to produce solutions that converge decreases. This result suggests that the MD-SLLM is computationally efficient. It implies that regardless of the terminal boundary point of ξ, the numerical scheme will require very few iterations to produce convergent solutions. The L_∞ –norms of the difference between two successive solutions are computed to measure the method's performance in terms of convergence, as follows:

$$\| \mathbf{F}_{r+1} - \mathbf{F}_r \|_\infty, \quad \| \mathbf{\Theta}_{r+1} - \mathbf{\Theta}_r \|_\infty, \quad \| \mathbf{\Gamma}_{r+1} - \mathbf{\Gamma}_r \|_\infty, \quad \| \mathbf{S}_{r+1} - \mathbf{S}_r \|_\infty .$$

Figure 13.2 depicts the error norms of the difference between successive solutions for increasing iterations. The results show that the errors diminish until further iterations no longer impact the errors. At this point, the iterative scheme is said to have reached full convergence. As seen, complete convergence occurs when the error norm is around 10^{-40} in Figure 13.2. This result shows that the iterative scheme produces convergent results with minimal error between successive solutions. However, the results in Figure 13.2 do not report the method's accuracy described in this chapter. The residual errors for each equation are calculated to verify the accuracy of the numerical solutions. To illustrate, consider the system

$$\mathbf{Ax} = \mathbf{b}, \tag{13.28}$$

where $\mathbf{x}$ is the exact solution to the system (13.28), such that

$$\mathbf{Ax} - \mathbf{b} = \mathbf{0}. \tag{13.29}$$

Now, consider that an approximation to Equation (13.29) is $\mathbf{y}$. In this case, we have

$$\mathbf{Ay} - \mathbf{b} = \epsilon, \tag{13.30}$$

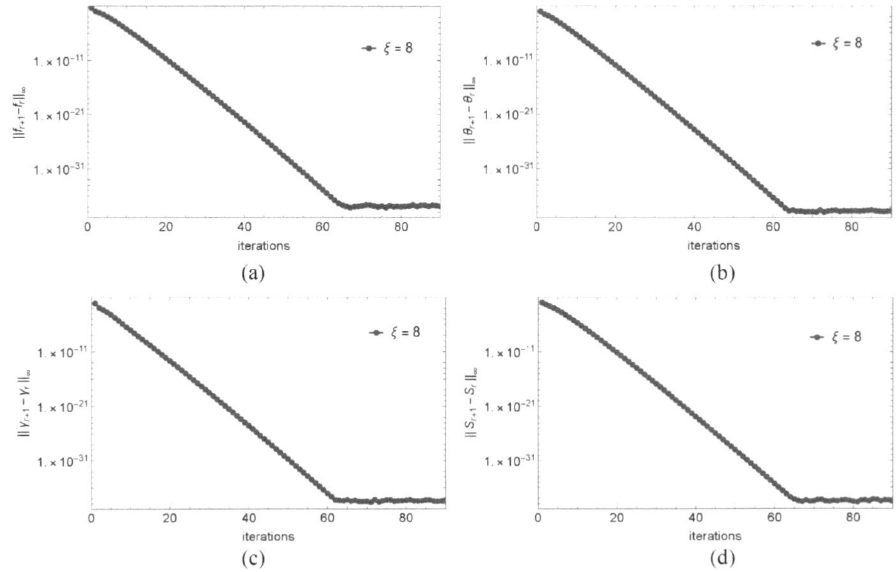

FIGURE 13.2 L_∞ –norm of the solution errors for: (a) equation (13.16), (b) equation (13.17), (c) equation (13.18), and (d) equation (13.19).

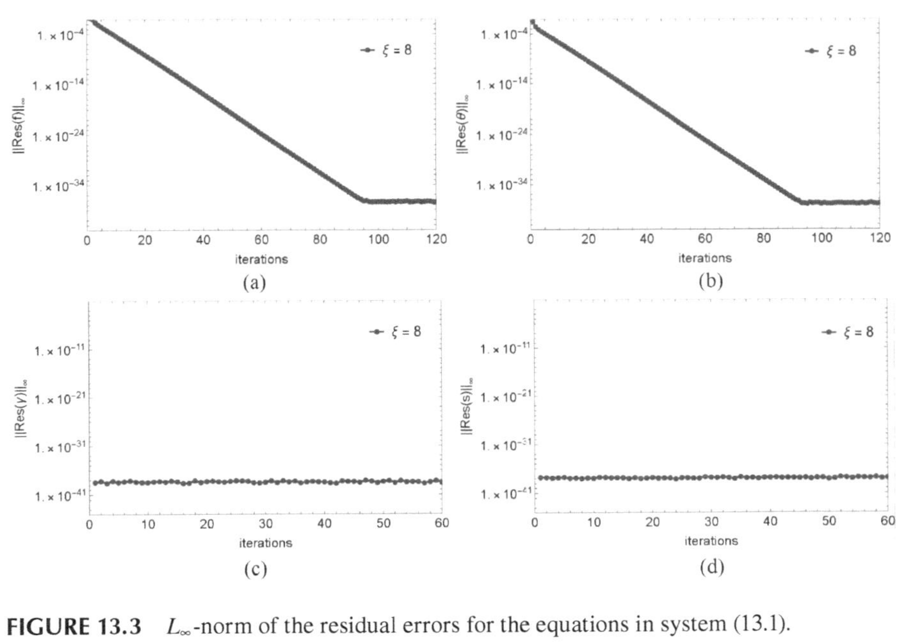

FIGURE 13.3 L_∞-norm of the residual errors for the equations in system (13.1).

where $\epsilon \neq 0$. This implies that the closer ϵ is to 0, the more accurate the method.

Figure 13.3 shows how the norm of the residuals, for each equation in system (13.1), changes as more iterations of the numerical scheme are performed.

We note that the solutions at the boundaries, corresponding to the boundary conditions, are excluded from the calculation, and the infinity norms are computed. These graphical representations of the residual norms, depicted in Figure 13.3, show that the error norms decrease as more iterations of the scheme are performed. The results also show that the errors are minimized until they reach between 10^{-24} and 10^{-40}, where further iterations do not affect the residual errors, thus indicating the high accuracy of the multi–domain spectral local linearization method.

13.5 CONCLUSION

This chapter has successfully described the multi-domain spectral local linearization method for a general nonlinear system of PDEs. The proposed method is applied to a system of PDEs that model a heat and mass transfer process. The simplicity of the implementation of the numerical scheme was noted. Based on the results presented in this study, it was shown that the method can solve problems defined on a large domain while requiring few collocation points and iterations. The solution error analysis shows that the MD-SLLM produces convergent solutions to high digit precision. The residual error analysis also indicates that the MD-SLLM is highly accurate. The analysis shows that the MD-SLLM is easy to implement and requires few grid points and iterations to produce highly accurate and convergent solutions. We can also conclude that the method can efficiently solve similar heat and mass transfer problems.

REFERENCES

[1] McLean, D. (2012). Continuum fluid mechanics and the Navier-Stokes equations. In McLean, D. (ed.), *Understanding Aerodynamics: Arguing from the Real Physics* (pp. 13–78). Wiley Online Library: West Sussex.

[2] McDavid, R. M. (2011). Macroscopic modeling. In *Encyclopedia of Materials:* Science and Technology (pp. 4720–4732). Elsevier, Boston.

[3] Baytas¸ A. C., & Baytas¸ A. F. (2005). Entropy generation in porous media. In D.B. Ingham and I. Pop (eds.), *Transport Phenomena in Porous Media III* (pp. 201–226). Pergamon, Oxford.

[4] Motsa, S. S., RamReddy, C., & Rao, C. V. (2017). Non-similarity solution for Soret effect on natural convection over the vertical frustum of a cone in a nanofluid using new bivariate pseudo-spectral local linearisation method. *Applied Mathematics and Computation*, 314, 439–455.

[5] Lewis, R. W., Morgan, K., Thomas, H. R., & Seetharamu, K. N. (1996). *The Finite Element Method in Heat Transfer Analysis.* John Wiley & Sons: New Jersey.

[6] Chau, K. V., & Gaffney, J. J. (1990). A finite-difference model for heat and mass transfer in products with internal heat generation and transpiration. *Journal of Food Science*, 55(2), 484–487.

[7] Chai, J. C., & Patankar, S. V. (2018). Finite-volume method for radiation heat transfer. In W.J. Minkowycz and E.M. Sparrow (eds.), *Advances in Numerical Heat Transfer* (pp. 109–141). CRC Press: Florida.

[8] Batool, S., Rasool, G., Alshammari, N., Khan, I., Kaneez, H., & Hamadneh, N. (2022) Numerical analysis of heat and mass transfer in micropolar nanofluids flow through lid driven cavity: Finite volume approach. *Case Studies in Thermal Engineering*, 37, 102233.

[9] Canuto, C., Husseini, M. Y., Quarteroni, A., & Zang, T. A. (1998). *Spectral Methods in Fluid Dynamics*. Springer-Verlag, Berlin.

[10] Jang, J., & Hong, S. Y. (2016). Comparison of nonhydrostatic and hydrostatic dynamical cores in two regional models using the spectral and finite difference methods: dry atmosphere simulation. *Meteorology and Atmospheric Physics*, 128(2), 229–245.

[11] Graham, N., Quandt, M., & Weigel, H. (2009). *Spectral Methods in Quantum Field Theory, Volume* 777. Springer, Berlin.

[12] Ascher, U. M., & Petzold, L. R. (1998). *Computer Methods for Ordinary Differential Equations and Differential-Algebraic Equations, Volume* 61. SIAM, Pennsylvania.

[13] Soomro, F. A., Usman, M, Haq, R. U., & Wang, W. (2018). Melting heat transfer analysis of Sisko fluid over a moving surface with nonlinear thermal radiation via collocation method. *International Journal of Heat and Mass Transfer*, 126, 1034–1042.

[14] Hernandez-Calderon, O. M., Rubio-Castro, E., & Rios-Iribe, E. Y. (2014). Solving the heat and mass transfer equations for an evaporative cooling tower through an orthogonal collocation method. *Computers & Chemical Engineering*, 71, 24–38.

[15] Li, B. W., Sun, Y. S., & Zhang, D. W. (2009). Chebyshev collocation spectral methods for coupled radiation and conduction in a concentric spherical participating medium. *Journal of Heat Transfer*, 131(6), 062701.

[16] Lekoko, M. L., Oloniiju, S. D., & Magalakwe, G. (2022). Analysis of buoyancy driven flow inside a vertical filter chamber. *Chaos, Solitons & Fractals*, 161, 112292.

[17] Oloniiju, S. D., Goqo, S. P., & Sibanda, P. (2020). Heat and mass transfer in an unsteady second grade nanofluid with viscous heating dissipation. *International Journal of Computational Methods*, 17(5), 1940005.

[18] Bellman, R. E., & Kalaba, R. E. (1965). *Quasilinearization and Nonlinear Boundary-Value Problems*. American Elsevier Publishing Co., Inc., New York.

[19] Motsa, S. S. (2013). A new spectral local linearization method for nonlinear boundary layer flow problems. *Journal of Applied Mathematics*, 2013, 423628. https://doi.org/10.1155/2013/423628.

[20] Otegbeye, O., Goqo, S. P., & Ansari, M. S. (2020). Comparative study of some spectral based methods for solving boundary layer flow problems. In *AIP Conference Proceedings* (Vol. 2253, No. 1, p. 020013). AIP Publishing LLC.

[21] Magagula, V. M., Motsa, S. S., & Sibanda, P. (2017) A multi-domain bivariate pseudospectral method for evolution equations. *International Journal of Computational Methods*, 14(4), 1750041.

[22] Agbaje, T. M., Mondal, S., Motsa, S. S., & Sibanda, P. (2017). A numerical study of unsteady non-Newtonian Powell-Eyring nanofluid over a shrinking sheet with heat generation and thermal radiation. *Alexandria Engineering Journal*, 56(1), 81–91.

[23] Goqo, S. P., Mondal, S., Sibanda, P., & Motsa, S. S. (2019). Efficient multi-domain spectral collocation solution for MHD laminar natural convection flow from a vertical permeable flat plate with uniform surface temperature and thermal radiation. *International Journal of Computational Methods*, 16(6), 1840029–1840046.

[24] Canuto, C., Hussaini, M. Y., Quarteroni, A., & Zang, T. A. (2007). *Spectral Methods: Evolution to Complex Geometries and Applications to Fluid Dynamics*. Springer Science & Business Media, Berlin.

[25] Trefethen, L. N. (2000). *Spectral Methods in MATLAB*. Society for Industrial and Applied Mathematics, Philadelphia.

14 An Improved Multi-Derivative Hybrid Block Method for Chaotic and Nonlinear Initial Value Problems

Uthman O. Rufai, Precious Sibanda, Sicelo Goqo, and Salma Ahmedai

14.1 INTRODUCTION

Consider the first-order (1st-order) ordinary differential equation (ODE)

$$y' = f(t, y), \qquad y(t_0) = y_0, \tag{14.1}$$

where the unique continuous solution y is assumed to be a differentiable function on the closed interval $[t_0, b]$. Equation (14.1) can be expressed in the form

$$y_r'(t) + \sum_{j=1}^{e} \lambda_{r,j} y_j(t) + f_r[y_1(t), y_2(t), \ldots, y_e(t)] = \Upsilon_r, y_r(0) = a_r, r = 1, 2, \ldots, e. \tag{14.2}$$

This represents an initial-value system described by e nonlinear 1st-order DEs, where $\lambda_{r,j}$ and Υ_r are known constant input parameters, y_r are the unknown variables, a_r are the corresponding initial conditions, f_r is the nonlinear (N) component of the r-th equation and the prime denotes differentiation with respect to time t [1].

Systems that exhibit rapidly fluctuating solutions, which are extremely sensitive to modest perturbations of the original data, are chaotic systems [1]. Chaos has numerous applications in various fields such as the study of turbulent flow of fluids, electrical circuits, chemical reactions, plasma physics, mechanical devices, population dynamics, and many other scientific areas. It was first demonstrated by Lorenz [2], who observed chaotic behavior in a system of ODEs that model meteorological occurrences. Several chaotic dynamical systems have been studied by various researchers (see references [3,4]). Chaotic dynamical systems have been solved using a variety of numerical techniques (see references [5,6]).

 DOI: 10.1201/9781003638414-14

For long time intervals, numerical methods like higher-order explicit Runge-Kutta methods are ineffective. This has given rise to the conception that chaotic systems can only be solved for short time intervals. To circumvent this restriction, researchers developed the concept of domain decomposition. In order to effectively solve the equations across each sub-interval, the time interval must be divided into several non-overlapping smaller intervals. The use of compact and well-conditioned matrices in multi-domain methods gives them an edge over single-domain computation. As a result, the approaches are more effective, and the results are more accurate. Examples of numerical approaches where multi-domain methods have been successfully applied can be found in references [1,7,8]. The downside of multi-domain methods is that analytical integration in each of the various subdomains can become monotonous and time-consuming, especially with multi-domain approaches that rely on analytical approximations.

Some researchers have proposed multi-derivative linear multistep methods [9–11] for the direct solution of IVPs. They concluded that multi-derivative methods provide high accuracy with good stability properties. In the past, hybrid methods were initially proposed to overcome the Dahlquist Barrier theorem [12] by imposing intra-step points during formulation (see [13,14]). There has also been interest in block hybrid approaches for the direct solution of IVPs [15,16].

In this study, we present a new IMDHBM implemented with a Gauss-Seidel type iterative method to solve N-systems and high-order chaotic systems. We examine the accuracy, stability, consistency properties, and the applicability of the new IMDHBM in N-IVPs and IVPs with chaotic behavior that include the RS and ACS. We show that the numerical simulations are precise, effective, reliable, and that the method is simple to use. The numerical outcomes are also compared with the multistage spectral relaxation method (MSRM) [1], the Matlab ode45 solver [1], the piece-wise successive linearization method (PSLM) [7], an optimized two-step hybrid block method (OTSHBM) [16], and continuous block backward differentiation formula (CBBDF) [17].

14.2 DERIVATION OF THE IMDHBM

A one-step second derivative (2-D) method in the form

$$y_{(n+p_i)} = y_n + h\beta_{i,0}f_n + h\left(\sum_{j=1}^{\Psi}\alpha_{ij}f_{n+p_j}\right) + h^2\tau_{i,0}g_n + h^2\left(\sum_{j=1}^{\Psi}\gamma_{ij}g_{n+p_j}\right), \tag{14.3}$$

for $i = 1, 2, \ldots, \Psi$, is developed for a 1st-order ODE on the interval from t_n to t_{n+1}. The power series polynomial is designed to locally approximate the solution on the interval $[t_n, t_{n+1}]$

$$Y(t) = \sum_{j=0}^{\Psi+\Lambda}\eta_{n,j}(t-t_n)^j, \tag{14.4}$$

where η_n, j are unknown coefficients in the interval $[t_n, t_{n+1}]$ to be deduced through a system of $\Psi+\Lambda$ equations with $\Psi+\Lambda$ unknowns, generated from

$$y'_{n+p_i} = f\left(t_{n+p_i}, y_{n+p_i}\right), \qquad i = 0,1,2, \ldots, \Psi,$$

$$y''_{n+p_i} = g\left(t_{n+p_i}, y'_{n+p_i}\right), \qquad i = 0,1,2, \ldots, \psi, \tag{14.5}$$

$$Y(t) = \eta_{n,0} = y_n, \qquad n = 0,1, \ldots, \Delta-1,$$

where $\Lambda=\Psi+2$ and Δ is the number of grid points. After algebraic simplification, the 2-D method in the form (14.3) is obtained. Here, α_{ij}, $\beta_{i,0}$, $\tau_{i,0}$ and γ_{ij} are continuous coefficients, h and n represent the step-length and grid index, respectively. In addition, $p_0=0$, $p_1=\varpi$, $p_2=\varrho$, $p_3=\varsigma$, and $p_4=1$, where

$$\varpi = \frac{1}{14}\left(7-\sqrt{21}\right), \quad \varrho = \frac{1}{14}\left(7+\sqrt{21}\right), \quad \varsigma = \frac{1}{2},$$

are the off-step points obtained from the zeros of the derivative of the S-L polynomial of degree four. Solving the equations that arise in (14.5) for all values of Ψ. We obtained the unknown coefficients $\eta_{n,j}$ and substituted them into (14.4) to obtain an improved multi-derivative block hybrid method equations. Evaluating the result at the collocation points t_{n+pj} for $i=1, 2, \ldots, \Psi$, we obtain

$$y_{n+p_i} = Y\left(t_{n+p_i}\right), \qquad i = 1,2,\ldots,\psi, \tag{14.6}$$

with $\Psi=4$. This leads to a matrix in the form:

$$\begin{bmatrix} y_{n+\varpi} \\ y_{n+\varrho} \\ y_{n+\varsigma} \\ y_{n+1} \end{bmatrix} = \begin{bmatrix} y_n \\ y_n \\ y_n \\ y_n \end{bmatrix} + h \begin{bmatrix} \frac{573}{6860} - \frac{153}{48020}\sqrt{21} \\ \frac{573}{6860} + \frac{153}{48020}\sqrt{21} \\ \frac{57}{640} \\ \frac{1}{10} \end{bmatrix} [f_n]$$

$$
+h\begin{bmatrix}
\left(\frac{49}{405}-\frac{67}{13230}\sqrt{21}\right) & \left(\frac{49}{405}-\frac{347}{13230}\sqrt{21}\right) & \left(\frac{64}{405}-\frac{1216}{36015}\sqrt{21}\right) & \left(\frac{133}{6860}-\frac{153}{48020}\sqrt{21}\right)\\
\left(\frac{49}{405}+\frac{347}{13230}\sqrt{21}\right) & \left(\frac{49}{405}+\frac{67}{13230}\sqrt{21}\right) & \left(\frac{64}{405}+\frac{1216}{36015}\sqrt{21}\right) & \left(\frac{133}{6860}+\frac{153}{48020}\sqrt{21}\right)\\
\left(\frac{49}{405}+\frac{343}{13824}\sqrt{21}\right) & \left(\frac{49}{405}-\frac{343}{13824}\sqrt{21}\right) & \frac{64}{405} & \frac{7}{640}\\
\frac{98}{405} & \frac{98}{405} & \frac{128}{405} & \frac{1}{10}
\end{bmatrix}
\begin{bmatrix} f_{n+\varpi}\\ f_{n+\varrho}\\ f_{n+\varsigma}\\ f_{n+1}\end{bmatrix}
$$

$$
+h^{2}\begin{bmatrix}
\frac{101}{49392}-\frac{27}{192080}\sqrt{21}\\
\frac{101}{49392}+\frac{27}{192080}\sqrt{21}\\
\frac{209}{92160}\\
\frac{1}{360}
\end{bmatrix}\left[g_{n}\right]
$$

$$
+h^{2}\begin{bmatrix}
\left(\frac{-221}{21168}+\frac{7}{6480}\sqrt{21}\right) & \left(\frac{461}{105840}-\frac{7}{6480}\sqrt{21}\right) & \left(\frac{-64}{46305}\right) & \left(\frac{-181}{246960}+\frac{27}{192080}\sqrt{21}\right)\\
\left(\frac{461}{105840}+\frac{7}{6480}\sqrt{21}\right) & \left(\frac{-221}{21168}-\frac{7}{6480}\sqrt{21}\right) & \left(\frac{-64}{46305}\right) & \left(\frac{-181}{246960}-\frac{27}{192080}\sqrt{21}\right)\\
\left(\frac{343}{276480}+\frac{7}{6480}\sqrt{21}\right) & \left(\frac{-221}{21168}-\frac{7}{6480}\sqrt{21}\right) & \frac{-17}{1080} & \frac{-47}{92160}\\
\frac{7\sqrt{21}}{3240} & -\frac{7\sqrt{21}}{3240} & 0 & \frac{-1}{360}
\end{bmatrix}
\begin{bmatrix} g_{n+\varpi}\\ g_{n+\rangle}\\ g_{n+\varsigma}\\ g_{n+1}\end{bmatrix} \quad (14.7)
$$

14.3 ANALYSIS OF THE IMDHBM

In this section, we present and analyze the local truncation error (LTE) and the stability properties of the IMDHBM.

14.3.1 Order of Accuracy

According to references [14] and [18], the local truncation error for equation (14.3) can be expressed in terms of a linear operator $\hat{L}$ as follows

$$\hat{\mathfrak{L}}[y(t_n);h]=\sum_{j=0}^{\psi}\begin{bmatrix}\alpha_{ij}y(t_n+p_j h)-h\beta_{ij}y'(t_n+p_j h)\\ -h^2\gamma_{ij}y''(t_n+p_j h)\end{bmatrix}. \tag{14.8}$$

Assuming that $y(t_n)$ is sufficiently many times differentiable, the terms $y(t_n+p_jh)$, $y'(t_n+p_jh)$ and $y''(t_n+p_jh)$ can be expanded using Taylor's series about t_n, to yield

$$\begin{aligned}\hat{\mathfrak{L}}[y(t_n);h]&=\widehat{C_0}\,y(t_n)+\widehat{C_1}\,h\,y'(t_n)+\widehat{C_2}\,h^2\,y''(t_n)\\&\quad+\ldots+\widehat{C_\psi}\,h^\psi\,y^\psi(t_n),\end{aligned} \tag{14.9}$$

where $\widehat{C_1}, \widehat{C_2}, \ldots, \widehat{C_\psi}$ are constants given as follows:

$$\widehat{C_0}=\sum_{j=1}^{\Psi}[\alpha_{ij}],\quad \widehat{C_1}=\sum_{j=1}^{\Psi}[p_j\alpha_{ij}-\beta_{ij}],$$

$$\widehat{C_2}=\frac{1}{2!}\left(\sum_{j=1}^{\Psi}p_j^2\alpha_{ij}\right)-\sum_{j=1}^{\Psi}p_j\beta_{ij}-\sum_{j=1}^{\Psi}\gamma_{ij},$$

$$\ldots,\ \widehat{C_\psi}=\frac{1}{\psi!}\left(\sum_{j=1}^{\Psi}p_j^\psi\alpha_{ij}\right)-\frac{1}{(\psi-1)!}\sum_{j=1}^{\Psi}p_j^{\psi-1}\beta_{ij}$$

$$-\frac{1}{(\psi-2)!}\sum_{j=1}^{\Psi}p_j^{\psi-2}\gamma_{ij},$$

According to reference [19], the method (14.3) has order ψ if

$$\hat{\mathfrak{L}}[y(t_n);h]=O(h^{\psi+1}),\qquad \widehat{C_0}=\widehat{C_1}=\ldots=\widehat{C_\psi}=0,\qquad \widehat{C_{\psi+1}}\neq 0.$$

Therefore, $\widehat{C_{\psi+1}}$ is the error constant and $\widehat{C_{\psi+1}}h^{\psi+1}y^{\psi+1}(t_n)$ is the principal local truncation error (LTE) at the point t_n. The LTE of the IMDHBM is given by

$$\hat{\mathfrak{L}}\left[y(t_n);h\right]=\left\{\frac{2401-405\sqrt{21}y^{(11)}(t_n)h^{11}}{4733740976025600}\right.$$

$$+O\left(h^{12}\right),\ \frac{2401+405\sqrt{21}y^{(11)}(t_n)h^{11}}{4733740976025600}$$

$$+O\left(h^{12}\right),\ \frac{y^{(11)}(t_n)h^{11}}{1971570585600}$$

$$\left.+O\left(h^{12}\right),\ \frac{y^{(11)}(t_n)h^{11}}{985785292800}+O\left(h^{12}\right)\right\},$$

We have

$$\widehat{C_0}=\widehat{C_1}=\ldots=\widehat{C_{10}}=0,$$

and the error constants are

$$\widehat{C_{11}}=\begin{pmatrix}\dfrac{2401-405\sqrt{21}}{4733740976025600},\dfrac{2401+405\sqrt{21}}{4733740976025600},\\ \dfrac{1}{1971570585600},\dfrac{1}{985785292800}\end{pmatrix}^T,$$

indicating that the IMDHBM has order $\psi=10$.

14.3.2 Stability Analysis

The IMDHBM can be expressed by a matrix finite difference equation in block form as:

$$E_\psi Y_{n+\psi}=F_\psi Y_n+h\,A_\psi\,F_{n+\psi}+h\,B_\psi\,F_n+h^2V_\psi G_{n+\psi}+h^2U_\psi G_n, \tag{14.10}$$

where

$$\begin{aligned}
Y_{n+\Psi}&=\left(y_{n+\varpi},y_{n+\varrho},y_{n+\varsigma},y_{n+1}\right)^T, && Y_n=\left(y_{n+\varpi-1},y_{n+\varrho-1},y_{n+\varsigma-1},y_n\right)^T,\\
F_{n+\Psi}&=\left(f_{n+\varpi},f_{n+\varrho},f_{n+\varsigma},f_{n+1}\right)^T, && F_n=\left(f_{n+\varpi-1},f_{n+\varrho-1},f_{n+\varsigma-1},f_n\right)^T,\\
G_{n+\Psi}&=\left(g_{n+\varpi},g_{n+\varrho},g_{n+\varsigma},g_{n+1}\right)^T, && G_n=\left(g_{n+\varpi-1},g_{n+\varrho-1},g_{n+\varsigma-1},g_n\right)^T.
\end{aligned} \tag{14.11}$$

The combined coefficients of the methods obtained above are given as follows:

$$A_\psi = \begin{bmatrix} \left(\frac{49}{405} - \frac{67}{13230}\sqrt{21}\right) & \left(\frac{49}{405} - \frac{347}{13230}\sqrt{21}\right) & \left(\frac{64}{405} - \frac{1216}{36015}\sqrt{21}\right) & \left(\frac{133}{6860} - \frac{153}{48020}\sqrt{21}\right) \\ \left(\frac{49}{405} + \frac{347}{13230}\sqrt{21}\right) & \left(\frac{49}{405} + \frac{67}{13230}\sqrt{21}\right) & \left(\frac{64}{405} + \frac{1216}{36015}\sqrt{21}\right) & \left(\frac{133}{6860} + \frac{153}{48020}\sqrt{21}\right) \\ \left(\frac{49}{405} + \frac{343}{13824}\sqrt{21}\right) & \left(\frac{49}{405} - \frac{343}{13824}\sqrt{21}\right) & \frac{64}{405} & \frac{7}{640} \\ \frac{98}{405} & \frac{98}{405} & \frac{128}{405} & \frac{1}{10} \end{bmatrix},$$

$$B_\psi = \begin{bmatrix} 0 & 0 & 0 & \left(\frac{573}{6860} - \frac{153}{48020}\sqrt{21}\right) \\ 0 & 0 & 0 & \left(\frac{573}{6860} + \frac{153}{48020}\sqrt{21}\right) \\ 0 & 0 & 0 & \frac{57}{640} \\ 0 & 0 & 0 & \frac{1}{10} \end{bmatrix}, \quad E_\psi = \begin{bmatrix} 1 & 0 & 0 & 0 \\ 0 & 1 & 0 & 0 \\ 0 & 0 & 1 & 0 \\ 0 & 0 & 0 & 1 \end{bmatrix}, \tag{14.12}$$

$$F_{\psi} = \begin{bmatrix} 0 & 0 & 0 & 1 \\ 0 & 0 & 0 & 1 \\ 0 & 0 & 0 & 1 \\ 0 & 0 & 0 & 1 \end{bmatrix}, \; U_{\psi} = \begin{bmatrix} 0 & 0 & 0 & \frac{101}{49392} - \frac{27}{192080}\sqrt{21} \\ 0 & 0 & 0 & \frac{101}{49392} + \frac{27}{192080}\sqrt{21} \\ 0 & 0 & 0 & \frac{209}{92160} \\ 0 & 0 & 0 & \frac{1}{360} \end{bmatrix},$$

$$V_{\psi} = \begin{bmatrix} \left(\frac{-221}{21168} + \frac{7}{6480}\sqrt{21}\right) & \left(\frac{461}{105840} - \frac{7}{6480}\sqrt{21}\right) & \frac{-64}{46305} & \left(\frac{-181}{246960} + \frac{27}{192080}\sqrt{21}\right) \\ \left(\frac{461}{105840} + \frac{7}{6480}\sqrt{21}\right) & \left(\frac{-221}{21168} - \frac{7}{6480}\sqrt{21}\right) & \frac{-64}{46305} & \left(\frac{-181}{246960} - \frac{27}{192080}\sqrt{21}\right) \\ \left(\frac{343}{276480} + \frac{7}{6480}\sqrt{21}\right) & \left(\frac{-221}{21168} - \frac{7}{6480}\sqrt{21}\right) & \frac{-17}{1080} & \frac{-47}{92160} \\ \frac{7\sqrt{21}}{3240} & -\frac{7\sqrt{21}}{3240} & 0 & \frac{-1}{360} \end{bmatrix}.$$

We explore the matrix (14.10) in the limit as $h \to 0$ in order to examine the zero stability of the IMDHBM. In light of this, the matrix system reduces to:

$$E_{\psi} Y_{n+\psi} = F_{\psi} Y_n. \tag{14.13}$$

Whose first characteristics polynomial $\aleph(R)$ is given as

$$\aleph(R) = \det\left(R\left(E_{\psi}\right) - F_{\psi}\right) = R^3(R-1). \tag{14.14}$$

This has solutions $R=0$ and $R=1$. For the roots with $|R_j|=1$, the multiplicity is not more than 1. According to reference [18], the block method (14.7) is zero stable. Since the order $\psi > 1$, the IMDHBM is consistent and converges.

14.3.3 Absolute Stability

The region of linear stability is considered to be the set of points $z \in \mathbb{C}$ such that the roots of the characteristic equation associated with the test equations

$$y' = \lambda y, \qquad y'' = \lambda^2 y, \qquad Re(\lambda) \le 0,.$$

lie within the unit circle. These test equations are applied on the matrix equation (14.10) to obtain

$$Y_{n+\psi} = \mathbf{H}(z) Y_n, \qquad z = \lambda h, \tag{14.15}$$

where the matrix $\mathcal{H}(z)$ is given by

$$\mathcal{H}(z) = (E_{\Psi} - zA_{\Psi} - z^2 V_{\Psi})^{-1}\left(F_{\Psi} + zB_{\Psi} + z^2 U_{\Psi}\right) \tag{14.16}$$

The matrix $\mathcal{H}(z)$ is known as the amplification matrix. By computing the eigenvalues of the matrix $\mathcal{H}(z)$, the dominant eigenvalue is the stability function $R(z)$. As illustrated in Figure 14.1, the IMDHBM is A-stable because the stability region for this scheme encompasses the entire left half of the complex plane.

14.4 APPLICATION OF THE IMDHBM

The IMDHBM for a system of 1st-order linear DEs is given by

$$y_{n+p_i} = y_n + h\beta_{i,0}\left(\kappa_n + \omega_n y_n\right) + h\left(\sum_{j=1}^{\Psi} \alpha_{ij}\left(\kappa_{n+p_j} + \omega_{n+p_j} y_{n+p_j}\right)\right) + h^2 \tau_{i,0}\left(p_n + q_n y_n + s_n y_n'\right).$$

$$+h^2\left(\sum_{j=1}^{\Psi} \gamma_{ij}\left(p_{n+p_j} + q_{n+p_j} y_{n+p_j} + s_{n+p_j} y_{n+p_j}'\right)\right), \tag{14.17}$$

for $i = 1, 2, \ldots, \Psi$, where

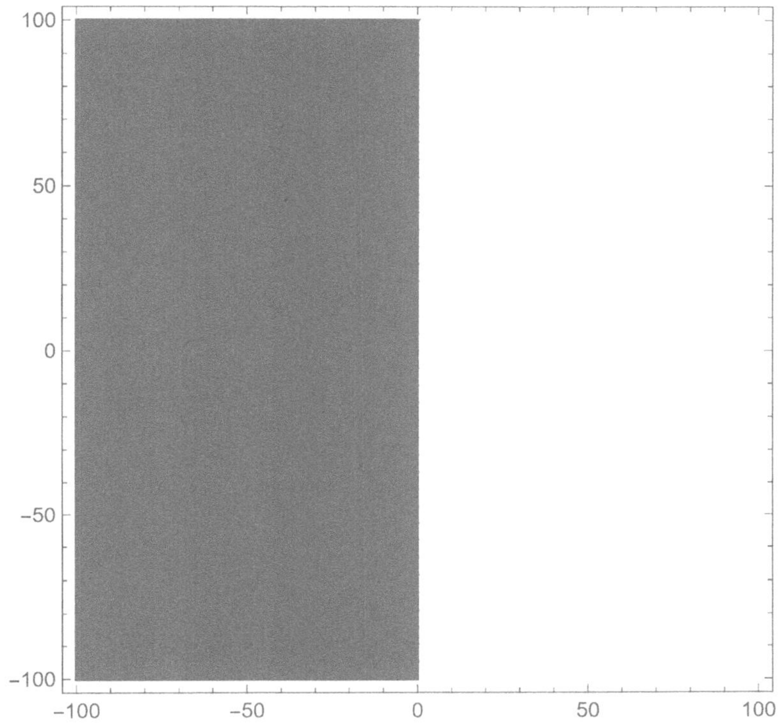

FIGURE 14.1 Stability region of the IMDHBM.

$$f(t,y)=\kappa(t)+\omega(t)y, \qquad g(t,y)=p(t)+q(t)y+s(t)y', \qquad t_0\le t\le t_T, \tag{14.18}$$

are the general linear first- and second-order equations, respectively, $\kappa(t)$, $\omega(t)$, $p(t)$, $q(t)$, and $s(t)$ are known functions of t. In matrix form, equation (14.17) takes the form

$$\begin{aligned}&\left(I-h\,A_\Psi W_{n+\Psi}-h^2V_\Psi\,Q_{n+\Psi}-h^2S_{n+\Psi}V_\Psi W_{n+\Psi}\right)Y_{n+\Psi}\\&=Y_n+h\,A_\Psi K_{n+\Psi}+h\,B_\Psi\,F_n+h^2V_\Psi P_{n+\Psi}+h^2S_{n+\Psi}K_{n+\Psi}V_\Psi+h^2U_\Psi G_n,\end{aligned}\tag{14.19}$$

or

$$A_nY_{n+\psi}=B_n, \tag{14.20}$$

where

$$A_n=I-h\,A_\psi W_{n+\psi}-h^2V_\psi\,Q_{n+\psi}-h^2S_{n+\psi}V_\psi W_{n+\psi},$$

$$B_n=Y_n+h\,A_\psi K_{n+\psi}+h\,B_\psi\,F_n+h^2V_\psi P_{n+\psi}+h^2S_{n+\psi}K_{n+\psi}V_\psi+h^2U_\psi G_n,$$

$Q_{n+\Psi}$, $W_{n+\Psi}$, and $S_{n+\Psi}$ are 4×4 diagonal matrices, Y_n, F_n, $K_{n+\Psi}$, $P_{n+\Psi}$, and $Y_{n+\Psi}$ are 4×1 matrices, A_Ψ and V_Ψ are 4×4 matrices in (14.12), B_Ψ and U_Ψ are 4×4 diagonal matrices, given by

$$B_\psi = \begin{bmatrix} \left(\frac{573}{6860} - \frac{153}{48020}\sqrt{21}\right) & 0 & 0 & 0 \\ 0 & \left(\frac{573}{6860} + \frac{153}{48020}\sqrt{21}\right) & 0 & 0 \\ 0 & 0 & \frac{57}{640} & 0 \\ 0 & 0 & 0 & \frac{1}{10} \end{bmatrix}, \tag{14.21}$$

$$U_\psi = \begin{bmatrix} \left(\frac{101}{49392} - \frac{27}{192080}\sqrt{21}\right) & 0 & 0 & 0 \\ 0 & \left(\frac{101}{49392} + \frac{27}{192080}\sqrt{21}\right) & 0 & 0 \\ 0 & 0 & \frac{209}{92160} & 0 \\ 0 & 0 & 0 & \frac{1}{360} \end{bmatrix},$$

14.4.1 Nonlinear System of IVPs

Consider an N-system of e 1st-order IVPs of the form

$$y_u' = f_u\left(t, y_1, y_2, \ldots, y_e\right), \text{ for } u = 1, 2, \ldots, e. \tag{14.22}$$

Differentiating (14.22), we have

$$y_u'' = \frac{\partial f_u}{\partial t}\left(t, y_1', y_2', \ldots, y_e'\right) + \sum_{j=1}^{e} \frac{\partial f_u}{\partial t}\left(t, y_1', y_2', \ldots, y_e'\right) f_j, \quad u = 1, 2, \ldots, e. \tag{14.23}$$

These equations can be solved using a Gauss-Siedel type iterative method, discussed in reference [20], with

$$y_{u,\imath}' = \kappa_u\left(t, y_{1,\imath}, \ldots, y_{u-1,\imath}, y_{u,r}, \ldots, y_{e,r}\right) + \omega_u\left(t, y_{1,\imath}, \ldots, y_{u-1,\imath}, y_{u+1,r}, \ldots, y_{e,r}\right) y_{u,\imath}, \tag{14.24}$$

and

$$y_{u,\imath}'' = p_u\left(t, y_{1,\imath}, \ldots, y_{u-1,\imath}, y_{u,r}, \ldots, y_{e,r}\right) + q_u\left(t, y_{1,\imath}, \ldots, y_{u-1,\imath}, y_{u+1,r}, \ldots, y_{e,r}\right) y_{u,\imath},$$
$$+ s_u\left(t, y_{1,\imath}, \ldots, y_{u-1,\imath}, y_{u+1,r}, \ldots, y_{e,r}\right) y_{u,\imath}', \tag{14.25}$$

where r and $\imath = r + 1$ are the previous and current iteration levels and $u = 1, 2, \ldots, e$. The term s_u are constant coefficients of y_u' in the second derivative, ω_u and q_u are the constant coefficients of y_u in the first and second derivative, respectively, in the u-th equation, κ_u and p_u are the remaining terms, including nonlinear terms in the u-th equation, also in the first and second derivative, respectively. In matrix form, (14.24) and (14.25) take the form:

$$A_{u,n}Y_{u,n+\Psi} = B_{u,n}.. \tag{14.26}$$

In the next examples, we illustrate the application of the IMDHBM to N-systems of 1st-order IVPs with chaotic behaviour.

14.5 NUMERICAL EXAMPLES

We conducted numerical experiments on systems of nonlinear IVPs and systems that exhibit chaotic behavior to examine the efficiency and effectiveness of the derived scheme. These numerical experiments fit the general governing equation profile (14.1). All the results were obtained using the number of partitions $\Delta = (b-a)/h$, where h is the step-size. Fifteen iterations of the new method in each interval $[a, b]$ are contrasted with the results derived using the MSRM [1], the Matlab ode45 solver [1], the PSLM [7], the OTSHBM [16], and CBBDF [17].

14.5.1 Example 1

Consider the nonlinear system as appeared in references [16] and [17].

$$y_1' = -1002y_1 + 1000y_2^2, \qquad y_1(0) = 1,$$

$$y_2' = y_1 - y_2(1 + y_2), \qquad y_2(0) = 1,$$

with exact solution given by $y_1(t) = \exp^{-2t}$ and $y_2(t) = \exp^{-t}$. The N-system is solved in the closed interval [0, 10].

The parameters are defined as follows

$$\kappa_1 = 1000y_2^2, \quad \omega_1 = -1002, \qquad \kappa_2 = y_1 - y_2^2, \quad \omega_2 = -1.$$

The second derivatives are given by

$$y_1'' = -1002y_1 + 2000y_2y_2' = (1004004 + 2000y_2)y_1 - (1004000 + 2000y_2)y_2^2,$$

$$y_2'' = y_1' - y_2' - 2y_2y_2' = -1003y_1 + 1003y_2^2 + 2y_2^3 + (1 - 2y_1)y_2,$$

The iteration scheme for the second derivatives are

$$p_1 = -(1004000 + 2000y_2)y_2^2, \quad q_1 = 1004004 + 2000y_2, \quad s_1 = 0,$$

$$p_2 = -1003y_1 + 1003y_2^2 + 2y_2^3, \quad q_2 = (1 - 2y_1), \quad s_2 = 0.$$

Table 14.1 shows the absolute error at the final points ($t = 1$, and 10) and computation time, both exact to four decimal places, at different grid points (Δ) and fifteen iterations obtained using the IMDHBM. The results were compared with references [16] and [17]. The numerical results show that the IMDHBM outperforms other

TABLE 14.1
Absolute Error for Example 1

Δ	t	Methods	y_1	y_2	CPU Time
50	1	IMDHBM	1.0490×10^{-29}	8.7481×10^{-30}	0.0731
		OTSHBM [16]	1.2258×10^{-13}	2.4555×10^{-15}	
		CBBDF [17]	9.1102×10^{-13}	1.2527×10^{-12}	
500	10	IMDHBM	5.7741×10^{-37}	5.6790×10^{-33}	0.7609
		OTSHBM [16]	2.1200×10^{-21}	3.0914×10^{-18}	
		CBBDF [17]	2.1977×10^{-20}	1.3542×10^{-15}	

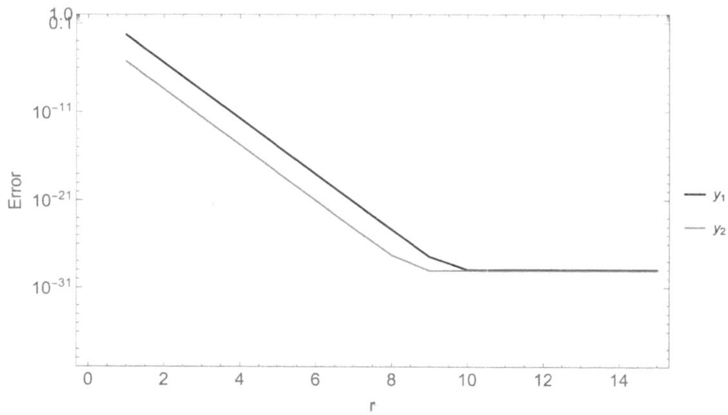

FIGURE 14.2 Convergence graph of y_1 and y_2 when $\Delta = 50$ on the closed interval [0, 1] for Example 1.

methods in terms of accuracy by returning smaller errors. In terms of method order, the IMDHBM is of order ten, whereas the OTSHBM [16] and CBBDF [17] are of order five and six, respectively.

Figures 14.2 and 14.3 provide the convergence graph of y_1 and y_2 when $\Delta = 50$ on the closed interval [0, 1] and when $\Delta = 500$ on the closed interval [0, 10], respectively, at fifteen iterations. In Figure 14.2, y_1 and y_2 converge after the 10th and 9th iterations, respectively, on the closed interval [0, 1]. In Figure 14.3, y_1 and y_2 converge after the 5th and 4th iterations, respectively, on the closed interval [0, 10].

The nonlinear system presented in references [16] and [17] was solved using the IMDHBM, demonstrating greater accuracy and efficiency, as shown by the smaller absolute errors and faster convergence in both tabular and graphical results.

14.5.2 Example 2

Consider the Rossler system (RS) [21] governed by the IVPs

$$y_1' = -y_2 - y_3, \qquad y_1(0) = 0,$$

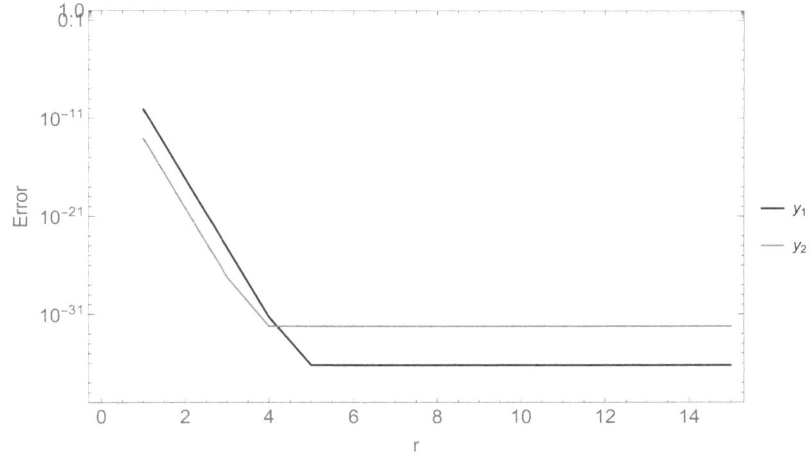

FIGURE 14.3 Convergence graph of y_1 and y_2 when $\Delta=500$ on the closed interval [0, 10] for Example 1.

$$y_2' = y_1 + a\,y_2, \qquad y_2(0) = 0,$$

$$y_3' = b + y_3(y_1 - c), \quad y_3(0) = 0.$$

A chaotic attractor with $a=0.2$, $b=0.2$ and $c=5.7$ was studied by [21]. For convenience, we define:

$$\kappa_1 = -y_2 - y_3, \quad \omega_1 = 0, \qquad \kappa_2 = y_1, \quad \omega_2 = a, \qquad \kappa_3 = b, \quad \omega_3 = y_1 - c.$$

The second derivatives of the Rossler system is given by

$$y_1'' = -y_2' - y_3' = -y_1(1 + y_3) - a\,y_2 - b + y_3 c,$$

$$y_2'' = y_1' + a\,y_2' = -y_2 - y_3 + a\,y_2',$$

$$y_3'' = y_3 y_1' + y_3'(y_1 - c) = -y_2 y_3 - y_3^2 + y_3'(y_1 - c),$$

and the parameters for the iteration scheme are

$$p_1 = -a\,y_2 - b - c\,y_3, \quad q_1 = -1 - y_3, \quad s_1 = 0,$$

$$p_2 = -y_3, \quad q_2 = -1, \quad s_2 = a,$$

$$p_3 = -y_3^2, q_3 = -y_2, \quad s_3 = y_1 - c.$$

Table 14.2 presents a comparison between the RS solutions obtained using IMDHBM with the MSRM [1] and PSLM [7]. The results were obtained using $h=0.025$,

TABLE 14.2

Numerical Solution for Rossler System Compared with MSRM and PSLM Results When $h = 0.025$

T	y_1			y_2			y_3		
	IMDHBM	MSRM [1]	PSLM [7]	IMDHBM	MSRM [1]	PSLM [7]	IMDHBM	MSRM [1]	PSLM [7]
10	0.05478	0.05478	0.05478	−0.11622	−0.11622	−0.11622	0.03534	0.03534	0.03534
20	−0.20350	−0.20350	−0.20350	0.09717	0.09717	0.09717	0.03403	0.03403	0.03403
40	−1.38930	−1.38930	−1.38930	−0.77840	−0.77840	−0.77840	0.02792	0.02792	0.02792
60	1.13402	1.13402	1.13402	−10.49973	−10.49976	−10.49976	0.03330	0.03330	0.03330
80	−1.84291	−1.84291	−1.84291	1.82289	1.82289	1.82289	0.73236	0.73236	0.73236
CPU Time	0.50606	0.73038	1.25968						

$\Delta = 3200$ and fifteen iterations on the closed interval [0, 80]. The computation time for producing results for y_1, y_2, and y_3 that are exact to five decimal places is also compared. The table shows that there is good agreement between the results of the IMDHBM and those of references [1] and [7]. In terms of computation speed, a comparison of the computation time shows that references [1] and [7] are less effective than the IMDHBM. This shows that the new method is accurate and useful for solving the RS.

Figure 14.4 shows the time series results and Figure 14.5 shows the corresponding phase portraits of the RS. Figure 14.4 was generated using $\Delta = 4000$ on the closed interval [0, 100] and Figure 14.5 was generated using $\Delta = 8000$ on the closed interval [0, 200]. The residual error in each block $[t_n, t_{n+1}]$ is defined as

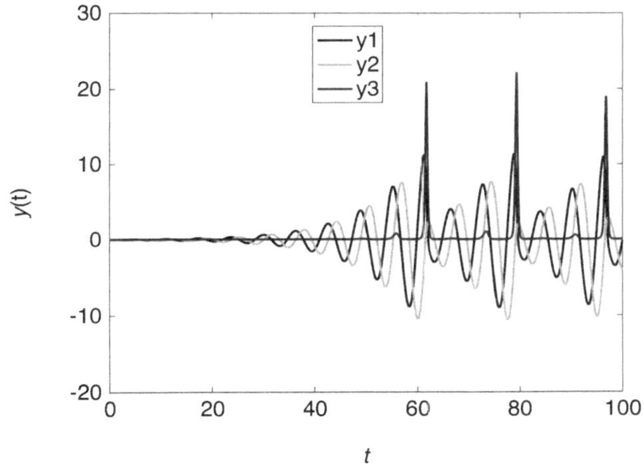

FIGURE 14.4 Time series solution for the Rossler equation when $h = 0.025$.

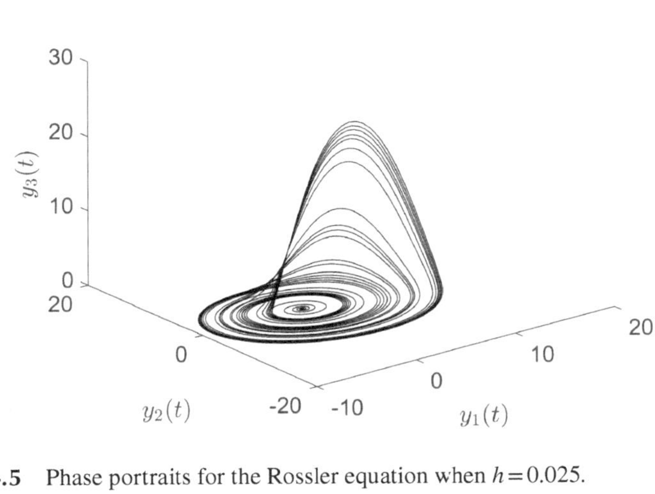

FIGURE 14.5 Phase portraits for the Rossler equation when $h = 0.025$.

$$\mathrm{RES}_n = \left\| \sum_{j=1}^{\psi+\Lambda} j\eta_{n,j}\left(t-t_n\right)^{j-1} - y'\left(t_{n+p_j}, y_{n+p_j}\right) \right\|. \tag{14.27}$$

To check that the IMDHBM is accurate, the RES and the iteration error were calculated with $\Delta = 4000$ and fifteen iterations on the closed interval [0, 100], we obtained Figures 14.6 and 14.7. The RES decreases as the number of iterations increases, and y_1, y_2, and y_3 converge after the 7th, 7th, and 8th iterations, respectively. This indicates that the IMDHBM is accurate, with RES of the order 10^{-10} to 10^{-15}.

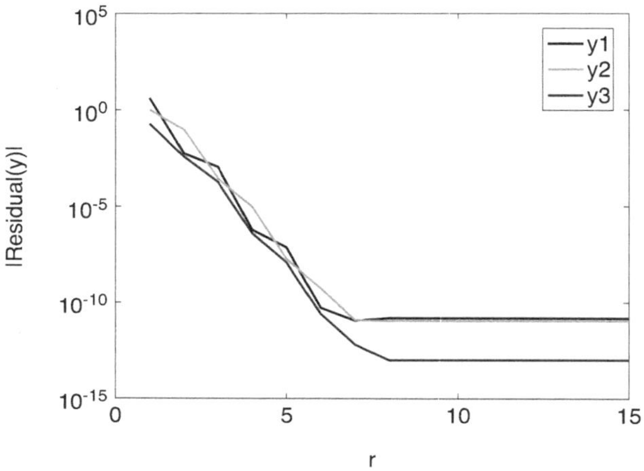

FIGURE 14.6 Residual error for the Rossler equation when $h = 0.025$.

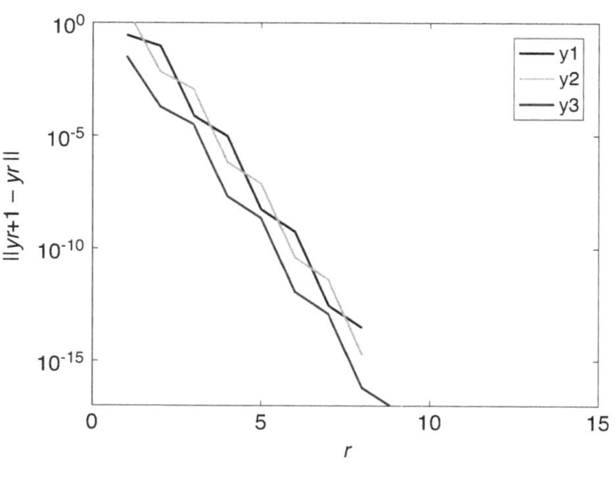

FIGURE 14.7 Iterative error for the Rossler equation when $h = 0.025$

The three-dimensional Rossler system, defined by a set of nonlinear differential equations has been solved using IMDHBM. The numerical solutions obtained via the IMDHBM method were compared with those from MSRM [1] and PSLM [7], showing good agreement and demonstrating the efficiency in terms of computation time and residual error.

14.5.3 Example 3

Consider the Arneodo-Coullet system [22] described by the following set of IVPs:

$$y_1' = y_2, \qquad y_1(0) = 0.21,$$

$$y_2' = y_3, \qquad y_2(0) = 0.22,$$

$$y_3' = a\,y_1 - by_2 - y_3 - y_1^3, \quad y_3(0) = 0.61.$$

The constants are $a = 5$ and $b = 3.8$. For the iteration scheme, we set

$$\kappa_1 = y_2, \quad \omega_1 = 0, \qquad \kappa_2 = y_3, \quad \omega_2 = 0, \qquad \kappa_3 = a\,y_1 - by_2 - y_1^3, \quad \omega_3 = -1.$$

The second derivatives of the Arneodo-Coullet system (ACS) are

$$y_1'' = y_2' = y_3,$$

$$y_2'' = y_3' = a\,y_1 - by_2 - y_3 - y_1^3,$$

$$y_3'' = a\,y_1' - by_2' - y_3' - 3y_1^2 y_1' = \left(a - 3y_1^2\right)y_2 - by_3 - y_3'.$$

The iteration scheme for the second derivatives are

$$p_1 = y_3, \quad q_1 = 0, \quad s_1 = 0,$$

$$p_2 = a\,y_1 - y_3 - y_1^3, \quad q_2 = -b, \quad s_2 = 0,$$

$$p_3 = \left(a - 3y_1^2\right)y_2, \quad q_3 = -b, \quad s_3 = -1.$$

The results of the MSRM and the Matlab solver ode45 from reference [1] are compared to the approximate IMDHBM solutions of the ACS with $h = 0.05$, $\Delta = 200$, and fifteen iterations on the closed interval [0, 10], as shown in Table 14.3. The results are in good agreement with one another, demonstrating the validity of the IMDHBM. The computation time for y_1, y_2, and y_3, exact to five decimal places, show that the IMDHBM is the fastest among the methods.

Figure 14.8 shows the time series results and Figure 14.9 shows the corresponding phase portraits of the ACS. The results were obtained using $\Delta = 1000$ on the closed interval [0, 50] and $\Delta = 2000$ on the closed intervals [0, 100], respectively.

TABLE 14.3
Numerical Solution for Arneodo–Coullet System Compared with MSRM and Ode45 Results of [1] When $h=0.05$

T	y_1			y_2			y_3		
	IMDHBM	**MSRM [1]**	**Ode45 [1]**	**IMDHBM**	**MSRM [1]**	**Ode45 [1]**	**IMDHBM**	**MSRM [1]**	**Ode45 [1]**
2	1.54703	1.54703	1.54703	1.13437	1.13437	1.13437	0.45461	0.45461	0.45461
4	1.74192	1.74192	1.74192	−1.58122	−1.58122	−1.58122	0.90371	0.90371	0.90371
6	3.22572	3.22572	3.22572	0.20200	0.20200	0.20200	−7.14070	−7.14070	−7.14070
8	−1.17969	−1.17970	−1.17970	0.52899	0.52899	0.52899	1.67458	1.67458	1.67458
10	−2.225058	−2.22506	−2.22506	−1.71510	−1.71510	−1.71510	0.65501	0.65501	0.65501
CPU Time	0.04181	0.08662	0.39798						

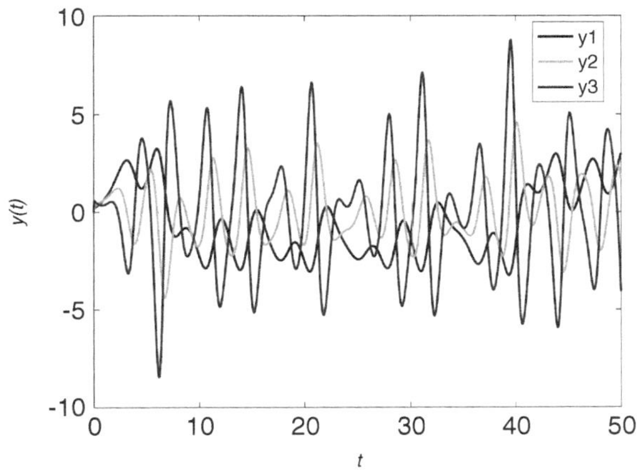

FIGURE 14.8 Time series solution for the Arneodo-Coullet equation when $h = 0.05$.

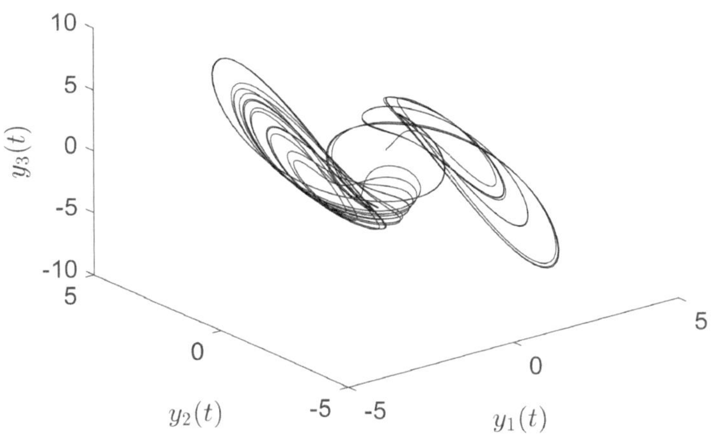

FIGURE 14.9 Phase portraits for the Arneodo-Coullet equation when $h = 0.05$.

The RES and iterative error are shown in Figures 14.10 and 14.11, respectively. Figure 14.10 was generated using $\Delta = 200$ and fifteen iterations on the closed interval [0, 10]. The results for y_1, y_2, and y_3 converged after the tenth iterations. This further demonstrates the numerical scheme's high level of accuracy, with residual errors of order 10^{-10} to 10^{-15}.

The Arneodo-Coullet system (ACS) was numerically solved using the IMDHBM, MSRM and Ode45 [1] methods, showing good agreement among them, with IMDHBM being the fastest. The results confirm the accuracy and efficiency of the IMDHBM method, as evidenced by the error analysis.

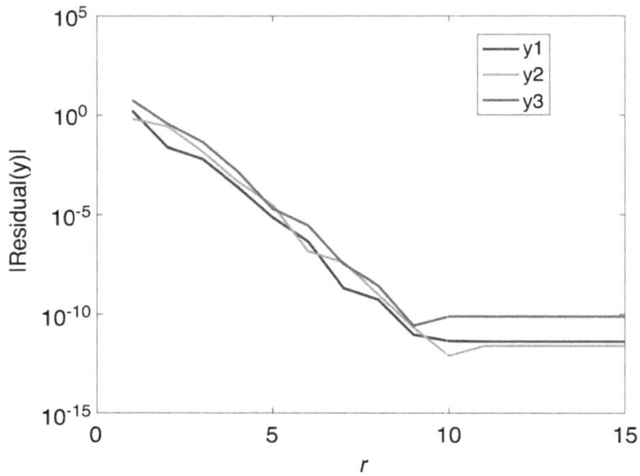

FIGURE 14.10 Residual error for the Arneodo-Coullet equation when $h = 0.05$.

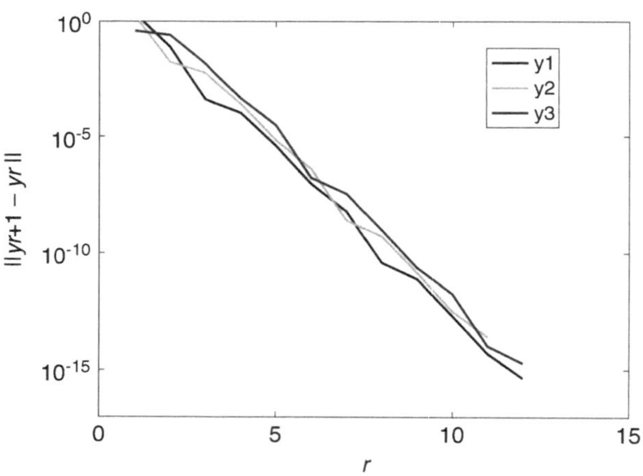

FIGURE 14.11 Iterative error for the Arneodo-Coullet equation when $h = 0.05$.

14.6 CONCLUSION

In this study, an IMDHBM has been developed to find solutions to nonlinear initial value problems (IVPs) and IVPs that exhibit general chaotic behavior. The IMDHBM was combined with a Gauss-Seidel type iterative method to simultaneously solve IVPs. The IMDHBM is zero-stable, consistent, convergent, and of order ten. A nonlinear system and two well-known chaos systems, namely, the Rössler and Arneodo-Coullet systems have been used to verify the method's accuracy. The results were compared with references [1], [7], [15], and [16] (see Tables 14.1–14.3 and Figures 14.2–14.11), and it was shown that the IMDHBM is more accurate and efficient in terms of error and computation time compared to the existing methods.

REFERENCES

[1] Motsa, S. S., Khumalo, M., Dlamini, P.G. (2018). A new multistage spectral relaxation method for solving chaotic initial value systems. *Nonlinear Dynamics*, 72(1), 265–283.

[2] Lorenz, E. N. (1963). Deterministic non-periodic flow. *Journal of Atmospheric Sciences*, 20(2), 130—141.

[3] Cheng, H., Zhou, J., Wu, Q. (2011). Adaptive synchronization of coupled hyperchaotic Chua systems. In *2011 Chinese Control and Decision Conference (CCDC)*, 143–148. IEEE.

[4] Rech, P. C., Albuquerque, H. A. (2009). A hyperchaotic Chua system. *International Journal of Bifurcation and Chaos*, 19(11), 3823–3828.

[5] Khan, M. S., Khan, M. I. (2018). A novel numerical algorithm based on Galerkin-Petrov time-discretization method for solving chaotic nonlinear dynamical systems. *Nonlinear Dynamics*, 91(3), 1555–1569.

[6] Motsa, S. S., Magagula, V. M., Sibanda, P. (2017). The multi-domain spectral relaxation method for chaotic system of ordinary differential equations. In *Chaotic Systems: Dynamics Algorithm and Synchronization*. Nova Science Publishers, Inc, Hauppauge, NY, 29–52.

[7] Motsa, S. S. (2012). A new piecewise-quasilinearization method for solving chaotic systems of initial value problems. *Central European Journal of Physics*, 10(4), 936–946.

[8] Motsa, S. S., Dlamini, P. G., Khumalo, M. (2012). Solving hyperchaotic systems using the spectral relaxation method. *Abstract and Applied Analysis*, 2012, Article ID 203461, 18 p.

[9] Akinnukawe, B. I., Akinfenwa, O. A., Okunuga, S. A. (2020). Third derivative block hybrid obrechkoff methods. *Nigerian Journal of Mathematics and Applications*, 30, 46– 60.

[10] Akinfenwa, O. A., Okunuga, S. A., Akinnukawe, B. I., Rufai, U. O., Abdulganiy, R. I. (2018). Multi-derivative hybrid implicit Runge-Kutta method for solving stiff system of a first order differential equation. *Far East Journal of Mathematical Sciences*, 106(2), 543–562.

[11] Rufai, U. O., Sibanda, P., Goqo, S. P. (2023). A one-step multi-derivative hybrid block method with modified-picard iteration for the solution of second order IVP. *Engineering Letters*, 31(4), 1–7.

[12] Dahlquist, G. (1963). A special stability problem for linear multistep methods. *BIT Numerical Mathematics*, 3(1), 27–43.

[13] Gear, C. W. (1965). Hybrid methods for initial value problems in ordinary differential equations. *Journal of the Society for Industrial and Applied Mathematics, Series B: Numerical Analysis*, 2(1), 69–86.

[14] Lambert, J. D. (1973). *Computational Methods in Ordinary Differential Equations*. John Wiley, New York.

[15] Jator, S. N., Adeyefa, E. O. (2019). Direct integration of fourth order initial and boundary value problems using Nystrom type methods. *IAENG International Journal of Applied Mathematics*, 49(4), 1–12.

[16] Ramos, H. (2017). An optimized two-step hybrid block method for solving first-order initial-value problems in ODEs. *Journal of Differential Geometry-Dynamical Systems*, 19, 107–118.

[17] Akinfenwa, O. A., Jator, S. N., Yao, N. M. (2013). Continuous block backward differentiation formula for solving stiff ordinary differential equations. *Computers and Mathematics with Applications*, 65(7), 996–1005.

[18] Fatunla, S. O. (1991). Block methods for second order IVPs. *International Journal of Computer Mathematics*, 42(9), 55–63.

[19] Henrici, P. (1962). *Discrete Variable Methods in ODE*. John Wiley and Sons, New York.

[20] Motsa, S. S. (2022). Overlapping grid-based optimized single-step hybrid block method for solving first-order initial value problems. *Algorithms*, 15(11), 427.
[21] Rossler, O. E. (1976). An equation of continuous chaos. *Physics Letters A*, 57(5), 397–398.
[22] Arneodo, A., Coullet, P., Tresser, C. (1981). Possible new strange attractors with spiral structure. Communications in Mathematical *Physics*, 79(4), 573–579.

15 Water Quality Predictive Analysis and Forecasting Dissolved Oxygen (DO) Using Neural Network Model

Sayantika Mukherjee, Srimanta Gupta, and Sumanta Nayek

15.1 INTRODUCTION

The role of water is crucial for maintaining the human health and wellbeing. The concentration of DO has a major role in maintaining the water quality. It is an indicator parameter of the health of a water body. The water temperature affects the concentration of DO in a water body. Low concentration of DO is responsible for water quality deterioration in term of loss of aquatic life, causing eutrophication. Surface water quality can be evaluated using various methods, which include the analysis of different water quality parameters and heavy metals [1]. To forecast water quality, applications of artificial intelligence are becoming more popular day by day [2,3]. In several studies, the model has been applied to water quality prediction in rivers [4,5]. Two ANN networks, in terms of multilayer perceptron (MLP) and radial basis function (RBF), were used to evaluate the total dissolved solid (TDS) content of the Jajrood River in Iran [4]. The computation of DO and BOD using three-layer feedforward neural networks with back propagation learning over the Gomti River in India has been done with precision [5]. The outcome confirms that both neural network models have the potential to forecast water parameters with higher precision (>90%). Furthermore, a successful modeling of BOD has been done using temperature, hardness, dissolved oxygen, and electrical conductivity data as input variables for the neural network.

The study is projected to evaluate the range of the thirty parameters to assess the overall water quality in Kolkata. For the purpose of DO modeling, an artificial neural network approach is used. Half yearly water quality data of BOD, COD, and DO from 2011 to 2019 are used to run the models. Furthermore, the present research work aims to evaluate the forecast error in predicting the DO level based on mean absolute error (MAE) and root mean square error (RMSE) [6].

DOI: 10.1201/9781003638414-15

15.2 METHODOLOGY

15.2.1 Study Location and Water Variables

In this study, seven different locations were selected all over Kolkata along the stretch of the River Ganga. These include Dakshineswar, Ganga at Garden Reach, Victoria (Pond No 1, 2, 3, and 4), and Rabindra Sarovar National Lake. The data used for the study were obtained from West Bengal Pollution Control Board (WBPCB), covering the years from 2011 to 2019. A total of 15 water parameters were investigated for the evaluation of water quality, with BOD and COD selected as input parameters for the proposed artificial neural network.

15.2.2 Statistical Methods

All statistical analysis and conventional graphical representation were performed using MS Excel, version 10.

15.2.3 Artificial Neural Network Model

Artificial intelligence is useful for functional prediction and system modeling where the physical processes are not understood or are highly complex [7,8]. ANN models, in terms of MLP, have been used to predict the DO from BOD and COD as input parameters over Kolkata, India. The following equation (15.1) is considered in this study.

Equation (15.1) is considered as:

$$y = \left(\sum_{i=1}^{n} \omega_i x_i + b \right) = \left(w^T x + b \right)$$

Where w = vector of weights, x = vector of inputs; b represents the bias, and T denotes the activation function.

15.3 RESULTS AND DISCUSSION

The dataset, consisting of fifteen physico-chemical parameters, was considered to evaluate the water quality of river Ganga from seven different sampling sites across Kolkata from 2011 to 2019. The variability of all the measured parameters is tabulated in Table 15.1.

Estimation of pH is useful for determining the nature of the solution (water sample). If the pH value remains within the range of 0–7, it indicates that the sample would be acidic, while a range of 7–14 indicates that the sample is alkaline, with a neutral value at 7. The pH values of the water sample ranged from 7.3 to 8.1. The electrical conductivity (EC) of the water samples ranged from 423.3 to 2000.3 µs/cm. Total dissolved solids (TDS) measures the dissolved materials in the water. Elevated concentration of TDS are of great concern due to their impact on water quality. The range in Kolkata was observed to be 10 to 2490 mg/L, with maximum variability observed between 143.8 and 1220.6 mg/L. Total suspended solids (TSS) can cause

TABLE 15.1
Physicochemical Parameters of River Water Samples

SL No	Parameter	Unit	Max	Min	Mean	St. dev
1	Temperature	°C	36	16	28.13	3.63
2	pH		9.83	6.85	7.69	0.43
3	EC	µs/cm	6070.00	376.00	1212.81	38.54
4	DO	mg/L	17.90	0.20	7.03	2.83
5	BOD	mg/L	65.04	0.05	8.08	10.49
6	COD	mg/L	131.99	1.02	27.53	15.77
7	Turbidity	NTU	463	10.14	52.39	23.05
8	TDS	mg/L	2490	310	682.20	28.39
9	TSS	mg/L	1672	286	79.06	29.25
10	Hardness	mg/L	950	64	328.75	21.47
11	Alkalinity	mg/L	490	55	255.53	13.57
12	Phosphate-P	mg/L	0.30	0.01	0.06	0.05
13	Nitrate-N	mg/L	0.78	0.004	0.20	0.19
14	Ammoniacal-N	mg/L	1.05	0.10	0.31	0.17
15	Chloride	mg/L	374.76	6.00	206.80	14.88

severe problems for water bodies and aquatic life too. These are the solid particles that are typically trapped by a filter. In this study, the TSS values were recorded within the range of 110.2 to 268.3 mg/L. Estimation of alkalinity is an indicator of the potential of stream water to resist alteration in pH. The alkalinity value varied from 55 to 490 mg/L, with maximum variability found to be between 142 and 369.1 mg/L. Water hardness ranged from 118.3 to 539.2 mg/L, with the maximum value recorded as 950 mg/L at the sampling stations. The chloride concentrations ranged from 6 to 775 mg/L, with a mean value of 208 mg/L. Nitrate-N is being released from fertilizers, animal residues, domestic, and industrial sources. Elevated nitrate levels can easily mobilize and leach into groundwater. The highest nitrate concentration recorded in the water sample was 0.8 mg/L, which can be harmful to humans and other life forms. A worsening situation can occur due to nitrogen contamination in water. The value of ammoniacal-N varied from 0.14 to 0.48 mg/L across the sampling sites in Kolkata. Phosphate-P is an essential plant nutrient, which is extensively used in fertilizer and other chemicals. The highest phosphate-P value recorded was 0.3 mg/L among the sampling stations.

DO is a water quality indicator parameter for any surface water system. The temperature of water affects the DO level. Low DO content is responsible for water quality deterioration particularly due to the loss of aquatic life. The range of DO concentrations varies from 0.2 to 17.9 mg/L. BOD and COD are key parameters used to examine surface water quality. High concentrations of BOD are a concern for plants and animals, as they result in less availability of dissolved oxygen. The measured values for BOD and COD values fall within the range of 2.4 to 18.6 mg/L, and 1.8 to 53.3 mg/L, respectively across the monitored stations. Chloride concentrations are recorded within a range of 6.0–411.7 mg/L. For further analysis, three water quality parameters – COD, BOD, and DO – are considered over seven sites (depending on the availability of data of the selected parameters). The study also aims to compare these three parameters site wise (Figures 15.1–15.3). Figure 15.1 depicts the average values of BOD across seven sampling stations. The minimum value is observed at Victoria Pond 4, and the maximum value at Rabindra Sarovar Lake. The variability of COD is shown in Figure 15.2. It reveals that the highest value was obtained at Rabindra Sarovar Lake. The lowest value of COD was observed at Victoria Pond 3. Figure 15.3 shows the schematic of the average values of DO across the seven sites, where Garden Reach reflects the minimum value. The maximum value of DO was obtained at Rabindra Sarovar over the study period.

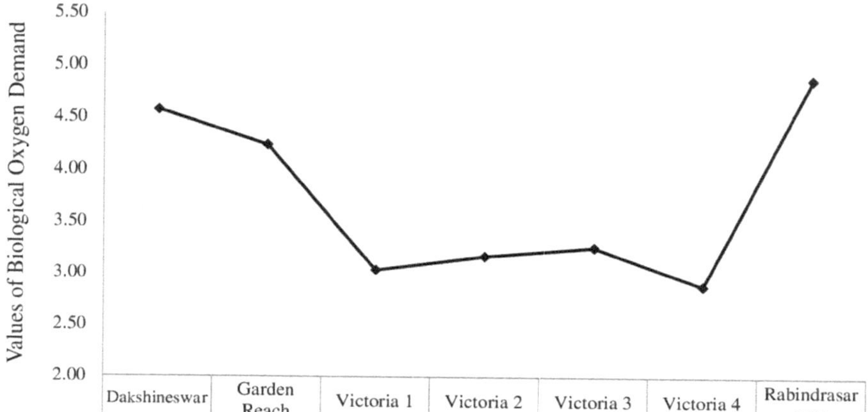

	Dakshineswar	Garden Reach	Victoria 1	Victoria 2	Victoria 3	Victoria 4	Rabindrasarovar
BOD	4.57	4.24	3.04	3.17	3.26	2.88	4.86

FIGURE 15.1 The average value of BOD over seven sites.

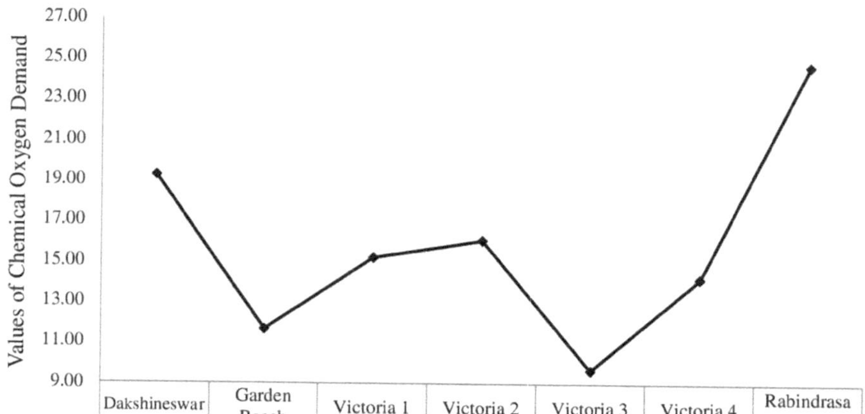

	Dakshineswar	Garden Reach	Victoria 1	Victoria 2	Victoria 3	Victoria 4	Rabindrasarovar
COD	19.26	11.64	15.19	16.05	9.65	14.18	24.68

FIGURE 15.2 The average value of COD over seven sites.

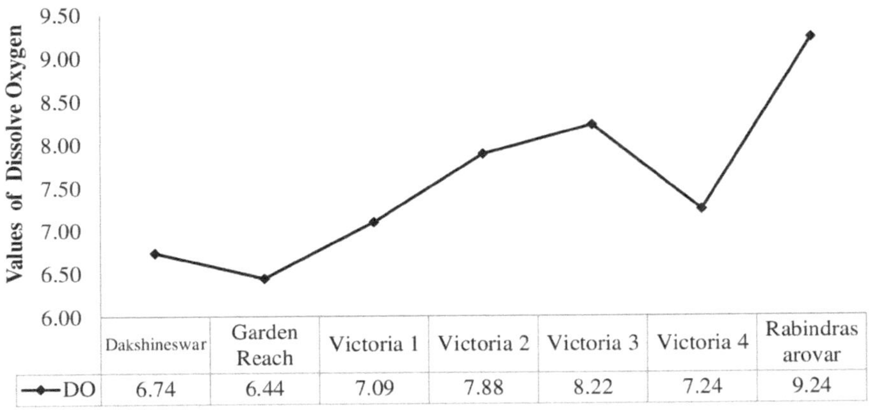

FIGURE 15.3 The average value of DO over seven sites.

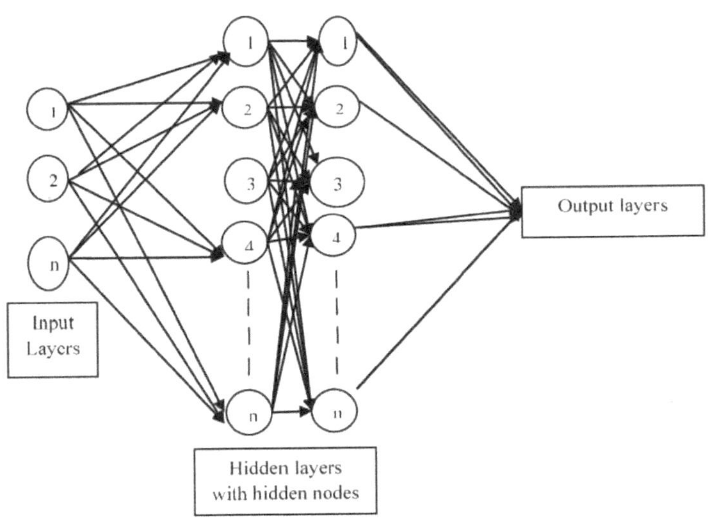

FIGURE 15.4 Schematic of ANN model having different input and output parameters.

The ANN models are developed by employing various numbers of neurons in the hidden layers (Figure 15.4). The effectiveness of the models is determined with respect to root mean square error and mean absolute error to assess the DO prediction potential. After calculating the forecast errors, three MLP models are selected using different network architecture. The abbreviations of the models showing the minimum error are MLP-1, MLP-2, and MLP-3. The construction of the model is given in Figure 15.5a–c, which illustrates the number of hidden layers and hidden nodes for each model. The forecast error has been calculated and is expressed as MAE and RMSE using the formula mentioned in the implementation procedure for the seven sites. For Dakshineswar, the mean absolute error values are 1.30, 1.10

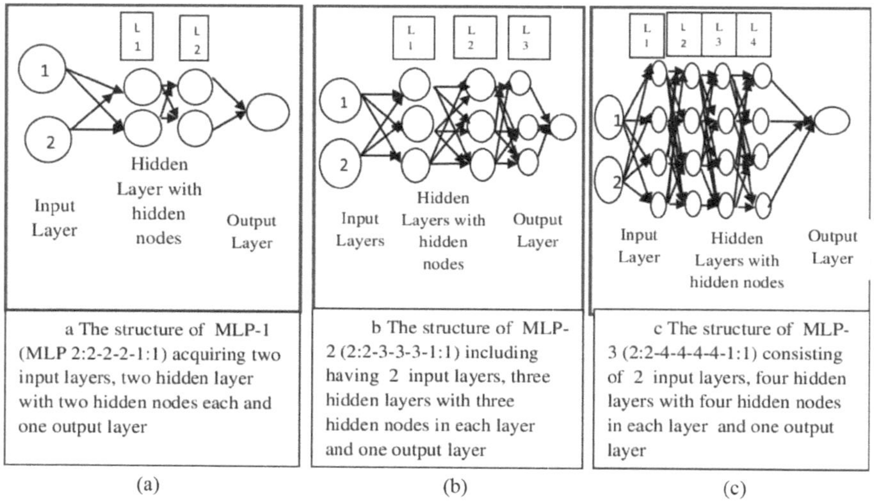

FIGURE 15.5 (a) The structure of MLP-1 (MLP 2:2-2-2-1:1) acquiring two input layers, two hidden layer with two hidden nodes each and one output layer, (b) The structure of MLP-2 (MLP 2:2-3-3-3-1:1) including having two input layers, three hidden layers with three hidden nodes in each layer and one output layer, (c) The structure of MLP-3 (2:2-4-4-4-4-1:1) consisting of two input layers, four hidden layers with four hidden nodes in each layer and one output layer.

and 0.20 for MLP-1, MLP-2 and MLP-3 respectively. The root mean square error values are 1.60, 1.10, and 0.78 for MLP-1, MLP-2, and MLP-3, respectively. From the schematic, it can be stated that the MLP-3 model shows the minimum error in predicting the DO value over Dakshineswar (Figure 15.6a). Figure 15.6 depicts the performance of the three ANN models in predicting the DO level. In this regard, it is observed that the MLP-3 model shows the minimum value of MAE and RMSE in estimating the DO value over Garden Reach using COD and BOD as input parameters (Figure 15.6b). Similarly, while evaluating the prediction accuracy of ANN models over Victoria (Pond1, Pond 2, Pond 3, and Pond 4); it is observed that MLP-3 model shows the lowest values of MAE and RMSE (Figure 15.6c–f). Figure 15.6g shows the comparison of the ANN models in estimating the dissolved oxygen. It depicts that the minimum error is obtained using MLP-3 model over Rabindra Sarovar.

15.4 CONCLUSIONS

In this study, artificial neural networks (ANNs) were used to develop models that can forecast and predict DO at various sampling stations along the stretch of River Ganga in Kolkata. The proposed model has shown good accuracy in predicting the DO levels in surface water. The multilayered perceptron model (MLP-3), based on BOD and COD as input parameters, has produced reliable results. Based on the observations,

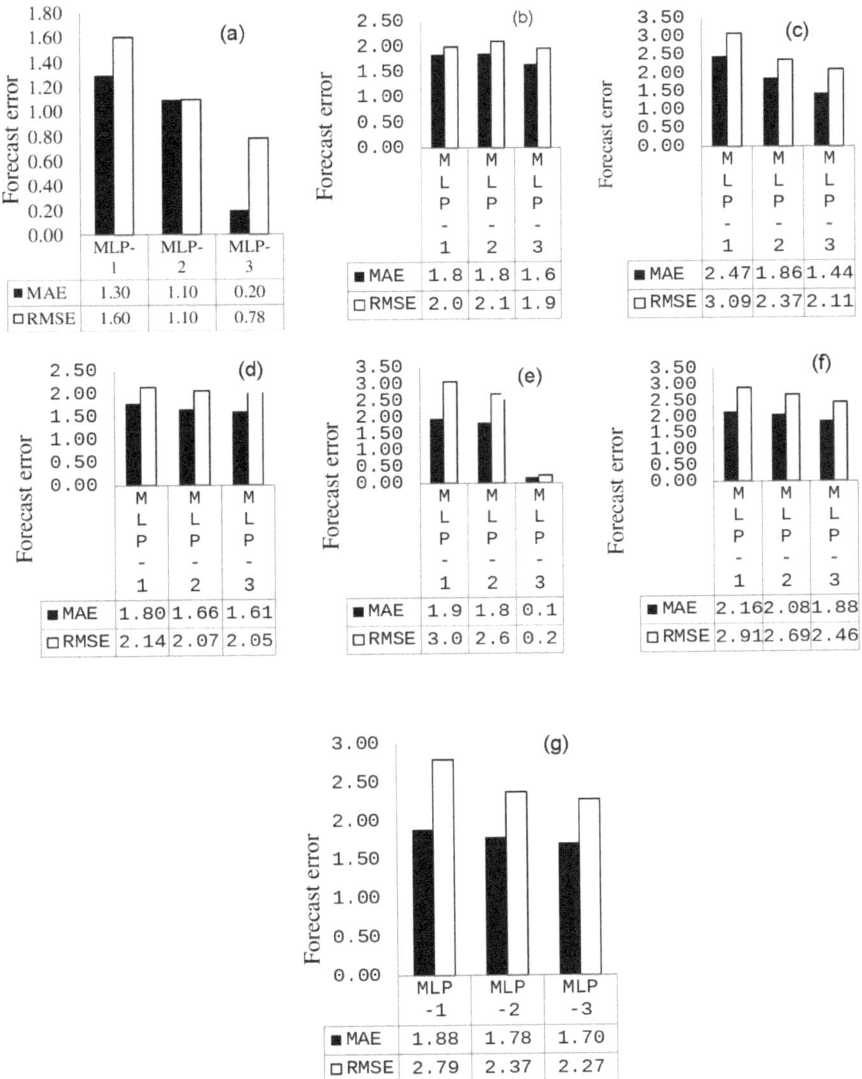

FIGURE 15.6 Forecast error in estimating the dissolved oxygen over (a) Dakshineswar (b) Garden Reach (c) Victoria Pond 1 (d) Victoria Pond 2 (e) Victoria Pond 3 (f) Victoria Pond 4 (g) Rabindra Sarovar Using MLP models.

it can be concluded that forecasting and prediction of DO level at the monitored stations of River Ganga in Kolkata can be effectively achieved with an acceptable range of accuracy using a small set of variables and the multilayered perceptron model. In this regard, the obtained results prove to be crucial for the evaluating water quality and managing surface water resources effectively.

ACKNOWLEDGMENTS

The authors wish to thank West Bengal Pollution Control Board for sharing the data with no cost, and Amity University Kolkata for providing infrastructure and facilities to conduct this research work.

REFERENCES

[1] Rahmanian, N., Ali, S. H. B., Homayoonfard, M., Ali, N. J., Rehan, M., Sadef, Y., Nizami, A. S. (2015). Analysis of physiochemical parameters to evaluate the drinking water quality in the State of Perak, Malaysia. *Journal of Chemistry*, 2015, 716125.
[2] French, M. N., Krajewski, W.F., Cuykendall, R. R. (1992). Rainfall forecasting in space and time using a neural network. *Journal of Hydrology*, 137(1–4), 1–31.
[3] Sakai, Y.M., Fässler, T., Ruoslahti, E. (2003). Antiangiogenic proteins require plasma fibronectin or vitronectin for in vivo activity. *Proceedings of the National Academy of Sciences of the United States of America*, 100(20), 11435–11438.
[4] Niroobakhsh, M.., Musavi-Jahromi, S. H., Manshouri, M., Sedghi, H. (2012). Prediction of water quality parameter in Jajrood River basin: Application of multi-layer perceptron (MLP) perceptron and radial basis function networks of artificial neural networks (ANNs). *African Journal of Agricultural Research*, 7(29), 4131–4139.
[5] Singh, P., Kunwar, P., Basant, A., Malik, A., Jain, G. (2009). Artificial neural network modeling of the river water quality-a case study. *Ecological Modelling*, 220(6), 888–895.
[6] Chaudhuri, S., Das, D, Sarkar, I, Goswami, S. (2015). Multilayer perceptron model for nowcasting visibility from surface observations: Results and sensitivity to dissimilar station altitudes. *Pure and Applied Geophysics*, 172(10), 2813–2829.
[7] Gardner, M.W., Dorling, S.R. (1998). Artificial neural networks (the multilayer perceptron)—a review of applications in the atmospheric sciences. *Atmospheric Environment*, 32(14–15), 2627–2636.
[8] Chaudhuri, S., Dutta, D., Goswami, S., Middey, A. (2015). Track and intensity forecast of tropical cyclones over the North Indian Ocean with multilayer feed forward neural nets. *Meteorological Applications*, 22(3), 563–575.

16 Understanding the Tropospheric Ozone Photochemistry in Gauteng, South Africa, Using Tropospheric Ultraviolet and Visible (TUV) Radiation Model

Sharadia Dey, Sabyasachi Mondal, and Precious Sibanda

16.1 INTRODUCTION

Tropospheric ozone plays a key role in atmospheric chemistry and climate change (Reddy et al. [1]). Understanding tropospheric ozone is essential not only because it is a greenhouse gas, but also because it is a precursor to the highly reactive hydroxyl radical, which determines the chemical composition of the troposphere (Lal et al. [2]). Variation in tropospheric ozone concentration is a complex function of surface emissions and deposition, photochemistry, horizontal transport, and interaction with the free atmosphere (Kuang et al. [3]). The mechanism of photochemical formation of ozone was first identified in California in the 1950s (Haagen-Smit [4]) and the role of photochemical oxidation of carbon monoxide and hydrocarbons catalysed by NO_x and HO_x was suggested in the 1970s (Crutzen [5], Chameides and Walker [6]). Surface ozone notably exceeds the permissible limit in different regions of the world due to increasing emissions of its precursors from anthropogenic sources. Several studies attempt to explore the sources and mechanisms of surface ozone formation. Oxides of nitrogen (NO_x) are known to play an important role in tropospheric ozone formation (David and Nair [7]). Durao et al. [8] employed a combination of Classification and Regression Trees techniques and the Multilayer Perceptron model for forecasting O_3 at an industrial site in Portugal. Abdallah et al. [9] used a chemical transport model (CTM) for simulating O_3 and $PM_{2.5}$ over Lebanon and investigated the influence of boundary conditions and

DOI: 10.1201/9781003638414-16

anthropogenic emission inventories on the simulated O_3 and $PM_{2.5}$ concentrations. Ghosh and Sarkar [10] focused on the relative importance of NO_x in the net production of ozone by using a chemical kinetics based model and used the Tropospheric Ultraviolet and Visible (TUV) Radiation Model for the computation of photolytic rate constants over Kolkata, India. The contribution of long-range horizontal transport of ozone precursors over continental air masses can be explored by air mass back-trajectory simulation (Srivastava et al. [11]). Few studies focus on the variation of tropospheric ozone and the occurrence of elevated ozone over different regions of South Africa. Diab et al. [12] reported the climatology of tropospheric ozone in Irene, South Africa and compared the results between two time periods, 1990–1994 and 1998–2002. Raghunandan et al. [13] studied tropospheric column ozone (TCO) variation over Johannesburg, South Africa, and identified the days with enhanced ozone as well as the sources of elevated ozone from the Measurement of Ozone and Water Vapor by Airbus In-Service Aircraft (MOZAIC) database for the period 1995 to 1998. Various studies conducted over South Africa confirm that both the dynamics of the region and multiple diverse sources play a crucial role in ozone buildup (Diab et al. [14], Thompson et al. [15]).

The present study identifies the days on which the 8-hourly average limit value of ozone (as per South African National Standards) was exceeded over the study site during January- December, 2015. The importance of the basic NO_2-NO-O_3 photochemical cycle in ozone production is analysed using a simple chemical kinetics based model. The Tropospheric Ultraviolet and Visible (TUV) Radiation Model Version 4.1 of UCAR is used for the computation of photolytic rate constants. The output of the model is evaluated based on different statistical tools, and the probable sources of elevated ozone concentrations on the identified days are illustrated. The roles of other ozone precursors (such as VOCs and carbon monoxide) in daytime ozone formation are examined. The influence of meteorological phenomena and long-range horizontal transport of ozone and its precursors is analysed to understand the temporal variation of ozone on the case study days over the chosen site. To the best of the authors' knowledge, such a comprehensive study on the role of different precursors, horizontal transport, and meteorology in the occurrence of elevated surface ozone levels is a novel contribution to the study of this region of African continent.

16.2 MATERIALS AND METHODS

16.2.1 Description of the Site

The site is located in Three Rivers, Gauteng province in South Africa. It is an air quality monitoring station of the South African Air Quality System, situated at 26.66 S and 28.00 E. This site is situated in the southern part of Gauteng province with an elevation of 1432 m (above sea level), and experiences warm, wind-free summers and chilly winter. Rainfall mostly occurs during the summer season, with occasional thunderstorms. Vereeniging CBD and Vanderbijlpark, which have significant industrial areas, are located to the northwest and southwest of this chosen site. The location of the study site is shown in Figure 16.1.

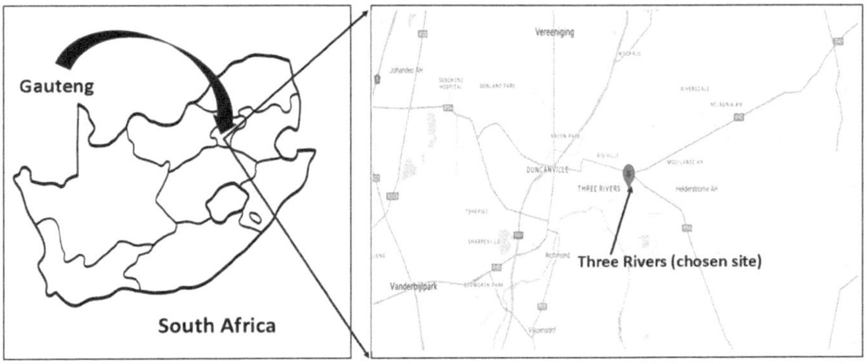

FIGURE 16.1 Location of the study site.

16.2.2 Data Collection

The data for different air pollutants (ozone, nitrogen dioxide, nitric oxide, carbon monoxide, benzene, toluene and xylene) and meteorological parameters (temperature, humidity, solar radiation, rainfall, wind direction, and wind speed) are collected from the South African Weather Service (SAWS) through the South African Air Quality System (SAAQIS). The data provider for this site (Three Rivers) is the DEA and this station falls within the Vaal Airshed Priority Area network. The data on cloud cover is collected from the NOAA Air Resources Laboratory (ARL) website. (https://www.ready.noaa.gov/READYcmet.php)

Air mass backward trajectory analysis is useful for ascertaining the sources and transport pathways of different trace species in the atmosphere. This simulation has been done using the National Oceanic and Atmospheric Administration (NOAA) HYbrid Single-Particle Lagrangian Integrated Trajectory (HYSPLIT) model (Draxler and Hess [16]), using GDAS Meteorological data. The 24-hour back-trajectories at three altitude levels (100, 500, and 1000 m AGL) are simulated.

16.3 THEORY AND CALCULATION

16.3.1 Daytime Photochemistry of Tropospheric Ozone

The concentration of tropospheric ozone is significantly influenced by the diurnal and seasonal variation of nitrogen oxide concentration. During the daytime, NO_x is quickly interconverted into two forms, NO_2 and NO. Tropospheric ozone chemistry is highly nonlinear and its production depends on a cycle of photochemical reactions involving NO_x and volatile organic compounds (VOC). According to Finlayson-Pitts and Pitts [17], a basic NO_2-NO-O_3 photochemical cycle always operates in the troposphere. In the troposphere, ozone formation results from the following reactions (Seinfeld and Pandis [18]):

$$NO_2 + hv \rightarrow NO + O\left({}^3P\right) \ (\text{where } hv < 420) \tag{16.1}$$

$$O(^3P) + O_2 + M \rightarrow O_3 + M \tag{16.2}$$

Here, '*M*' represents a molecule (e.g. N_2 or O_2) that absorbs energy from the reaction as heat and thereby stabilises the O_3 formed.

Once formed, ozone reacts with NO to regenerate NO_2.

$$O_3 + NO \rightarrow NO_2 + O_2 \tag{16.3}$$

Applying steady state approximation on all the reacting species, the steady state formation of O_3 based on only NO_x cycle can be computed (Ghosh and Sarkar [10]) as follows

$$\begin{aligned}[O_3] = &-\frac{1}{2}\left([NO]_0 - [O_3]_0 + \frac{j_{NO_2}}{k}\right) \\ &+\frac{1}{2}\left\{\left([NO]_0 - [O_3]_0 + \frac{j_{NO_2}}{k}\right)^2 + 4\frac{j_{NO_2}}{k}\left([NO_2]_0 + [O_3]_0\right)\right\}^{1/2}\end{aligned} \tag{16.4}$$

where, $[NO]_0$ and $[O_3]_0$ represent their respective initial concentrations of NO and O_3 respectively, j_{NO_2} is the photolytic rate constant for photolysis of NO_2 and k is the rate constant for the reaction of NO with O_3.

Leighton [19] introduced a parameter, Leighton ratio (Φ), which can be used to explore the chemical processes leading to the conversion of NO to NO_2. The Leighton ratio (Φ) can be defined as

$$[O_3] = \frac{j_{NO_2}[NO_2]}{k[NO]} \tag{16.5}$$

where $[NO_2]$ and [NO] represent the concentration of NO_2 and NO respectively, j_{NO_2} is the photolytic rate constant for photolysis of NO_2 and k is the rate constant for the reaction of NO with O_3.

Numerous studies have shown that Leighton ratios are consistently unity in areas with high ambient NO_x concentrations (Calvert and Stockwell [20], Carpenter et al. [21], and Yang et al. [22]), whereas in rural or isolated areas with high sun conditions, the value of the Leighton ratio ranges from approximately 1.2 to 3 (Ridley et al. [23], Hauglustaine et al. [24], and Volz-Thomas et al. [25]). Higher averages and ranges of Φ are reported for measurements performed at higher elevations (Davis et al. [26], Mannschreck et al. [27]).

When a chemical process other than the reaction between NO and O_3 converts NO to NO_2, the value of **Φ** shows a positive deviation from unity (Griffin et al. [28]). Other chemical processes leading to the daytime interconversion of NO_x generally involve hydroperoxy radical (HO_2) and organic peroxy radicals (RO_2), which are mainly produced in the troposphere as intermediates in the photochemical oxidation of carbon monoxide (CO) and Volatile Organic Compounds (VOCs).

16.3.2 TUV Model and Calculation of j_{NO_2}

The Tropospheric Ultraviolet and Visible (TUV) Radiation Model (https://www.acom.ucar.edu/Models/TUV/Interactive_TUV/) has been used for the calculation of the photolytic rate constant for the photolysis of NO_2 (j_{NO_2}) (Vuilleumier et al. [29]). This interactive model computes molecular photolysis frequencies over the visible and ultraviolet spectral range in the atmosphere under clear sky conditions by using radiative transfer scheme of pseudo spherical discrete ordinate four streams. The solar zenith angle (degree), overhead ozone column (du), surface albedo, ground elevation (km asl), measurement altitude (km asl) are used as the input parameters for the simulation. The chosen site is a residential area, so the albedo is considered as 0.16 (Sailor [30]). The ground elevation and the measurement altitude are 1.450 km asl and 1.455 km asl, respectively. The data for the daily overhead ozone column has been collected from NASA's Ozone and Air quality website.

16.4 STATISTICAL TECHNIQUES

- **Root Mean Square Error (RMSE)**: It is an effective statistical tool for measuring the difference between the predicted values by a model and the actually observed values. RMSE can be calculated as follows:

$$\text{RMSE} = \sqrt{\frac{1}{n}\sum_{i=1}^{n}(P_i - O_i)^2} \quad (16.6)$$

  Where O_i is the actually observed value and P_i is the value predicted by the model.

  RMSE can be used for evaluating the performance of a model.
- **Pearson's Correlation Coefficient (*r*)**: It is a statistical measure of the strength of a linear association between two variables and can be calculated as follows

$$r = \frac{\sum_{i-1}^{x}(X_i - \overline{X})(Y_i - \overline{Y})}{(n-1)\,S_x\,S_y} \quad (16.7)$$

  Where X and Y are two variables, with means $\overline{X}$ and $\overline{Y}$, respectively, and with standard deviations S_x and S_y, respectively.

16.5 RESULTS AND DISCUSSION

The National Framework for Air Quality Management in South Africa has made provisions for the establishment of standard limits for common air pollutants. As per the South African National Standard, the 8-hour average concentration limit for ambient ozone is 120 $\mu g/m^3$ and the number of permissible exceedances for ozone is 11. During our study period, i.e., from January 2015 to December 2015 (1 year), 14 such occasions were found on which the prescribed permissible limit for ambient ozone was exceeded. The following dates were identified as the days on which the ozone concentration at the Three Rivers site was higher than the national permissible limit: 19.01.15, 20.01.15, 10.02.15, 21.02.15, 09.10.15, 10.10.15, 11.10.15, 12.10.15, 13.10.15, 14.10.15, 15.10.15, 19.10.15, 18.12.15 and 20.12.15. The sources of ground-level ozone are explored in the subsequent sections.

16.5.1 Photochemical Ozone Production

Tropospheric ozone shows a distinct diurnal and seasonal variation. Generally, ozone concentration begins to increase gradually after sunrise, reaches its maximum around noontime, and starts decreasing after late afternoon or evening. The present work attempts to find out the amount of ozone formed by the basic photochemical cycle involving NO, NO_2 and O_3 by using the TUV radiation model.

Figure 16.2 shows the variation of observed and model-predicted concentrations of surface ozone over the study site from 6 to 18 hours on all the above-mentioned days except 21.02.15. Due to the unavailability of NO and NO_2 concentration data,

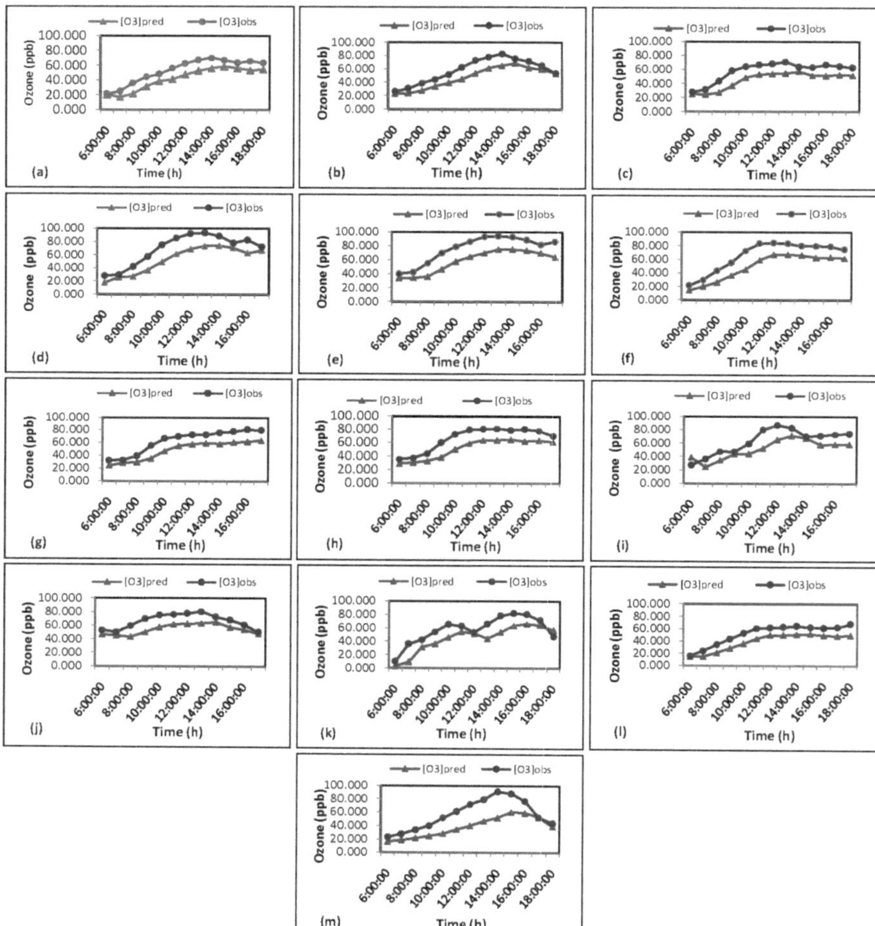

FIGURE 16.2 Variation of observed and predicted concentration of surface ozone over the study site during 06:00:00 to 18:00:00 hours on (a) 19.01.15, (b) 20.01.15, (c) 10.02.15, (d) 09.10.15, (e) 10.10.15, (f) 11.10.15, (g) 12.10.15, (h) 13.10.15, (i) 14.10.15, (j) 15.10.15, (k) 19.10.15, (l) 18.12.15 and (m) 20.12.15. The time period of this study is from 6 hours in the morning to 18 hours in the evening.

the computation of predicted ozone values was not possible. The plots show that the predicted and observed values of ozone are in fair agreement during the morning hours (i.e. 6 to 9 hours) and to some extent during the afternoon and evening (i.e. 13 to 18 hours). However, the predicted values deviate considerably during the late morning and afternoon (i.e., 10 to 14 hours). A similar result was also reported over the Asian mega city of Kolkata by Ghosh and Sarkar (2015). The chosen residential site has power a station to the south and a steel industry to the west. As a result, industrial emissions remain fairly constant throughout the day but domestic fuel burning emissions peak in the morning and evening hours. The fair agreement between the modelled and observed ozone values during the morning and evening might be attributed to the predominant NO_x cycle due to domestic fuel burning emission peaks.

16.5.2 Evaluation of Model Output

In general, the model provides under-predicted values in comparison to the actual ozone concentration. The interactive TUV radiation model does not take cloud cover into consideration while performing the simulation for computing j_{NO_2}. Therefore, the model simulated j_{NO_2} are the ideal values, assuming clear sky conditions. However, the presence of clouds over the study site (Table 16.1) might have introduced some error in the simulation.

Moreover, albedo, which is one of the input parameters in the TUV model, has been considered constant throughout the year. The underprediction of ozone concentration might be a manifestation of these assumptions.

The performance of the model can be evaluated by calculating the root mean square error. The values of RMSE in Table 16.2 show that the deviation remains minimum during the morning hours (i.e. 6–9 hours) and afternoon to evening hours (i.e., 15–18 hours), whereas the maximum deviation occurs during the late morning to afternoon hours (i.e., 10–14 hours). Ghosh and Sarkar (2015) also reported a similar trend in RMSE over Kolkata. The values of RMSE suggest that the basic NO_x cycle (NO_2-NO-O_3) remains the dominant source of tropospheric ozone during the morning hours but other ozone precursors also play a vital role in ozone production during the late morning to afternoon hours. The contribution of other ozone precursors to ozone production leads to a high deviation between the observed ozone level and the model-predicted ozone level.

TABLE 16.1
Variation of Cloud Cover (%) Over The Study Site on the Case Study Days

Time	Cloud Cover (%) On												
(h)	19.01	20.01	10.02	09.10	10.10	11.10	12.10	13.10	14.10	15.10	19.10	18.12	20.12
5	0.3	2	0	52	44	2.2	0.7	1	89.6	100	26.9	33.6	67.4
8	0	1	0	29.2	26.8	22.1	31.5	6.1	60.4	78.8	15.2	41.2	43
11	22.1	2.7	0	12.4	1.4	62.7	83.7	38.8	86.1	1.3	3.8	39.6	2.4
14	14.6	1	17	39.6	0.7	52.8	80.3	34.1	84.8	10.6	24	45.2	25.2
17	8.8	5.1	52.3	69	20.5	61.5	34	16.2	91.9	38.4	89.8	48	83.8

TABLE 16.2
Summary of Root Mean Square Error (RMSE) at Three Different Time Duration on the Case Study Days

Dates↓/Time (h)→	6–9	10–14	15–18
19.01.15	10.416	13.781	9.597
20.01.15	8.577	16.829	6.589
10.02.15	13.846	13.770	12.809
09.10.15	13.582	21.663	12.516
10.10.15	15.915	20.745	16.771
11.10.15	13.997	20.480	15.411
12.10.15	12.204	16.183	17.518
13.10.15	13.264	18.480	14.328
14.10.15	10.712	18.009	14.739
15.10.15	13.279	14.992	7.708
19.10.15	17.556	17.451	13.160
18.12.15	11.117	14.655	14.079
20.12.15	11.248	31.024	16.533

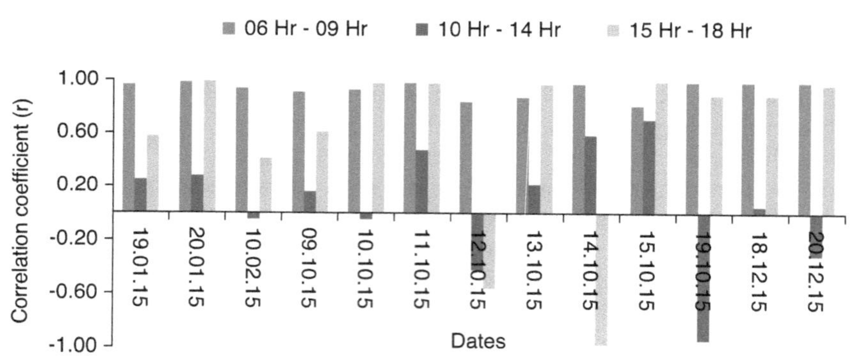

FIGURE 16.3 Variation of correlation coefficient of observed ozone concentration with j_{NO_2} at three different durations of the case study days.

The correlation coefficient (*r*) of observed ozone concentration and the photolytic rate constant for the photolysis of NO_2 (j_{NO_2}) shows a high positive correlation (close to 1) during the morning hours (6–9 hours) and evening hours (15–18 hours), as shown in Figure 16.3. However, during the late morning to afternoon hours (10–14 hours), the observed ozone concentration either shows a very low positive correlation or a negative correlation. Such an observation further supports the predominance of the NOx based cycle for the production of surface ozone during the morning and evening hours.

16.5.3 Role of Other Ozone Precursors

RMSE values (Table 16.2) and the values of the correlation coefficient (Figure 16.3) suggest that ozone photochemistry is not solely NO_x-driven during the late morning to afternoon hours. The computation of the Leighton Ratio is performed to examine the contribution of other photochemical cycles (apart from basic NO_x cycle) in formation of ozone.

Figure 16.4 shows that the Leighton ratio exhibits positive deviation from unity during the late morning to afternoon hours. This observation suggests the occurrence of some chemical processes other than the reaction between NO and O_3 leading to the conversion of NO to NO_2. Other daytime chemical processes responsible for the interconversion of NOx species involve free radicals like the hydroperoxy radical (OH_2) and organic peroxy radicals (RO_2). These free radicals are mainly produced in the troposphere as intermediates in the photochemical oxidation of CO and VOCs. Such type of reactions lead to the formation of photochemical O_3 as a tropospheric pollutant (Seinfeld and Pandis [31]). The hydroperoxy radical (OH_2) and organic peroxy radical (RO_2) provide additional NO to NO_2 conversion routes. The variation in the ambient concentrations of some VOCs (such as Benzene, Toluene and Xylene) and carbon monoxide (CO) on the case study days is examined (Figure 16.5). It is observed in Figure 16.4 that the concentrations of Benzene, Toluene, Xylene and CO remain high during the morning and evening hours, but the concentrations of these precursors of ozone reach their minimum around noon.

Along with the NOx-based cycle, the photochemical oxidation of VOCs and CO also contributes to tropospheric ozone formation during the late morning to afternoon

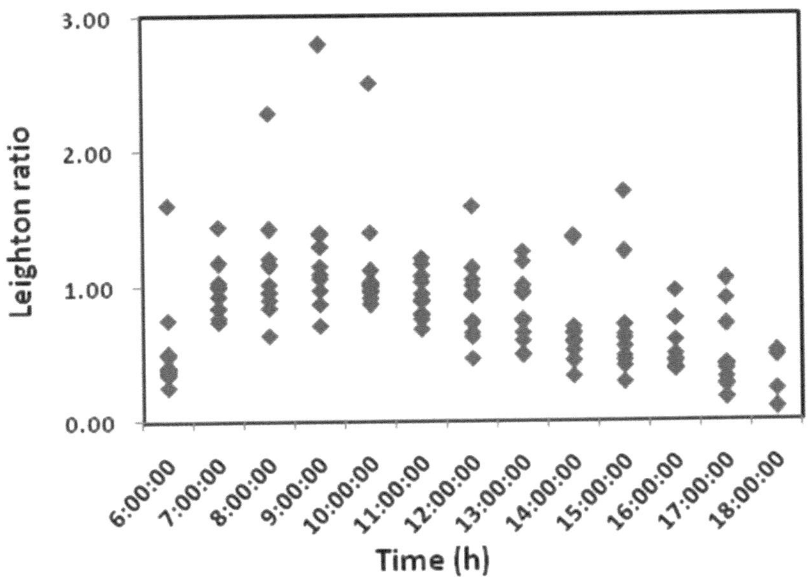

FIGURE 16.4 Variation of Leighton Ratio on the dates of case study.

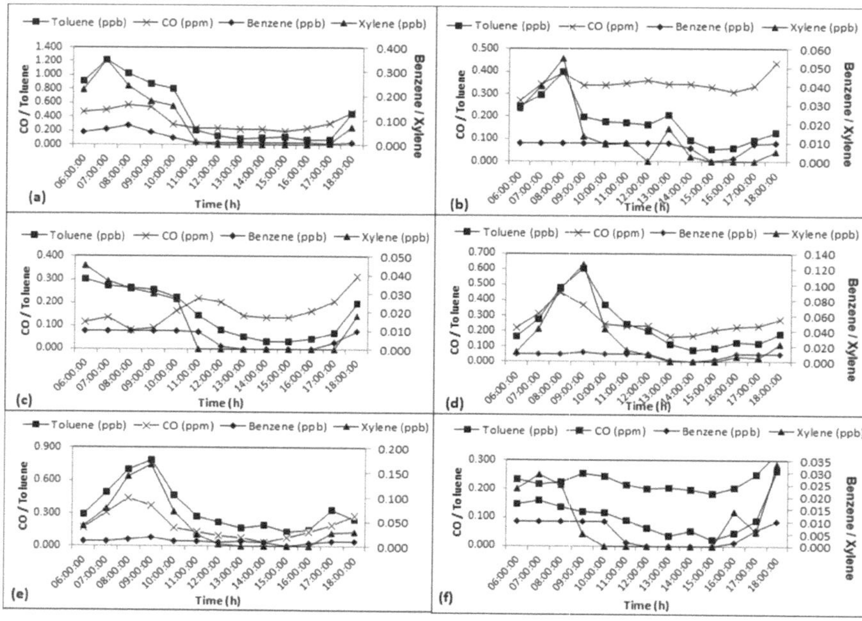

FIGURE 16.5 Variation of the ambient concentration of Benzene, Toluene, Xylene and CO from 06:00:00 to 18:00:00hours on (a) 09.10.15, (b) 10.10.15, (c) 11.10.15, 9 (d) 12.10.15, (13) 13.10.15 and (e) 18.12.15.

hours. Therefore, the concentrations of precursors decrease during this time of the day, supporting the high surface ozone concentration.

16.5.4 Influence of Advection and Meteorology

Horizontal transport (advection) of ozone and its precursors also influences the concentration of ozone over a region. An air mass containing the precursors of ozone, under suitable meteorological conditions, can form ozone during its transit from one place to another.

A detailed study of the pathways of the air masses reaching the site has been performed using the National Oceanic and Atmospheric Administration (NOAA) HYbrid Single-Particle Lagrangian Integrated Trajectory (HYSPLIT) model (Draxler and Hess [16]), with GDAS Meteorological data. These back trajectory simulations end at the local time when the deviation between the predicted and observed ozone values is maximum.

On 09.10.15 and 13.10.15 (Figures 16.6 and 16.10, respectively), air masses reside over continental areas and accumulate a considerable amount of ozone and its precursors. These air masses traverse different iron and steel industries, oil refineries and other industries located in Johannesburg before reaching the study site. Therefore, advection increases the concentration of tropospheric ozone over the chosen site, which is not considered by the TUV model. Such an incursion of ozone leads to a deviation between the observed and predicted values of ozone (Figure 16.7).

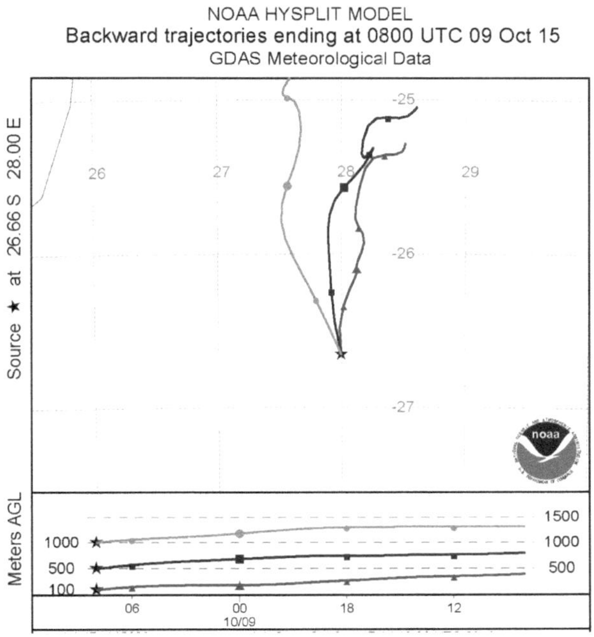

FIGURE 16.6 24 hours backward trajectory simulation at the chosen site on 09.10.15.

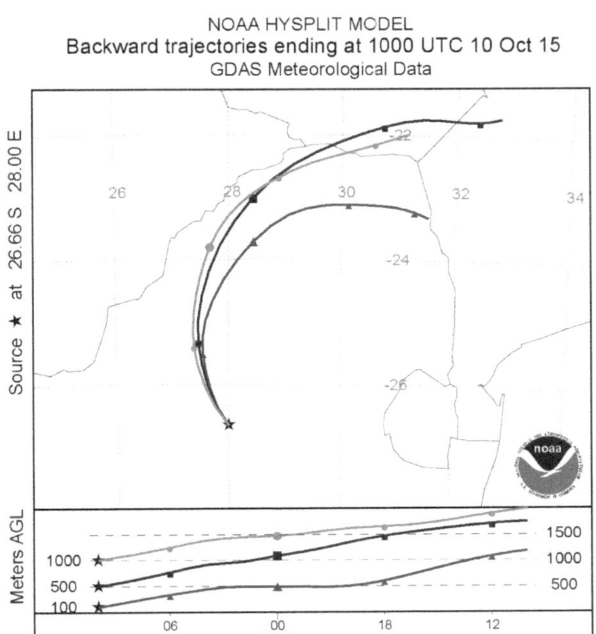

FIGURE 16.7 24 hours backward trajectory simulation at the chosen site on 10.10.15.

On 11.10.15, the air masses at 100m and 500m travel through polluted regions (such as the Lethabo power plant area), where ozone precursors accumulate, which, in turn, produce ozone under suitable weather conditions (Figure 16.8). Moreover, the air mass at 1000m travels through the polluted industrial regions of Vereeniging, where a substantial amount of ozone and its precursors accumulate. Horizontal transport of ozone increases the concentration of ozone over the study site and also results in large deviation between the observed and predicted value.

A similar kind of horizontal transport is observed on 12.10.15 and 18.12.15 in Figures 16.9 and 16.11, respectively. In addition to the polluted industrial regions of Johannesburg, the air mass (at 1000m on 18.12.15) also travels through the industrial regions of Vereeniging and transports ozone and its precursors to the study site (Figures 16.10 and 16.11).

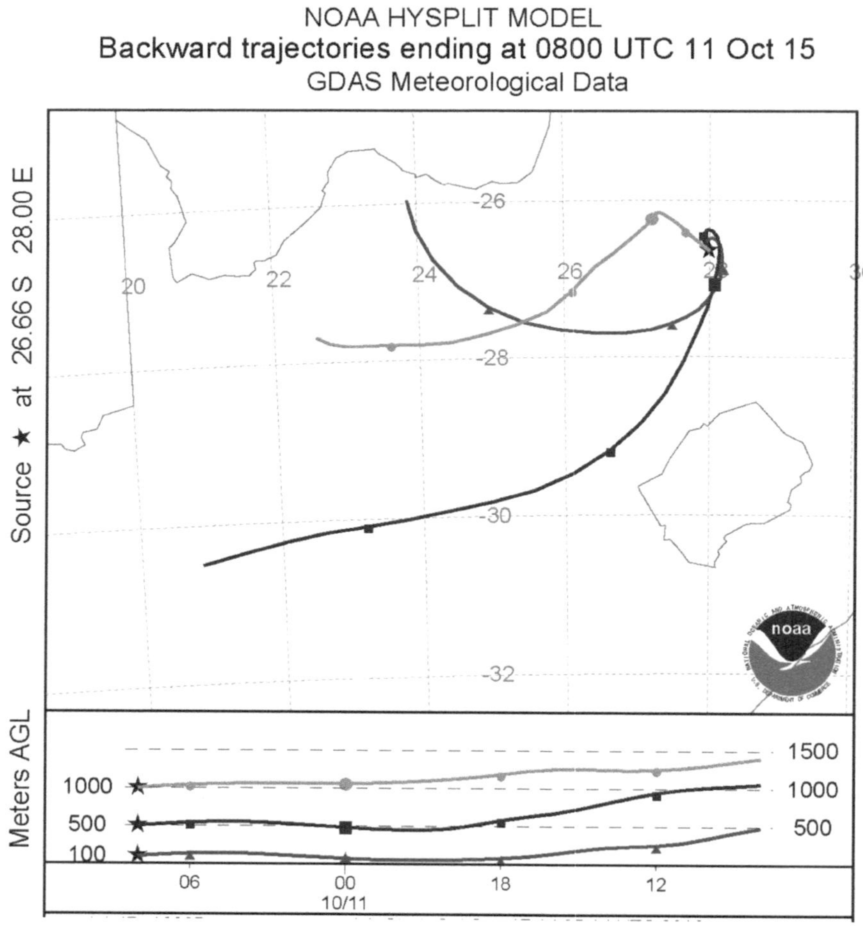

FIGURE 16.8 24 hours backward trajectory simulation at the chosen site on 11.10.15.

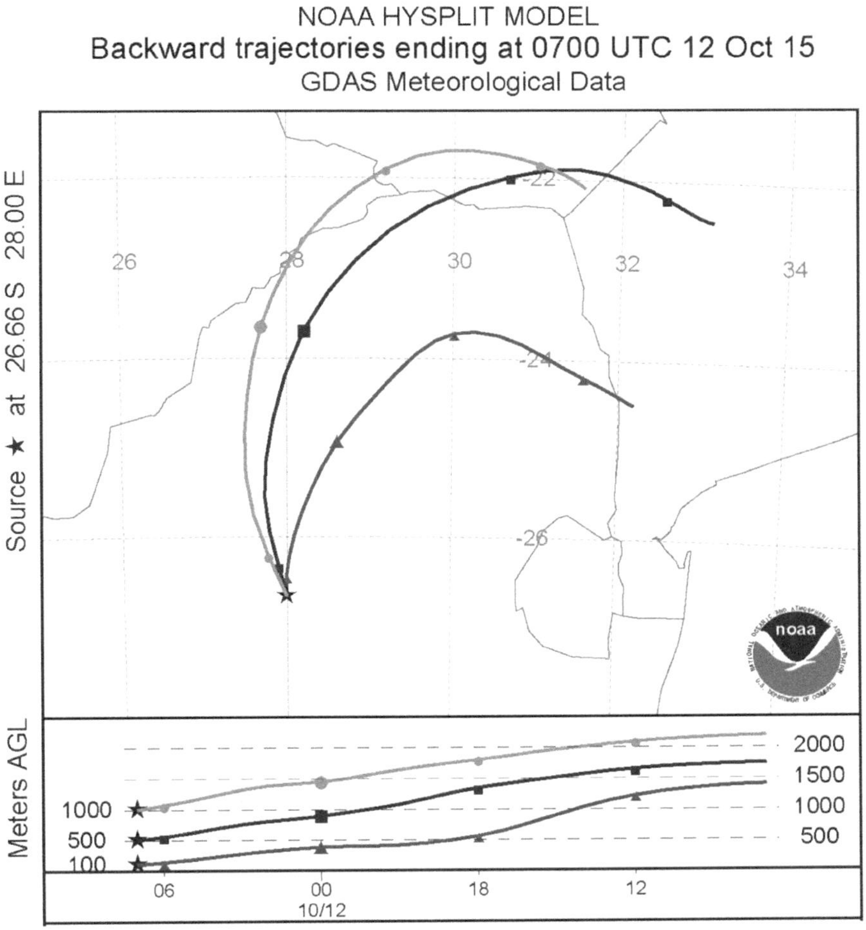

FIGURE 16.9 24 hours backward trajectory simulation at the chosen site on 12.10.15.

Meteorology plays a crucial role in tropospheric ozone formation. Temperature is found to have a significant positive correlation with ozone concentration throughout the day (Table 16.3). High temperatures enhance the in-situ photochemical ozone production rate, thereby increasing the concentration of ozone near the Earth's surface. Table 16.4 shows that humidity is inversely proportional to ozone concentration. Low ozone concentrations are actually associated with wet and rainy weather with high humidity (Tarasova and Karpetchko [32]). It is observed from Table 16.4 that there is either no rainfall or very little rainfall on the case study days. The absence of precipitation reduces the wet deposition rate of ozone, thereby leading to high ambient ozone concentrations. It is interesting to note that solar radiation shows a significant positive correlation with ozone concentration during the morning (6–9 hours) and late afternoon-evening (13–15 hours), but a negative correlation with ozone during

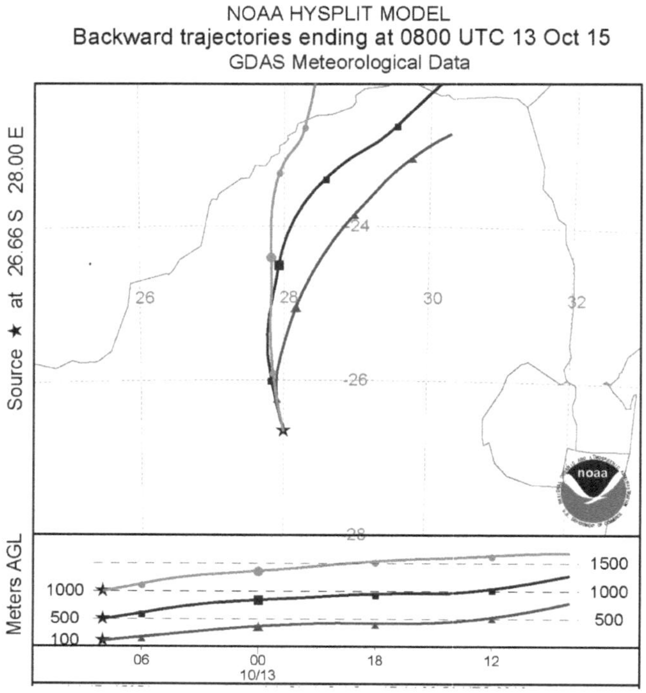

FIGURE 16.10 24 hours backward trajectory simulation at the chosen site on 13.10.15.

late morning-afternoon (10–12 hours). The model-predicted and observed ozone values show the maximum deviation during this late morning-afternoon period (10–12 hours). This finding further supports the fact that photochemical production of ozone is not the dominant source of ozone during the late morning-afternoon period (10–12 hours).

Where Par. is parameters; j_{NO_2} is photolytic rate constant for the photolysis of NO_2; SR is solar radiation; Temp. is temperature; Hum is humidity and WS is wind speed

16.6 CONCLUSION

This work identifies the occurrence of elevated ozone levels at the Three Rivers site in Gauteng province and examines the role of different precursors (NO_x and VOCs) in photochemical ozone production. The basic NO_x cycle (NO-NO_2-O_3) is used to model the photochemical formation of tropospheric ozone. During this study, it is found that the observed and model-predicted ozone values are in fair agreement during the morning hours (6–9 hours) and evening hours (15–18 hours), but these values deviate the most during the late morning to afternoon hours (10–14 hours). Such an observation suggests that although the basic NO_x cycle predominates photochemical

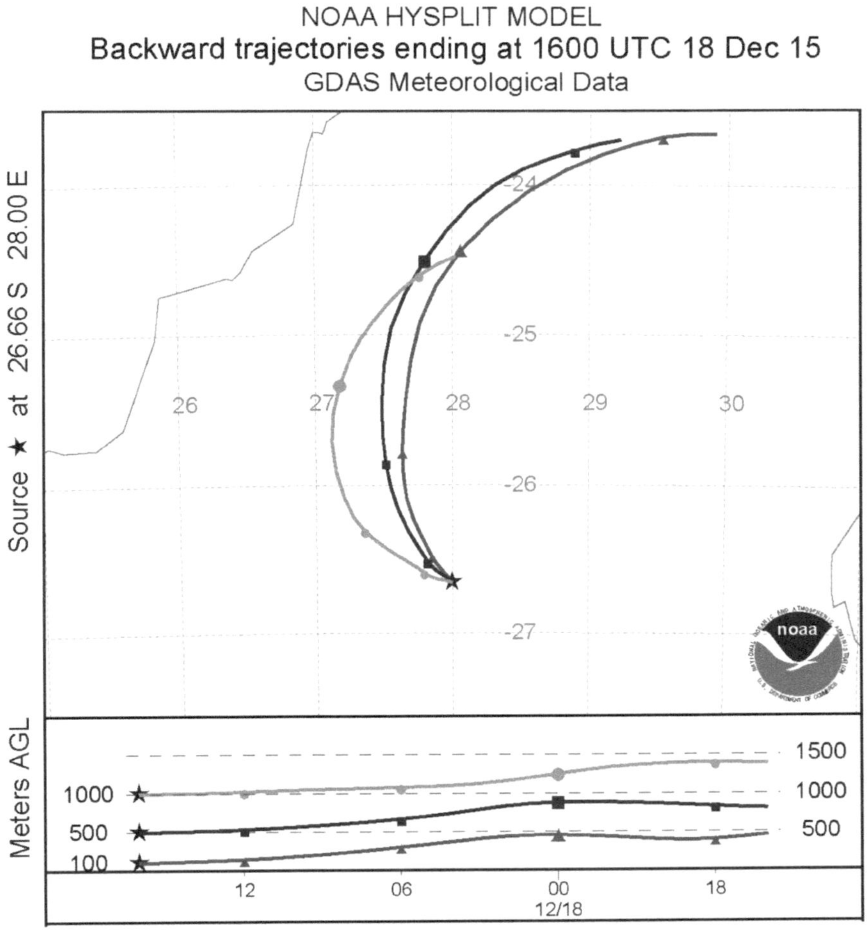

FIGURE 16.11 24 hours backward trajectory simulation at the chosen site on 18.12.15.

ozone formation during the morning and evening hours, the photochemical oxidation of other precursors (CO and VOCs) contributes considerably to ozone production during the late morning to afternoon hours. NO_x driven photochemistry predominates with enhanced domestic fuel-burning emissions during the morning and evening hours. The positive deviation of the Leighton ratio from unity and the very low concentrations of benzene, xylene, toluene and CO during the late morning to afternoon hours also support the crucial contribution of these precursors in photochemical ozone production. Temperature, solar radiation, and the rate constant for the photolysis of $NO_2(j_{NO_2})$ show a high positive correlation coefficient during the morning and evening hours but the reverse trend is observed during the late morning to afternoon hours. Advection also transports ozone and its precursors from polluted

TABLE 16.3

Correlation Coefficients (*r*) of Observed Ozone Concentration with Different Meteorological Parameters

Par.	Time (h)	19.01	20.01	10.02	09.10	10.10	11.10	12.10	13.10	14.10	15.10	19.10	18.12	20.12
j_{NO_2}	6–9	0.963	0.988	0.942	0.908	0.923	0.978	0.832	0.869	0.976	0.809	0.981	0.984	0.984
	10–14	0.253	0.273	−0.045	0.161	−0.045	0.477	−0.427	0.217	0.587	0.698	−0.935	0.051	−0.310
	15–18	0.581	0.995	0.416	0.617	0.978	0.979	−0.556	0.977	−0.970	0.999	0.897	0.897	0.975
SR	6–9	0.844	0.952	0.918	0.928	0.948	0.974	0.880	0.577	0.868	0.914	0.952	0.999	0.999
	10–14	−0.567	-0.274	−0.330	−0.072	−0.330	0.100	−0.136	−0.396	0.494	0.178	−0.741	0.543	−0.315
	15–18	0.780	0.886	0.427	0.372	0.974	0.990	−0.626	0.947	−0.644	1.000	0.891	0.891	0.954
Temp.	6–9	0.928	0.978	0.914	0.994	0.961	0.979	0.918	0.937	0.976	0.932	0.995	0.992	0.992
	10–14	0.966	0.997	0.960	0.865	0.960	0.617	0.915	0.675	0.856	−0.013	0.188	0.936	0.967
	15–18	0.901	0.986	0.569	0.525	0.979	0.999	−0.530	0.989	−0.934	0.907	0.997	0.997	0.953
Hum.	6–9	−0.945	−0.980	−0.967	−0.995	−0.975	−0.966	−0.931	−0.945	−0.991	−0.936	−0.994	−0.986	−0.986
	10–14	−0.984	−0.340	−0.949	−0.988	−0.949	−0.286	−0.881	−0.840	−0.631	0.281	−0.260	−0.924	−0.982
	15–18	−0.856	−0.997	−0.593	−0.527	−0.966	−0.979	0.667	−0.950	0.972	0.941	−0.903	−0.903	−0.899
WS	6–9	−0.416	0.864	0.605	−0.601	0.850	0.987	0.604	0.522	0.992	0.793	0.101	0.758	0.758
	10–14	0.878	−0.148	0.346	−0.616	0.346	0.870	0.667	0.790	0.646	0.070	−0.615	−0.916	−0.822
	15–18	−0.988	0.703	0.603	0.722	0.987	0.868	−0.332	0.999	0.674	0.921	0.972	0.972	−0.684

TABLE 16.4
Variation of Rainfall (mm) Over the Study Site on the Case Study Days

Time	Precipitation (mm) On												
(h)	19.01	20.01	10.02	09.10	10.10	11.10	12.10	13.10	14.10	15.10	19.10	18.12	20.12
5	0	0	0	0	0	0	0	0	0	0.64	0	0	0
8	0	0	0	0	0	0	0	0	0	0	0	0	0
11	0	0	0	0	0	0	0	0	0	0	0	0	0
14	0	0	0	0	0	0	0	0	0	1.29	0	0	0
17	0	0	0	0	0	0	0	0	0	0	0	0	0

industrial regions, contributing to ozone buildup in the ambient atmosphere of the study site. Incorporating the role of different VOCs in photochemical ozone production may improve the prediction accuracy of the model.

ACKNOWLEDGEMENT

The authors are grateful to South African Weather Service (SAWS) and data providers of DEA (Vaal Airshed Priority Area network) for providing the data of different air pollutants and meteorological parameters required for performing the research study. The authors are thankful to University of KwaZulu-Natal for providing all the facilities and necessary support for successful completion of this work. The authors also gratefully acknowledge National Center for Atmospheric Research (NCAR) for the interactive TUV model, NOAA Air Resources Laboratory (ARL) for the provision of the HYSPLIT transport and dispersion model and READY website (http://www.ready.noaa.gov) used in this work.

REFERENCES

[1] Reddy, K.K., Naja, M., Ojha, M., Mahesh, P. & Lal, P. (2012). Influences of the boundary layer evolution on surface ozone variations at a tropical rural site in India. *J. Earth Syst. Sci.,* 121 (4), 911–922.

[2] Lal, S., Naja, M. & Subbaraya, B.H. (2000). Seasonal variations in surface ozone and its precursors over an urban site in India. *Atmos. Environ.*, 34, 2713–2724.

[3] Kuang, S., Newchurch, M.J., Burris, J., Wang, L., Buckley, P.I., Johnson, S., Knupp, K., Huang, G., Phillips, D. & Cantrell, W., (2011). Nocturnal ozone enhancement in the lower troposphere observed by lidar. *Atmos. Environ.*, 45, 6078–6084.

[4] Haagen-Smit, A.J. (1952). Chemistry and physiology of Los Angeles Smog. *Indust. Eng. Chem.*, 44, 1342–1346.

[5] Crutzen, P.J. (1973). Photochemical reactions initiated by and influencing ozone in the unpolluted troposphere. *Tellus*, 26, 47–57.

[6] Chameides, W.L. & Walker, J.C.G. (1973). A photochemical theory for tropospheric ozone. *J. Geophys. Res.*, 78, 8751–8760.

[7] David, L.M. & Nair, P.R. (2011). Diurnal and seasonal variability of surface ozone and NOx at a tropical coastal site: Association with mesoscale and synoptic meteorological conditions. *J. Geophys. Res.,* 116, D10303.

[8] Durao, R.M., Mendes, M.T. & Pereira, M.J. (2016). Forecasting O_3 levels in industrial area surroundings up to 24h in advance, combining classification trees and MLP models. *Atmos. Pollut. Res.,* 7, 961–970.

[9] Abdallah, C., Sartelet, K. & Afif, C. (2016). Influence of boundary conditions and anthropogenic emission inventories on simulated O_3 and $PM_{2.5}$ concentrations over Lebanon. *Atmos. Pollut. Res.,* 7, 971–979.

[10] Ghosh, D. & Sarkar, U. (2015). Analysis of the photochemical production of ozone using tropospheric ultraviolet-visible (TUV) radiation model in an Asian megacity. *Air Qual. Atmos. Health,* 9, 367–377.

[11] Srivastava, S., Lal, S., Naja, M., Venkataramani, S. & Gupta, S. (2012). Influence of regional pollution and long range transport over western India: analysis of ozonesonde data. *Atmos. Environ.*, 47, 174–182.

[12] Diab, R.D., Thompson, A.M., Mari, K., Ramsay, L. & Coetzee, G.J.R. (2004). Tropospheric ozone climatology over Irene, South Africa, from 1990 to 1994 and 1998 to 2002. *J. Geophy. Res.,* 109, D20301.

[13] Raghunandan, A., Mahumane, G. & Diab, R. (2007). Elevated ozone events over Johannesburg based on analysis of tropospheric ozone partial columns. *S. Afr. J. Sci.,* 103, 248–252.

[14] Diab, R.D., Thompson, A.M., Zunckel, M., Coetzee, G.J.R., Combrink, J., Bodeker, G.E., Fishman, J., Sokolic, F., McNamara, D.P., Archer, C.B. & Nganga, D. (1996). Vertical ozone distribution over southern Africa and adjacent oceans during SAFARI-92. *J. Geophys. Res.,* 101(D19), 23823–23833.

[15] Thompson, A.M., Pickering, K.E., McNamara, D.P., Schoeberl, M.R., Hudson, R.D., Kim, J.H., Browell, E.V., Kirchoff, V.W.J.H. & Nganga, D. (1996). Where did tropospheric ozone over southern Africa and the tropical Atlantic come from in October 1992? Insights from TOMS, GTE TRACE A, and SAFARI 1992. *J. Geophys. Res.,* 101, 24251–24278.

[16] Draxler, R.R. & Hess, G.D. (1998). An overview of the HYSPLIT_4 modelling system for trajectories, dispersion, and deposition. *Aust. Meteorol. Mag.*, 47, 295–308.

[17] Finlayson-Pitts, B.J. & Pitts, J.N. (1997). Tropospheric air pollution: Ozone, airborne toxics, polycyclic aromatic hydrocarbons, and particles. *Science*, 276, 1045–1051.

[18] Seinfeld, J.H. & Pandis, S.N. (2006). *Atmospheric Chemistry and Physics: From Air Pollution to Climate Change.* John Wiley and Sons, New York.

[19] Leighton, P.A. (1961). *Photochemistry of Air Pollution.* Elsevier, New York.

[20] Calvert, J.G. & Stockwell, W.R. (1983). Deviations from the O_3-NO-NO_2 photostationary state in tropospheric chemistry. *Can. J. Chem.*, 61, 983–992.

[21] Carpenter, L.J., Klemitshaw, K.C., Burgess, R.A., Penkett, S.A., Cape, J.N. & McFadyen, G.G. (1998). Investigation and evaluation of the NOx/O_3 photochemical state. *Atmos. Environ.,* 32, 3353–3365.

[22] Yang, J., Honrath, R.E., Peterson, M.C., Parrish, D.D. & Warshawsky, M. (2004). Photostationary state deviation – estimated peroxy radicals and their implications for HO_x and ozone photochemistry at a remote Northern Atlantic coastal site. *J. Geophys. Res.*, 109, D02312, doi:10.1029/2003JD003983.

[23] Ridley, B.A., Madronich, S., Chatfield, R.B., Walega, G., Shetter, R.E., Carroll, M.A. & Montzka, D.D. (1992). Measurements and model simulations of the photostationary state during the Mauna Loa Observatory Photochemical Experiment: Implications for radical concentrations and ozone production and loss rates. *J. Geophys. Res.,* 97, 10375–10388.

[24] Hauglustaine, D.A., Madronich, S., Ridley, B.A., Walega, J.G., Cantrell, C.A., Shetter, R.E. & Hübler, G. (1996). Observed and model calculated photostationary state at Mauna Loa Observatory duringMLOPEX2. *J. Geophys. Res.,* 101, 14681–14696.

[25] Volz-Thomas, A., Pätz, H.W., Houben, N., Konrad, S., Mihelcic, D., Klüpfel, T. & Perner, D. (2003). Inorganic trace gases and peroxy radicals during BERLIOZ at Pabstthum: An investigation of the photo stationary state of NOx and O_3. *J. Geophys. Res.,* 108(D4), 8248, doi:10.1029/2001JD001255.
[26] Davis, D.D., et al. (1993). A photo stationary state analysis of the NO_2-NO system based on airborne measurements from the subtropical/tropical north and South Atlantic, *J. Geophys. Res.,* 98(D12), 23501–23523.
[27] Mannschreck, K., Gilge, S., Plass-Duelmer, C., Fricke, W. & Berresheim, H. (2004). Assessment of the applicability of NO-NO_2-O_3 photostationary state to long-term measurements at the Hohenpeissenberg GAW Station, Germany. *Atmos. Chem. Phys.,* 4, 1265–1277.
[28] Griffin, R.J., Beckman, P.J., Talbot, R.W., Sive, B.C. & Varner, R.K. (2007). Deviations from ozone photostationary state during the International Consortium for Atmospheric Research on Transport and Transformation 2004 campaign: Use of measurements and photochemical modeling to assess potential causes. *J. Geophys. Res.,* 112, D10S07.
[29] Vuilleumier, L., Bamer, J.T., Harley, R.A. & Brown, N.J. (2001). Evaluation of nitrogen dioxide photolysis rates in an urban area using data from the 1997 Southern California Ozone Study. *Atmos. Environ.,* 35, 6525–6537.
[30] Sailor, D.J. (1995). Simulated urban climate response to modifications in surface albedo and vegetative cover. *J Appl. Meteorol.*, 34, 1694–1704.
[31] Seinfeld, J.H. & Pandis, S.N. (1998). *Atmospheric Chemistry and Physics.* Wiley-Interscience, Hoboken, N. J.
[32] Tarasova, O.A. & Karpetchko, A.Y. (2003). Accounting for meteorological effects of ozone time-series of Lovozero (Kola Peninsula). *Atmos. Chem. Phys.*, 3, 941–949.

17 A Comprehensive Analysis of the Mismatch between Satellite and Ground-Based Particulate Matter Concentration over Gangetic West Bengal, India

Debjani Dutta, Srimanta Gupta, and Sharadia Dey

17.1 INTRODUCTION

The Indo-Gangetic plain (IGP) has been demarcated as one of the most polluted regions on Earth in terms of air quality [1]. The studies supporting this assertion mostly include observations with satellite data as it is easy to access at a low cost and in a short time, covering a large area, with some thorough analysis over a few urban stations like Delhi, Kanpur, and Kolkata. Mostly, CAMS-Chem output, MODIS AOD, MERRA, S5/TROPOMI data are used as satellite inputs in recent studies [2]. There are some studies based on simulations with mesoscale models, but they lack accuracy in the end result due to their coarse resolution. Although the high concentration of air pollutants over IGP cannot be denied, it is important to check the accuracy of satellite data, which is the primary basis for this postulation, with ground station data for precise measurement. Only data with high precision level can fabricate a successful mitigation planning. AERONET [3] and National Ambient Air Quality Monitoring (NAAQM) data [1] are frequently used to develop the accuracy of satellite data over some areas of upper Gangetic plain, but the lower Gangetic plain (LGP) lacks this methodology.

As with the upper part, the LGP also is considered a high-pollution region [4], especially in its southern part according to satellite observations. This portion, which falls in the states of West Bengal and Bangladesh, contains some urban-industrial pockets with a rapid urbanization rate, but mostly, the area is covered with agricultural fields. According to the West Bengal Pollution Control Board (WBPCB), 70%

 DOI: 10.1201/9781003638414-17

of the pollutants over LGP in West Bengal result from long-range transport [5], and the winter pollution over India is generally presumed to be the result of long-range transport from western deserts. However, some recent studies are pointing to localized reasons and emphasizing the reduction of indigenous pollution [6].

This study focuses on the dissimilarities between satellite data and in-situ data of suspended particulate matter (PM_{10}) over the lower Gangetic plain (LGP) in the state of West Bengal with the identification of pollution pockets in that area (Figure 17.1). It also examines the seasonal variation of PM_{10} and its possible causes over the study area, where the average annual level of PM_{10} is higher than the World Health Organization (WHO) and National Ambient Air Quality Standard (NAAQS) limits [7].

17.2 METHODS AND MATERIALS

17.2.1 Retrieval of Data

CAMS EAC4 (ECMWF Atmospheric Composition Reanalysis 4th generation) monthly average fields are used for the retrieval of PM_{10} and $PM_{2.5}$ data as modeled input at the surface level. The spatial resolution of this data is 0.75×0.75 degrees. CAMS reanalysis is the newest global reanalysis dataset of atmospheric composition (AC), processed by the Copernicus Atmosphere Monitoring Service. It consists of three-dimensional and time-accordant AC fields, which include greenhouse gases, chemical species, and aerosols [8]. These reanalysis data are produced using 4DVar data assimilation in CY42R1 of ECMWF's Integrated Forecast System. It has 60 hybrid sigma levels in the vertical (the top level at 0.1 hPa).

PM_{10} and $PM_{2.5}$ data are also collected from 40 stations of WBPCB distributed over 14 districts in a 3-year timespan (01.01.2019–01.01.2022).

TROPOMI/S5P aerosol index (AI) or Ultra Violet aerosol index (UVAI) is used to detect smoke and dust transport (spatial resolution 7 km × 3.5 km). The calculation of AI is based on wavelength-dependent changes in Rayleigh scattering in the ultraviolet spectral range where ozone absorption is very small (retrieved from https://www.tropomi.eu on 21.01.2022). TROPOMI radiances at 354 and 388 nm are used to calculate the aerosol index. Mathematically, it is as follows –

$$\text{UVAI} = -\ 100 \log_{10}\left[\ I_{354}^{\text{obs}} \,/\, I_{354}^{\text{cal}}\ \right] \tag{17.1}$$

where I is the observed and calculated radiances at 354 nm.

Radiance measurement at 388 nm is used to acquire a wavelength-independent cloud fraction, which is necessary to calculate the radiance at 354 nm [9].

The hybrid Single Particle Lagrangian Integrated Trajectory (HYSPLIT) model is used for back trajectory analysis of the particulate matter in the lower Gangetic plain of West Bengal, fitted with GFS (Global Forecast System) climate data of 0.25×0.25 degrees resolution.

The mixed layer height is also calculated using the HYSPLIT model.

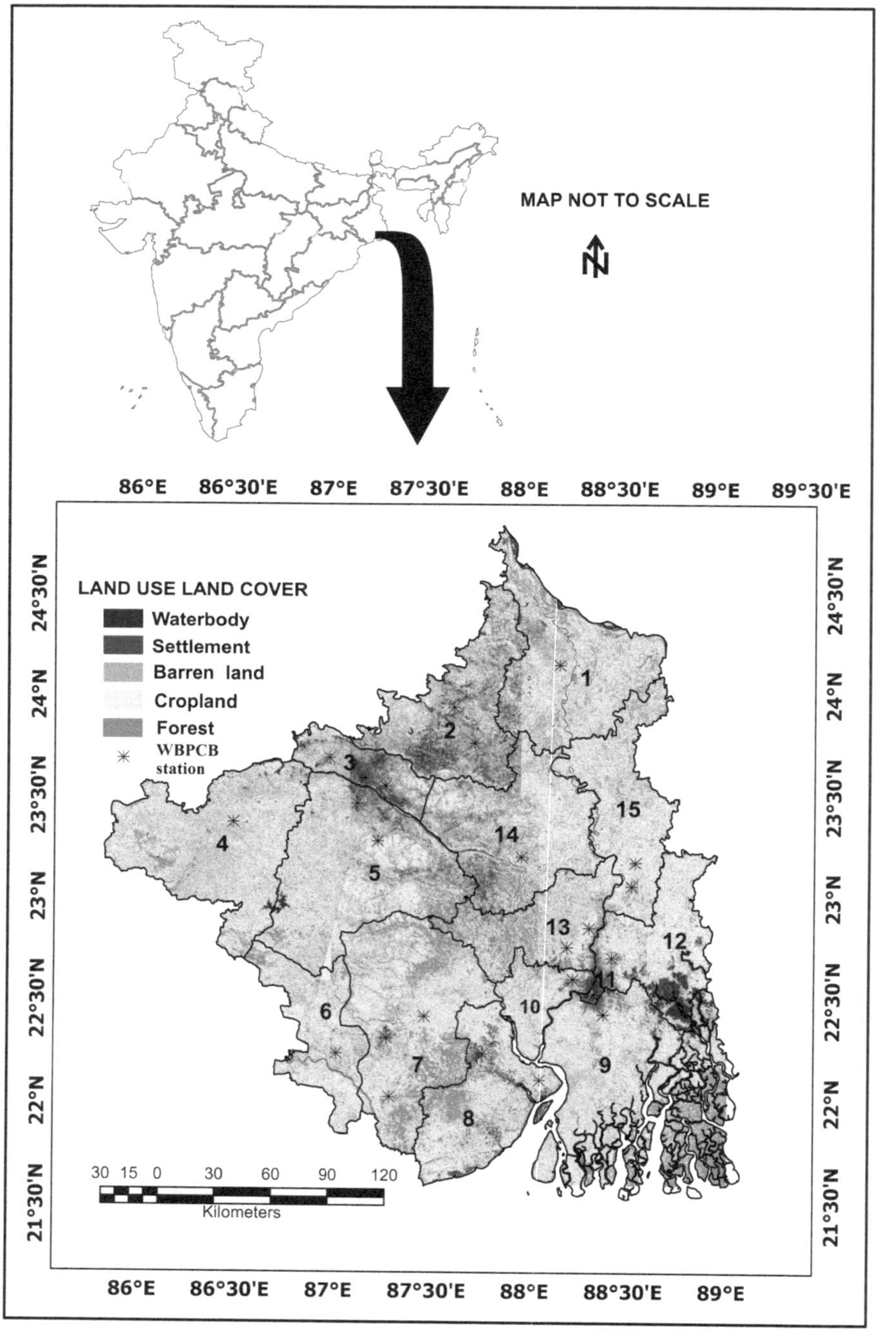

FIGURE 17.1 Lower Gangetic plain in the state of West Bengal, India with district annotations; 1 – Murshidabad, 2 – Birbhum, 3 – Paschim Bardhaman, 4 – Purulia, 5 – Bankura, 6 – Jhargram, 7 – Paschim Medinipur, 8 – Purba Medinipur, 9 – South 24 Parganas, 10 – Howrah, 11 – Kolkata, 12 – North 24 Parganas, 13 – Hooghly, 14 – Purba Bardhaman, 15 – Nadia.

17.2.2 Analysis of the Accuracy of Satellite Data on the Portrayal of Particulate Matter Concentration over LGP

The analysis is done over four seasons in 3 years. Daily and weekly data are collected first and then the average value for each season is calculated. All of the satellite dataset are resampled using bilinear interpolation technique [9] to the same spatial resolution of 0.75 × 0.75 degrees.

Back trajectories are calculated for each week and then are averaged to a season.

Lastly, the satellite and ground observed data are correlated to see their association using Pearson's correlation coefficient and student's *t*-test.

17.3 RESULTS AND DISCUSSIONS

17.3.1 Comparative Analysis of Particulate Matter Concentration from Different Dataset

In this study, the SPM value from the EAC4 reanalysis seemed to overestimate the particulate matter level in the atmosphere (Figure 17.2), as it showed much higher values than the ground stations (300 μg/m³ in annual average) and the average SPM values of the Indian subcontinent. However, the seasonal trend matches with the ground station data and follows the common idea of seasonal pollutant variation in India. The values are highest in winter, followed by pre-monsoon, post-monsoon, and monsoon. The EAC4 analysis shows PM_{10} concentrations as high as 400–450 μg/m³ over Kolkata, Howrah, and Hooghly district (middle part), North 24 Parganas and Nadia (eastern part) and South 24 Parganas district (south-eastern part), and around 225–300 μg/m³ over Paschim Bardhaman, Bankura, Purulia, Paschim Medinipur district (western part). The northern part (Murshidabad, Birbhum districts) shows lower concentration of PM_{10} compared to other areas (200–250 μg/m³), while Purba Medinipur district in the south shows a variation from 300 to 400 μg/m³. This spatial trend remains same throughout the year with variations in concentration, depending on the season. The PM_{10} concentration remains within 390–420 μg/m³ over the middle, eastern and south-eastern part of the LGP in West Bengal. The value ranges between 150 and 270 μg/m³, 100 and 125 μg/m³, and 250 and 275 μg/m³ over the western, northern, and southern parts, respectively, during the pre-monsoon season. In the monsoon season, the concentrations reduce to 150–180 μg/m³ over the middle, east, and south-eastern parts, 110–140 μg/m³ over the western part, 100–110 μg/m³ over the northern part and 115–135 μg/m³ over the southern part, due to the prevalence of rainfall throughout the season. In the post-monsoon season, the concentrations again increase to 370–420 μg/m³ in the middle, east and south-eastern parts, 100–210 μg/m³ in the north, 120–150 μg/m³ in the west, and 150–200 μg/m³ in the southern part.

This trend shows a pocket of pollution over the middle, east, and south-eastern part of LGP over the years. In the middle part, there is the metro city of Kolkata with a huge population and many anthropogenic activities, such as transportation, industries, and household emissions; Howrah and parts of North 24 Parganas

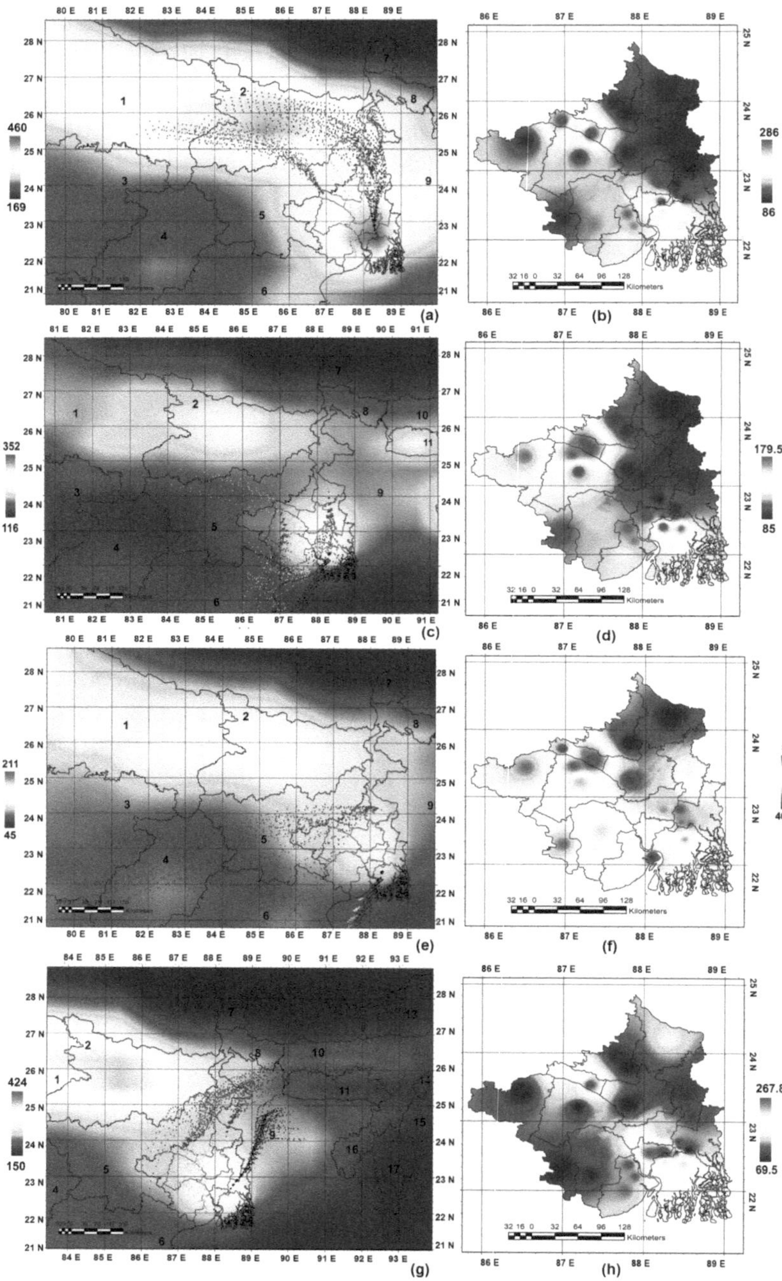

FIGURE 17.2 Spatio-temporal variation of PM_{10} distribution (μg/m³) using CAMS data (left column) and WBPCB station data (right column) with backward wind trajectories; (a, b) winter; (c, d) pre-monsoon; (e, f) monsoon; (g, h) post-monsoon.

also have a cluster of industries in that region. Nadia and most parts of South 24 Parganas, including Sundarban region, however, lack a significant number of industries and transportation activities. The probable cause of high particulate matter in these regions may be the presence of sea salt and organic aerosols in the air. The interpolation technique of CAMS model data could also be a cause, as it may have distributed the high values of the industrial clusters over the whole region due to the coarse resolution of this data. The high particulate matter concentration in the northern part, covering agricultural land use in the post-monsoon season, may have been caused by the crop-residue burning activities practiced during the season. The cause of the yearlong high concentration over Purba and Paschim Medinipur is quite surprising. It is also surprising that the western part shows lower values than the eastern, southern, and south-eastern parts over the year, despite containing a handful of industries and thermal power stations. Three cities from Paschim Bardhaman district (Asansol, Raniganj, and Durgapur) have been selected under the National Clean Air Programme (NACP) due to their high air pollution rates. Being a part of Chhotanagpur plateau, the western part of LGP in West Bengal also experiences recurrent dust episodes, which are omitted from the CAMS analysis results. The spatial trend of MODIS (Moderate resolution spectroradiometer) AOD (aerosol optical depth) matches the spatial variation of the CAMS analysis result, with a Pearson correlation coefficient of 0.78. The high AOD in the middle, southern, and south-eastern parts was previously accounted to the high water vapor content in that area because of its proximity to the Bay of Bengal (BoB), but this analysis shows the high concentration of particulate matter itself.

17.4 ANALYSIS OF THE TRANSPORTATION OF PARTICULATE MATTER BY WIND

Backward wind trajectory analysis showed the transportation of air mass from the northern Indo-Gangetic plain (IGP), i.e., mostly from Delhi, Uttar Pradesh, Bihar and Jharkhand during the winter seasons. Four cities from four zones were selected for the analysis, namely, Asansol from the western region, Baharampur from the northern region, Kolkata from the middle region and Tamluk from the southern region. All of the backward wind trajectories ending at the mixed layer height of these four cities were found to be coming from the same region with nearly the same pollution levels, yet the four cities did not show the same pollution level (Figure 17.2). In pre-monsoon season, the trajectories come from the southern India, where the total pollution level is lower than in those four cities. In the post-monsoon season, the wind trajectories started from northeastern India and Bangladesh, where episodic crop-residue burning activities are found, but the concentration of particulate matter remained higher at the end points. In the monsoon season, most of the wind trajectories started flowing from the south, west and southwest directions with wide variation, and none of the places consists of high pollution load (Figure 17.2).

The UV aerosol index (AI) shows dust and smoke transport to Kolkata and its adjacent regions in winter from Northern India (Figure 17.3), as well as a smoke patch over the ADDA region. Talukdar et al. [7] also found long-range dust transport over Kolkata in the winter season at 2000 m above ground level. A dust patch is found

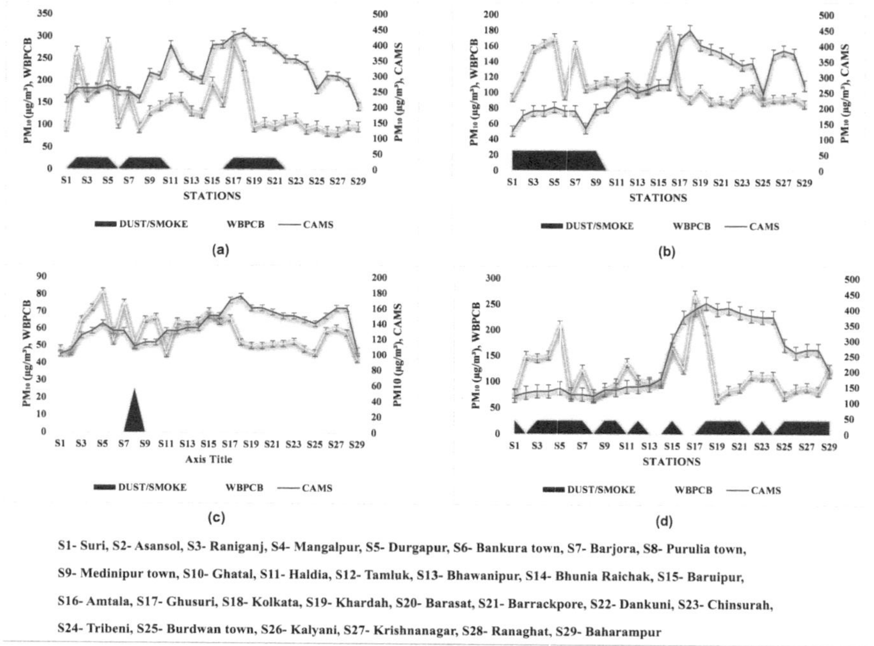

FIGURE 17.3 Graphical representation of variations in PM_{10} distribution over WBPCB station using CAMS and WBPCB data on (a) winter; (b) pre-monsoon; (c) post-monsoon; (d) winter seasons.

to linger over the upper part of the western region in winter. Small patches of smoke are found around the LGP, but no way they cover the whole region. Most of the area is dominated by anthropogenic smaller particulate matter. In the post-monsoon season, patches of smoke are found to linger over agricultural fields in the LGP - probably a result of crop-residue burning. Smoke is also found to be transported from the upper parts of Bangladesh, which is dominated by agricultural land use. However, the wind trajectories from regions with smoke in their atmosphere do not always result in high pollution concentrations in the destination regions. In pre-monsoon season, dust is found to cover the upper atmosphere of the western part.

17.5 ACCURACY OF SATELLITE DATA IN IDENTIFICATION OF POLLUTION POCKETS IN LGP REGION

According to the data from ground stations, there are several pollution pockets spread over the LGP. One is Kolkata and its adjacent regions, and the other is ADDA (Asansol Durgapur Development Authority) region. Like Kolkata, ADDA and its adjacent regions also consist of urban cities, industries, open-cast coal mines and thermal power plants, all of which contribute to the yearlong high PM_{10} rates. Some stations in West Medinipur, like Medinipur city, also show moderate to high PM_{10} levels at times, probably due to sand mining activity. The Baharampur station from Murshidabad district in the northern part exceeds the NAAQS standard in post-monsoon season due to crop-residue burning activity. Most of the stations in the western and middle parts remain above the NAAQS standard yearlong, except during the monsoon season. However, unlike CAMS analysis, stations in the southern and eastern parts do not show higher values than those in the western part. Student's t-test shows a significant difference between the CAMS and WBPCB data at 95% confidence level. The Pearson correlation coefficient of PM_{10} concentration between CAMS analysis and WBPCB station data over 42 stations in the LGP is 0.22 in winter, −0.43 in pre-monsoon, 0.09 in monsoon, and 0.15 in post-monsoon season. The correlation coefficient between AI and WBPCB station data of PM_{10} is above 0.65 for the whole year, and the correlation coefficient of AI and CAMS PM_{10} data is above 0.35 for the whole year. This clearly declares a strong mismatch between the satellite and ground station data. The lower values of PM_{10} in WBPCB stations may have resulted from the low pollution rate during the 2020 lockdown, as the graphs here show the seasonal average from 3 years (Figure 17.3). With respect to ground station data, CAMS underestimates the PM_{10} concentration in western part and overestimates it in eastern and southern parts, and vice versa.

This analysis clearly portrays that there are permanent pollution pockets in the LGP region of West Bengal. Though the long-range transport of air pollutants in this region cannot be denied, it can be said that even without transportation, the pollution level would have been high here. During the 2020 lockdown, long-range dust transport did not reduce much, but the reduction in indigenous pollution had pulled down the pollution level by more than 50%. Long-range transport increases the pollutant load during certain periods of the year, but the base level of pollution is found to always be above the standards in most of the urban-industrial and town regions in LGP.

17.6 CONCLUSION

From the previous discussions, it is quite clear that to accurately depict the air pollution sources over a large area, the area should be scanned minutely. Thus, it is of uttermost priority to monitor the air quality in the LGP by installing more ground stations, including sea-side and semi-urban or agricultural regions. Source apportionment and the identification of aerosols characteristics at ground stations are also important factors for accurately portraying the pollution conditions in the LGP. These insights can be used to modify the satellite data and to develop a proper pollution reduction plan.

REFERENCES

[1] Roozitalab B, Carmichael GB, Guttikunda, SK. Improving regional air quality predictions in the Indo-Gangetic Plain – case study of an intensive pollution episode in November 2017. *Atmos. Chem. Phys.* (2021) 21:2837–2860. https://doi.org/10.5194/acp-21-2837-2021

[2] Dutta V, Kumar S, Dubey D. Recent advances in satellite mapping of global air quality: evidences during COVID-19 pandemic. *Environ. Sustain.* (2021) 4:469–487. https://doi.org/10.1007/s42398-021-00166-w

[3] Vohra K, Marais EA, Suckra S, Kramer L, et al. Long-term trends in air quality in major cities in the UK and India: a view from space. *Atmos. Chem. Phys.* (2021) 21:6275–6296. https://doi.org/10.5194/acp-21-6275-2021
[4] Ojha N, Sharma A, Kumar M, Girach I. et al. On the widespread enhancement in fine particulate matter across the Indo-Gangetic Plain towards winter. *Sci. Rep.* (2020) 10:5862. https://doi.org/10.1038/s41598-020-62710-8
[5] CSIR-National Environmental Engineering Research Institute (CSIR-NEERI), Nagpur, INDIA. *PM_{10} and $PM_{2.5}$ Source Apportionment Study and Development of Emission Inventory of Twin Cities Kolkata and Howrah of West Bengal.* West Bengal Pollution Control Board (2019), Kolkata, West Bengal, India.
[6] Dey S, Purohit B, Balyan P, Dixit KA. A satellite-based high-resolution (1-km) ambient $PM_{2.5}$ database for India over two decades (2000–2019): applications for air quality management. *Remote Sens.* (2020) 12:3872. https://doi.org/10.3390/rs12233872
[7] Talukdar S, Jana S, Maitra A. Dominance of pollutant aerosols over an urban region and its impact on boundary layer temperature profile. *JGR Atmos.* (2017) 122(2):1001–1014. https://doi.org/10.1002/2016JD025770
[8] Inness A, Ades M. The CAMS reanalysis of atmospheric composition. *Atmos. Chem. Phys.* (2019) 19:3515–3556. https://doi.org/10.5194/acp-19-3515-2019
[9] Torres O, Jethva H, Ahn C, Jaross G, Loyola DG. TROPOMI aerosol products: evaluation and observations of synoptic-scale carbonaceous aerosol plumes during 2018–2020. *Atmos. Meas. Tech.* (2020) 13:6789–6806. https://doi.org/10.5194/amt-13-6789-2020

18 Variation of Ambient $PM_{2.5}$ and Associated Deposition in the Respiratory Tract of Children during the Festive Seasons over Kolkata, India

Bidisha Chatterjee, Somangee Chakrabarti, Sharadia Dey, Arup Kumar Mitra, Sabyasachi Mondal, and Jyoti Sekhar Banerjee

18.1 INTRODUCTION

One of the top environmental concerns with respect to air quality is particulate matter (PM). It has various sources, mainly the sudden short span of fireworks, various types of vehicular emissions and factory emissions. These displays of fireworks are generally carried out in various religious and cultural functions [1]. Diwali is the most widely celebrated festival in India and in West Bengal, Kali Puja is celebrated as the festival of light. It is usually celebrated in the months of October and November. The bursting of a large number of firecrackers is a common occurrence during this festival. This leads to the sudden deterioration of air quality, resulting in severe air pollution over a short span of time [2]. According to Chatterjee et al. [3], the firecrackers emit various kinds of gases and particulate matter into the air and they mainly contain nitrate, sulphur, potassium and trace elements, affecting both the environment and human beings. Studies have been carried out all over the world to understand the effect of firecrackers on the atmosphere because of the release of trace elements and their various components in the atmosphere. Such studies, reported from different places around the world like in Oahu [4], Germany during New Year [5], Italy [6] during the FIFA World Cup [6], Spain during "Las Fellas" [7] and China during Lantern Days [8], show a higher amount of air pollution due to firework displays compared to normal days.

DOI: 10.1201/9781003638414-18

In India, air pollution has been monitored over a long period of time. According to the National Ambient Air Quality Standard (NAAQS) given by CPCB, the 24 hours average permissible limit for $PM_{2.5}$ is 60 μg/m^3 [9]. Especially during the Diwali period it has been observed that there is a sudden, drastic deterioration of air quality over various cities. Reports of studies from Hyderabad, Lucknow and Delhi during this period also support this finding, and such drastic deterioration of air quality has led to various health effects [10]. The air quality during the pre- and post-Diwali and Dusshera period in North India was studied by Daga et al. [11] to find out the exposure effect of poor air quality on human health and the associated health risks. The study reported a significant increase in PM10 and PM 2.5 levels in the ambient atmosphere. One of the crucial factors for health issues in India is $PM_{2.5}$, which has a comparatively higher ratio of surface-to-volume, prevails in the atmosphere for a longer time, and contains a higher percentage of toxic compounds compared to PM_{10}. These particles also have the potential to reach distal regions in the lungs, produce a high amount of reactive oxygen species, and thus can be very harmful to human lungs [12].

A 30%–40 % increase in the cases of wheezing, respiratory infections, exacerbation of bronchial asthma, and chronic obstructive pulmonary disease (COPD) has been reported in India during Diwali by Gouder and Montefort [13]. According to the study by Shah et al. [12], there was an increase in the rate of hospitalization in both urban and rural areas, and the majority of the patients were children because their lungs are growing and thus were most affected during the post-Diwali period. Children were found to be the most vulnerable population to the harmful health impacts of firecrackers since their lungs are growing and guided by a complex and precise time sequence of chemical messages. These air pollutants make children more susceptible to asthma and bronchial diseases, which ultimately lead to poor lung development. This is mainly because their respiratory epithelium is easily penetrable by the pollutants, they have an underdeveloped defence mechanism to particulate matter relative to adults, and a contrasting ability to assimilate and remove harmful environmental agents. Health issues associated with elevated concentrations of ambient particulate matter include aggravated cardiovascular and respiratory illness and added stress to the heart and lungs. According to Daga et al. [11], longterm exposure to polluted air can have permanent health effects, such as accelerated aging of the lungs, loss of lung capacity and decreased lung function, and can also result in the development of diseases or rapid worsening of conditions such as asthma, bronchitis, emphysema and cause a shortened lifespan. Although there has been large-scale education of the general public along with the spreading of public awareness, child health in both urban and rural areas continues to be threatened by firework pollutants [11]. Exposure to and inhalation of high concentrations of particulate matter results in deposition in the lower respiratory tract, which adversely affects human health [14]. The variation in particle number concentration and mass concentration during the pre-Diwali, Diwali and post-Diwali periods over Kanpur, India, was studied and also reported significant deposition of the particles in the pulmonary region using the MPPD model [15]. A study conducted in Roorkee city, Uttarakhand, India to understand the deposition and retention of ambient particulate matter in the human respiratory tract across different age groups concluded that lobar deposition of $PM_{2.5}$ was maximum in children, and the fine fraction predominantly got deposited in the pulmonary region [16]. The age specific deposition of PM in the human airways was studied and shown to reveal the regional and lobar deposition of PM using the MPPD model [17]. The results obtained from this study can be further used for human health risk assessments.

Kolkata, situated on the banks of the river Ganga, is a metropolitan city in Eastern India. The air quality over the city has degraded as a result of urbanization and industrialization. Diwali is celebrated in Kolkata as it is in the rest of India. Firework displays are a major attraction that stretch over a few days, including Diwali. There have been very few studies on the effect of firework displays on air quality in Kolkata [3]. However, a recent study showing the air quality of Kolkata during this festive season and the effects of the particulate matter on the respiratory health of children has not been found. Understanding the air quality over Kolkata during this festive season, the influence of meteorology and the deposition of $PM_{2.5}$ in different parts of the respiratory system of children is essential for assessing the health risks associated with this exposure.

The aim of this study is to analyse the variation of ambient $PM_{2.5}$ over five consecutive years (2018–2022), investigate the deposition of $PM_{2.5}$ in different parts of the respiratory tract of children using the MPPD model, and understand the influence of meteorology on the $PM_{2.5}$ levels during the festive season, i.e., pre-Diwali, Diwali, and post-Diwali periods, over Kolkata, India. The results of such a study will help in evaluating the adverse health impacts of ambient particulate matter on children. To the best of the authors' knowledge, such deposition modelling and analysis have not been done over Kolkata before.

18.2 MATERIALS AND METHOD

18.2.1 Study Area

The data were collected from the meteorological station of the Central Pollution Control Board located at Victoria Memorial (22.5448° N, 88.3426° E) in Kolkata, West Bengal, India (Figure 18.1). This area experiences a tropical wet-and-dry climate (Köppen climate classification) with an annual mean temperature of 26.8°C. Victoria Memorial is a tourist spot surrounded by Queens Way in the north, Hospital Road in the west, Acharya Jagadish Chandra Bose Road in the south, and Cathedral Road in the east.

18.2.2 Data Collection

The data for $PM_{2.5}$, wind speed, relative humidity and temperature were collected for the pre-Diwali, Diwali and post-Diwali seasons. The Diwali data consist of the data from the day before Diwali, Diwali and the day after Diwali, i.e., three days. The pre-Diwali data comprise the ten days before Diwali, and the post-Diwali data comprise the ten days after Diwali. The data for $PM_{2.5}$ and meteorological parameters have been collected for five consecutive years, i.e., 2018 to 2022.

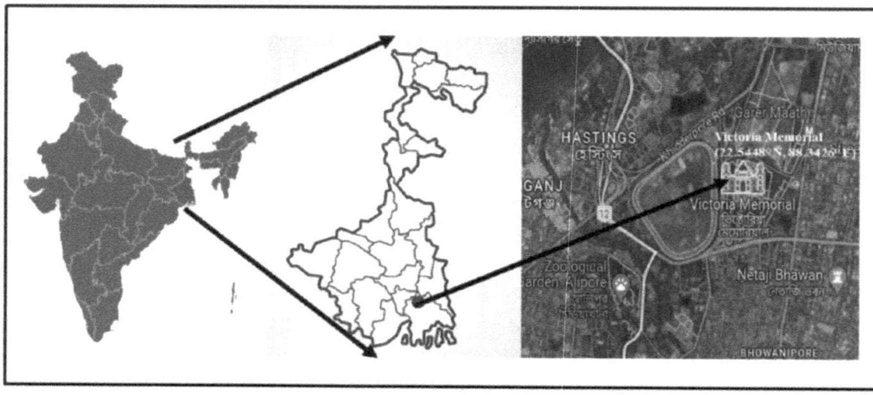

FIGURE 18.1 Location of the study area.

18.2.3 MPPD Model

In this work, the deposition of $PM_{2.5}$ has been studied using ARA's (Applied Research Associates) Multiple-Path Particle Dosimetry (MPPD) model, version 3.04. The MPPD model replicates the human lung through eight different options that the model offers. An age-specific five -lobe model has been used. Age-specific human lung geometries, representing ten distinct ages from 3 months old to 21 years old, are provided in this five-lobe symmetric multiple-path model. The age-specific lung structures are generated based on the morphometric measurements given by Mortensen et al. [18–20]. This model also requires input data for airway morphometry such as functional residual capacity in mL (FRC), upper respiratory tract volume (URT), exposure conditions such as aerosol concentration in mg/m^3, breathing frequency per minute (BF), tidal volume in mL (TV), inspiratory fraction, pause fraction, breathing scenario and particle properties such as density, size distribution and shape factor. This study has been carried out for children aged 8 years. In this model, the PM properties play a crucial role in estimating lung depositions. It was assumed that all the particles entered the respiratory tract through the nose. A density of 1.66 g/cm^3 has been used in the estimation. The other input data like airway morphometry, PM properties, and exposure conditions, have been provided by default. The deposited mass rate is the rate at which the aerosol mass is deposited in the head region, TB region and pulmonary region. It is derived from the product of the deposition fraction, the mass concentration of the inhaled aerosol, the tidal volume and the breathing frequency. The units for this quantity are micrograms per minute. The values of functional residual capacity in mL (FRC), upper respiratory tract volume (URT), breathing frequency per minute (BF), tidal volume in mL (TV), inspiratory fraction and pause fraction have been provided by the model by default.

18.3 RESULTS

18.3.1 Variation of $PM_{2.5}$ in Pre-Diwali, Diwali and Post Diwali Days

Figure 18.2a–c shows the exposure of 8-year-old children to $PM_{2.5}$ during the pre-Diwali, Diwali and post-Diwali periods, respectively, in the years 2018, 2019, 2020, 2021 and 2022. During the Pre-Diwali period, the highest mean concentration and highest concentration variation of $PM_{2.5}$ were observed in the year 2020.The mean concentration of $PM_{2.5}$ in 2020 was 77.26 µg/m^3. The lowest mean concentration (39.98 µg/m^3) and lowest concentration variation was recorded in 2018. During the Diwali period, the highest mean concentration (203 µg/m^3) and highest concentration variation were $PM_{2.5}$ were observed in 2018, whereas the year 2022 showed the lowest mean concentration (24.01 µg/m^3) and lowest concentration variation.

The mean concentration of $PM_{2.5}$ in the post-Diwali period was found to be highest in 2019, with a value of 97.761 µg/m^3 but the highest concentration variation was observed in 2018. Both the mean concentration and the variation in the concentration of $PM_{2.5}$ were found to be minimum in the year 2021.It has been observed that, in most years, the mean concentration of $PM_{2.5}$ was higher than the 24-hour average CPCB permissible limit of 60 µg/m^3 [9].

Figure 18.3 shows the 24 hours average $PM_{2.5}$ concentration (µg/m^3) during Pre-Diwali, Diwali and Post-Diwali periods over the study area. It is observed that the 24 hours average $PM_{2.5}$ concentration was below the NAAQS by CPCB only during the pre-Diwali period of 2018, both the pre-Diwali and post-Diwali periods of 2021, and much below the prescribed limit during the pre-Diwali and Diwali periods of 2022. During 2019 and 2020, the 24 hours average $PM_{2.5}$ concentration was above the permissible limit during the pre-Diwali, Diwali and post-Diwali periods, and much above the permissible limit during the Diwali and post-Diwali periods of 2018.

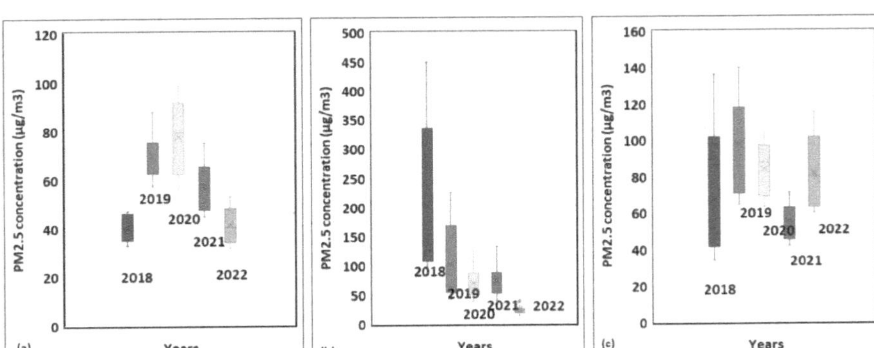

FIGURE 18.2 Exposure of children to $PM_{2.5}$ concentration (a) during pre-Diwali period, (b) Diwali period and (c) post-Diwali period.

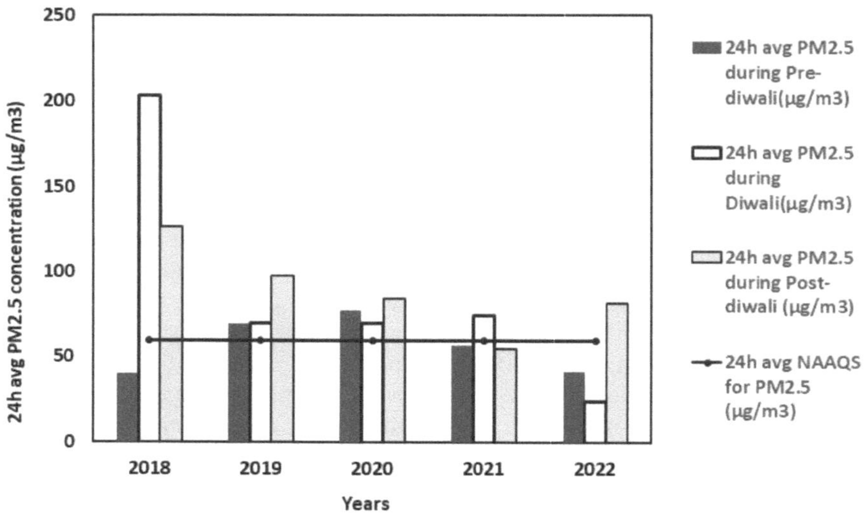

FIGURE 18.3 24 hours average $PM_{2.5}$ concentration (µg/m³) during pre-Diwali, Diwali and post-Diwali periods over the study area.

18.3.2 Deposited Mass Rate of $PM_{2.5}$ in Respiratory Tract

Figure 18.4a–c shows the deposition mass rate of $PM_{2.5}$ in the head, TB and pulmonary regions during the pre-Diwali, Diwali and post-Diwali periods.

It is observed that the year 2018 had the highest total deposited mass rate during the Diwali and post-Diwali periods, whereas 2020 had the highest deposited mass rate during the pre-Diwali period. Although 2022 showed a low deposited mass rate in the pre-Diwali and Diwali periods, the post-Diwali period of 2022 witnessed a high deposited mass rate. This highlights the fact that although the ambient $PM_{2.5}$ concentration of the study area was moderate during the pre-Diwali and Diwali periods of 2022, the ambient $PM_{2.5}$ concentration sharply increased during the post-Diwali period, which in turn drastically increased the deposited mass rate of $PM_{2.5}$ in different parts of the respiratory system. The pulmonary deposited mass rate was found to be the highest, followed by the head deposited mass rate and TB deposited mass rate. A similar finding of the predominance of fine fraction region in the pulmonary deposition was reported in Roorkee city, Uttarakhand, India, by Khan et al. [16].

18.3.3 Diurnal Variation of Deposited Mass Rate

Figure 18.5 shows the diurnal variation of the deposited mass rate in the respiratory tract of the 8-year-old children during the (a) pre-Diwali, (b) Diwali and (c) post – Diwali periods. No significant change is noticed in the deposited mass rate during the pre-Diwali period and it tends to attain its minimum value during the afternoon. However, the deposited mass rate remains high in the morning, attains a minimum during the afternoon and again increases in the evening. During the post-Diwali period, it remains very high until the afternoon, remains low for a very short duration, and then increases sharply in the evening.

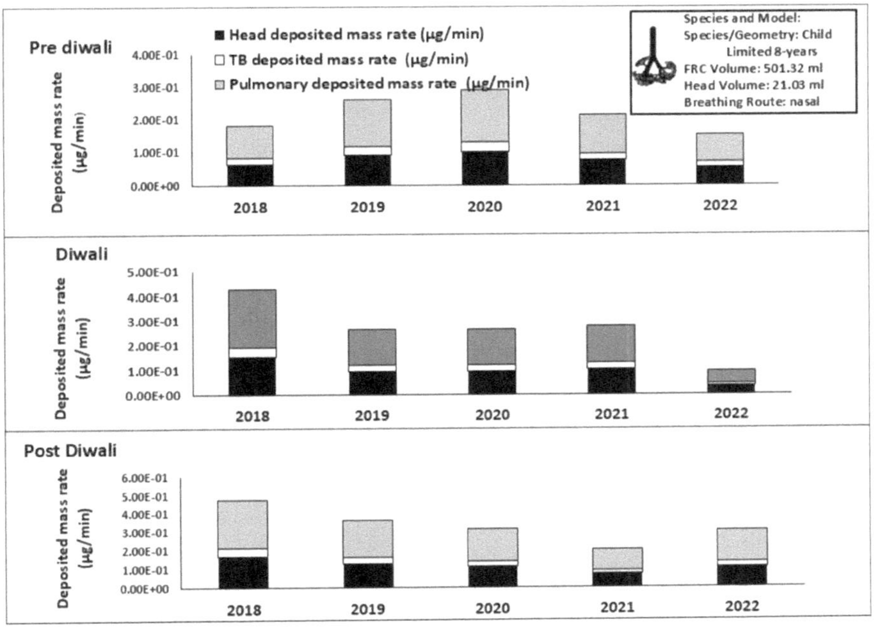

FIGURE 18.4 Deposited mass rate (µg/min) in the head, TB and pulmonary region of children of age 8 years during (a) pre- Diwali, (b) Diwali and (c) post - Diwali periods and the insert gives the geometry visualization of the respiratory tract of a child of 8 years.

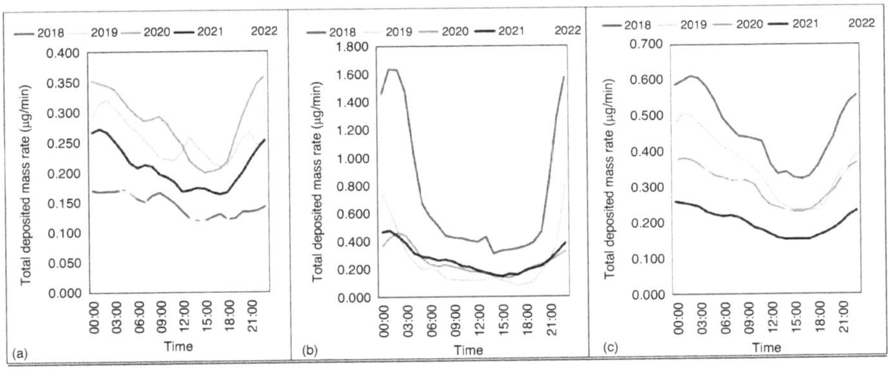

FIGURE 18.5 Diurnal variation of total deposited mass rate (µg/min) during (a) pre-Diwali, (b) Diwali and (c) post-Diwali.

18.4 DEPOSITED MASS RATE OF $PM_{2.5}$ AND METEOROLOGY

The relation of the deposited mass rate of PM2.5 with the meteorological parameters is shown in Table 18.1. The relation of the deposited mass rate with the meteorological parameters is shown in terms of the coefficient of correlation (*r*), along with the *p*-value and coefficient of determination (R^2). It is observed that temperature mostly holds a significant inverse relationship with the deposited mass rate of $PM_{2.5}$,

TABLE 18.1
Relation of Deposited Mass Rate (μg/min) with Temperature, Relative Humidity and Wind Speed

	Temp			RH			WS		
	r	*Pr*	R^2	*r*	Pr	R^2	*r*	*Pr*	R^2
Pre-Diwali 2018	−0.581	0.003	0.338	0.576	0.003	0.332	−0.468	0.021	0.219
Pre-Diwali 2019	−0.639	0.001	0.408	0.660	0.0004	0.436	−0.625	0.001	0.391
Pre-Diwali 2020	−0.753	<0.0001	0.567	0.793	<0.0001	0.629	−0.436	0.033	0.190
Pre-Diwali 2021	−0.691	0.0002	0.477	0.705	0.0002	0.498	−0.720	<0.0001	0.518
Pre-Diwali 2022	−0.814	<0.0001	0.663	0.840	<0.0001	0.706	−0.664	0.0004	0.440
Diwali 2018	−0.622	0.001	0.386	0.698	0.0001	0.487	−0.157	0.465	0.025
Diwali 2019	−0.564	0.004	0.318	0.587	0.003	0.345	−0.452	0.026	0.205
Diwali 2020	−0.776	<0.0001	0.602	0.816	<0.0001	0.666	−0.537	0.007	0.289
Diwali 2021	−0.660	0.0004	0.436	0.677	0.0003	0.458	−0.467	0.021	0.218
Diwali 2022	−0.394	0.057	0.155	0.476	0.019	0.227	−0.261	0.217	0.068
Post-Diwali 2018	−0.788	<0.0001	0.620	0.828	<0.0001	0.685	−0.324	0.123	0.105
Post-Diwali 2019	−0.639	0.001	0.409	0.651	0.001	0.424	−0.355	0.088	0.126
Post-Diwali 2020	−0.774	<0.0001	0.598	0.835	<0.0001	0.698	−0.319	0.128	0.102
Post-Diwali 2021	−0.714	<0.0001	0.510	0.713	<0.0001	0.508	−0.447	0.029	0.199
Post-Diwali 2022	−0.633	0.001	0.030	0.711	<0.0001	0.506	−0.444	0.030	0.197

and relative humidity holds a significant positive correlation with the deposited mass rate. Wind speed also holds significant negative correlation with deposited mass rate, which infers that the higher the wind speed, the lower the ambient $PM_{2.5}$ load, and hence the lower the deposited mass rate. It is observed that a higher percentage of the variability in the deposited mass rate is explained by the meteorological parameters (temperature, relative humidity and wind speed) during the post-Diwali periods than during the pre-Diwali and Diwali periods.

79.4%, 85.7%, 91.9%, 72.6% and 71.8% of the variability in the deposited mass rate of $PM_{2.5}$ is explained by the three meteorological parameters during the post-monsoon periods of 2018, 2019, 2020, 2021 and 2022 respectively.

18.5 DISCUSSION

The data on the variation of $PM_{2.5}$ concentration suggests that the ambient concentration of $PM_{2.5}$ was mostly above the permissible limit of the 24 hours average NAAQS value given by CPCB during the in Diwali and post – Diwali seasons of the years (2018–2022). A higher concentration of ambient $PM_{2.5}$ results in a higher deposited mass rate in different parts of the respiratory tract of 8-year-old children. The pulmonary region is found to have the maximum deposited mass rate, followed by the head and TB regions. Bronchioles and alveolar region of the lungs constitute the pulmonary region. The clearance mechanism in this region is very slow. A high residence time of $PM_{2.5}$ leads to the highest deposited mass rate in the pulmonary region [19]. The lowest deposited mass rate in the tracheobronchial (TB) region is due to the faster clearance rate by mucociliary activity [21]. Although the deposited mass rate remains lower during the daytime compared to the early morning, late evening and night during the Diwali period, the deposited mass rate remains quite high until noon during the post – Diwali period. Children burn a lot of firecrackers during the evening and night and get exposed to the high concentration of ambient $PM_{2.5}$. Children, in particular, burn firecrackers while holding them in their hands, which is barely a foot away from their breathing zone. The $PM_{2.5}$ generated from the smoke of these firecrackers mostly gets deposited in their pulmonary region, posing numerous respiratory and allergic issues in children. Children also get exposed to high ambient $PM_{2.5}$ load during the post-Diwali period, as it remains high until noon during that period.

18.6 CONCLUSION

The present study focuses on the variation of ambient $PM_{2.5}$ during the pre-Diwali, Diwali and post- Diwali periods, as well as the deposited mass rate of $PM_{2.5}$ in different parts of the respiratory tract of 8-year-old children over the Victoria Memorial area of Kolkata. This study shows that the concentration of $PM_{2.5}$ remains above the permissible limit during the Diwali period, especially in the evening and night due to fireworks and crackers, and it remains high even during the daytime in the post Diwali period. Ambient temperature, relative humidity and wind speed play a vital role in ambient $PM_{2.5}$ concentration as low temperature, high relative humidity and low wind speed favours the accumulation of particulate matter. The ambient $PM_{2.5}$ gets deposited in different parts of the respiratory tract of the children. The deposited mass rate of $PM_{2.5}$ is found to be highest in the pulmonary region of the children, followed by the head region and tracheobronchial (TB) region. Children are most vulnerable to health issues related to air pollution exposure, as they are mostly involved in fireworks and bursting crackers during Diwali and stay outdoors. A detailed personal exposure study of children may further help in understanding the health impacts of elevated ambient air pollution during the festive seasons over Kolkata.

ACKNOWLEDGEMENT

The authors are grateful to St. Xavier's College (Autonomous), Kolkata, West Bengal, India and North - Eastern Hill University, Shillong, Meghalaya, India for providing the necessary support for this book chapter.

REFERENCES

[1] Chandu, K., and Dasari, M. (2020). A case study on air pollution during Diwali festival at Visakhapatanam, India. *Ecology, Environment and Conservation Journal* 26(4), 1754–1759.

[2] Thakur, B., Chakraborty, S., Debsarkar, A., Chakraborty, S., and Srivastava, R.C. (2010). Air pollution from fireworks during festival of lights (Deepawali) in Howrah, India- A case study. *Atmosfera* 23(4), 347–365.

[3] Chatterjee, A., Sarkar, C., Adak, A., Mukherjee, U., Ghosh, S.K., and Raha, S. (2013). Ambient air quality during Diwali festival over Kolkata- A mega-city in India. *Aerosol and Air Quality Research* 13(13), 1133–1144.
[4] Bach, W., Danies, A., Dickinson, L., Hertlein, F.; Morrow, J., Margolis, S., and Dinh, D.V. (1975). Fireworks, pollution and health. *International Journal of Environmental Studies* 7, 183–192.
[5] Drewnick, F., Hings, S.S., Curtius, J., Eerdekens, G., and Williams, J. (2006). Measurement of fine particulate and gas-phase species during the New Year's fireworks 2005 in Mainz, Germany. *Atmosphere Environment* 40, 4316–4327.
[6] Vecchi, R., Bernardoni, V., Cricchio, D., D'Alessandro, A., Fermo, P., Lucarelli, F., Nava, S., Piazzalunga, A., and Valli, G. (2008). The impact of fireworks on airborne particles. *Atmospheric Environment* 42, 1121–1132.
[7] Moreno, T., Querol, X., Alastuey, A., Minguillon, M.C., Pey, J., Rodriguez, S., Miro, J.V., Felis, C., and Gibbons, W. (2007). Recreastional atmospheric pollution episodes: inhalable metalliferous particles from firework displays. *Atmosphere Environment* 41, 913–922.
[8] Wang, Y., Zhuang, G., Xu, C., and An, Z. (2007). The air pollution caused by the burning of fireworks during the lantern festival in Beijing, China. *Atmospheric Environment* 41, 417–431.
[9] Central Pollution Control Board. (2009). *National Ambient Air Quality Standards*. New Delhi: Central Pollution Control Board.
[10] Ghei, D., and Sane, R. (2018). Estimates of air pollution in Delhi from the burning of firecrackers during the festival of Diwali. *PLoS One* 13(8), e0200371.
[11] Daga, M.K., Mawari, G., Bharali, D., Banker, H., Mehar, P., Saluja, S., Kumar, D., Kumar, D., Jha, M.K., and Gargava, P. (2019). Assessment of the air quality and its impact on health and environment in India. *Indian Journal of Medical Specialities* 10(3), 117–125.
[12] Shah, R., Limaye, S., Ujagare, D., Madas, S., and Salvi, S. (2019). Personal exposures to particulate matter <2.5 µm in mass median aerodynamic diameter (PM2.5) pollution during the burning of six most commonly used firecrackers in India. *Lung India* 36(4), 324–329.
[13] Gouder, C., and Montefort, S. (2014). Potential impact of fireworks on respiratory health. *Lung India* 31, 375–379.
[14] Passi, A., Shiva Nagendra, S.M., and Maiya, M.P. (2021). Assessment of exposure to airborne aerosol and bio-aerosol particles and their deposition in the respiratory tract of subway metro passengers and workers. *Atmospheric Pollution Research* 12(11), 101218.
[15] Izhar, S., Rajput, P., and Gupta, T. (2017). Variation of particle number and mass concentration and associated mass deposition during Diwali festival. *Urban Climate* 2018, 1027–1036.
[16] Khan, S., Gurjar, B.M., and Sahu, V. (2022). Deposition modeling of ambient particulate matter in the human respiratory tract. *Atmospheric Pollution Research* 13(10), 101565.
[17] Manojkumar, N., Srimuruganandam, S., and Shive Nagendra, S.M. (2019). Application of multiple path particle dosimetry model for quantifying age specified deposition of particulate matter in human airway. *Ecotoxicology and Environmental Safety* 168, 241–248.
[18] Mortensen, J., Schaap, R., Bagley, B., Stout, L., Young, J., Stout, A., Burkart, J., and Baker, C. (1983a). *Final Report: A Study of Age Specific Human Respiratory Morphometry*. University of Utah Research Institute, UBTL Division, University of Utah, Salt Lake City, UT 01525–01010.

[19] Mortensen, J., Young, J., Stout, L., Stout, A., Bagley, B., and Schaap, R. (1983b). A numerical identification system for airways in the lung. *Anatomical Record* 206, 103–114.

[20] Mortensen, J., Stout, L., Bagley, B., Burkart, J., Schaap, R., Crapo, J., Smolko, E., Miller, F., Graham, J., and Hayes, A. (1988). Age related morphometric analysis of human lung casts. In *Extrapolation of Dosimetric Relationships for Inhaled Particles and Gases*, Crapo, J., Smolko, E., Miller, F., Graham, J. and Hayes, A. (Eds.), CA Academic Press, San Diego, pp. 59–68.

[21] Manojkumar, N., and Srimuruganandam, B. (2022). Age-specific and seasonal deposition of outdoor and indoor particulate matter in human respiratory tract. *Atmospheric Pollution Research* 13(2), 101298.

19 Book Recommender System Using Collaborative Filtering

Sukanya Kool, Abhijit Paul, and Tamoghna Mukherjee

19.1 INTRODUCTION

"Machine learning (ML) is the field of study that gives computers the ability to learn without being explicitly programmed," as said by Arthur Samuel (1959) [1]. It is a branch of statistics and artificial intelligence that deals with programming computers to optimize the performance criterion using our previous experience. The following are the types of machine learning.

Supervised machine learning is a type of machine learning where we have an input variable (x) and an output variable (Y) and we can use an algorithm to map the function from input to output (see Figure 19.1):

$$Y = f(x)$$

In this case, the system mainly analyses the current data, and whenever a new dataset and algorithm are fed to the system, it generates the correct solution by matching it with the previously trained dataset. Regression and classification come under supervised learning.

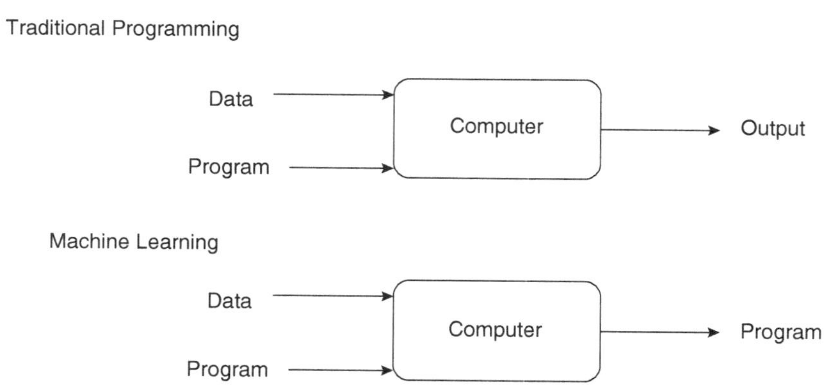

FIGURE 19.1 Difference b/w Traditional Programming and ML.

DOI: 10.1201/9781003638414-19

- **Classification**: Classification problems require that the algorithm predict a discrete value, identifying the input as a member of a specific class or group. In a training dataset of animal pictures, this would mean each photo is pre-labeled as a cat, koala, or turtle. The algorithm is then evaluated by how accurately it can classify new pictures of different koalas and turtles.
- **Regression**: A regression problem is one where the output variable is a real or continuous value, such as "salary" or "weight." A wide range of models can be used, with the simplest being linear regression. It attempts to fit the data with the best hyperplane that passes through the points.

Unsupervised learning can generally be termed as "learning without a teacher." In this case, the training data does not include targets, so we don't tell the system where to go; the system has to understand it itself from the data we provide. Clustering and association come under unsupervised learning.

- **Clustering**: Without being an expert ornithologist, it's possible to look at a collection of bird photographs and separate them generally by species, based on cues like feather color, size, or beak shape. That is how the most common application for unsupervised learning, clustering, works.
- **Association**: Fill an online shopping basket with diapers, fruit juice, and sippy cups, and the website might recommend that you add a kiddie apron and a baby monitor to your order. This is an example of association, where certain features of a data sample correlate with other features. By examining a few key characteristics of a data point, an unsupervised learning model can predict other traits with which they are commonly associated.

Semi-supervised learning is, generally, exactly what it sounds like: a training dataset with both labeled and unlabeled information. This technique is especially valuable when extracting significant features from the data is difficult, and labeling models is a time-intensive task for experts. Common scenarios for this type of learning are medical images, like CT scans or MRIs. A trained radiologist can review and label a small subset of scans for tumors or diseases. It would be too time-consuming and expensive to manually label all the scans, but the deep learning system can still benefit from the small portion of labeled data and improve its accuracy compared to a fully unsupervised model.

Reinforcement learning is a particular kind of machine learning paradigm that teaches an agent to make decisions by interacting with its surroundings. Based on the decisions it makes, the agent is rewarded or penalized, and its objective is to eventually figure out a policy that maximizes the total reward.

Use cases:

- **Self-Driving Cars**: A standout among the most obvious applications promised by the modern resurgence in AI is self-driving vehicles. Automobiles are likely the most dangerous modern technology to be accepted and embraced as a regular necessity, with annual road traffic deaths estimated at 1.25 million worldwide by the World Health Organization. Quantifying the economic effects of autonomous vehicles is challenging, but conservative estimates range from $190 billion to $642 billion per year in the U.S. Modern autonomous driving has roots in projects like ALVINN, which controlled the Carnegie Mellon autonomous driving testbed NAVLAB, and Ernst Dickmann's work at Bundeswehr University during the 1990s. Many aspects of these projects may seem familiar to modern self-driving engineers: ALVINN used a neural network to predict steering commands from a 30×32 video feed and laser rangefinder inputs. To overcome processing bottlenecks, Dickmann's dynamic vision framework concentrated computation on predefined areas of images based on predicted significance, a concept very similar to neural attention in modern systems. Many of the innovations from the self-driving cars of the 1990s were crucial for overcoming bottlenecks in data throughput or processing. More recently, the availability of computing power is no longer the bottleneck, and the primary challenges in overcoming the final 10% or so of obstacles for full vehicle autonomy are related to ensuring training data reliability and effectively handling rare edge cases. The success of neural networks in computer vision tasks makes them an obvious choice for deep learning, and a typical engineering approach may combine deep learning for vision with other learned or hard-coded modules to handle all aspects of driving. Gradually, rather than designing and training separate individual models that must be forced to cooperate by many skilled engineers, autonomous driving engineers are relying on intelligent models with high-quality training data and carefully considered target functions to learn a more comprehensive set of skills for driving.

- **Computer Games (AlphaGo)**: Reinforcement learning made a breakthrough in March 2016 when DeepMind's AlphaGo, trained using RL, defeated 18-time world champion Go player Lee Sedol 4-1. It turns out the game of Go was difficult to master for the machine, more so than games like chess, primarily because there are just too many possible moves and an overwhelming number of states the game can be in. However, how did AlphaGo beat the best in the world? Much like Mario, AlphaGo learned through experimentation over many cycles. AlphaGo doesn't know the best strategy, but it knows whether it won or lost. AlphaGo uses a tree search to examine each possible move it can make and see which is better. On a 19×19 Go board, there are 361 possible moves. For each of these 361 moves, there are 359 possible second moves, and so on. Altogether, there are about 4.67×10^385 possible moves—that's far too many. Even with its advanced hardware, AlphaGo cannot try every move. Therefore, it uses a different kind of tree search called the Monte Carlo Tree Search. In this search, only those moves that are most promising are explored.

A recommendation system is an information filtering system that primarily seeks to predict the 'preference' and 'rating' a user might give to an item. In the later part of this report, we discuss the Book Recommendation System. Recommendation systems are broadly used to recommend items to the most appropriate end users. Online book-selling sites nowadays are competing with each other using various methods. The recommendation system is one of the stronger tools to increase profit and retain customers. The book recommendation system must suggest books that align with the buyer's interests.

19.2 LITERATURE REVIEW

Why Recommendation System is required for e-commerce websites? The simplest answer to this question is that recommender systems introduce users to a wide range of items that they are likely to be interested in. Nowadays, the recommendation system is a fundamental component of any e-commerce website [2,3].

A couple of decades prior, individuals regularly depended on their companions, family, or specialists to advise them on what books they should read, films they should watch, or eateries they should try. However, nowadays, we are considerably more prone to depend on taste-production algorithms to help guide which purchases we make. Also, we depend on these algorithms so regularly that, most of the time, we may not even realize we're doing it anymore. We add a book to our cart on Amazon and then, with a click, another recommended book pops up on the screen. We think, "Okay, that book looks interesting, too," and we add it to our cart as well, without the slightest hesitation.

Paradigms of Recommendation System:
The two basic paradigms involved in any recommendation system are

- Content-based Filtering
- Collaborative Filtering

Content-based Filtering:
In this algorithm, we try to find items that look alike. Once we have an item look-alike matrix, we can easily recommend similar items to a customer who has purchased an item from the store.

It is a recommendation system technique that suggests items to users based on the characteristics or content of the items and the preferences of the users. This approach involves analysing the features of items and creating user profiles to recommend items that match the users' preferences. In Figure 19.2, we can see that content-based filtering mainly works by predicting similar types of books and then recommending them to the user. Here's a mathematical explanation of content-based filtering:

1. **Item Representation**:
   - Let I be the set of items, and each item $i \in I$ is represented by a feature vector $v_i = [v_{i1}, v_{i2}, ..., v_{in}]$, where n is the number of features.
2. **User Preferences**:
   - Each user u is associated with a preference vector $p_u = [p_{u1}, p_{u2}, ..., p_{un}]$ indicating the user's preferences for different features.

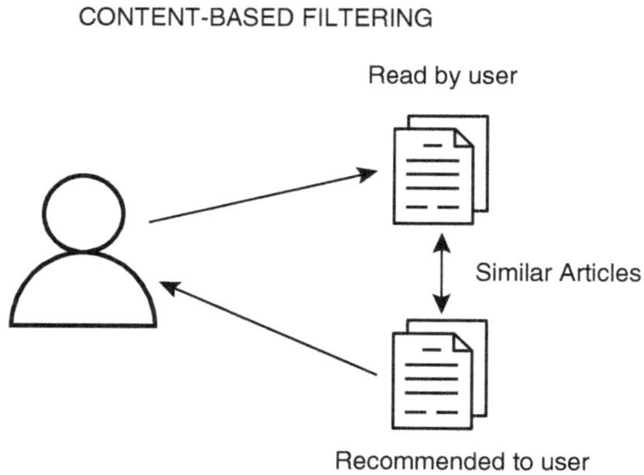

FIGURE 19.2 Content-based filtering approach.

3. **Item-User Compatibility**:
   - The compatibility between a user and an item can be measured using a similarity function, such as cosine similarity or Euclidean distance. The similarity score s_{ui} between user u and item i can be computed as:

$$s_{ui} = \frac{p_u . v_i}{|p_u| . |v_i|} \quad (19.1)$$

4. **Recommendation Score**:
   - To generate recommendations for a user, we compute a recommendation score for each item based on the user's preferences and the item's features. A common approach is to use a weighted sum of the features:

$$r_{ui} = p_u . v_i = \Sigma_{j=1}^{n} p_{uj} . v_{ij} \quad (19.2)$$

5. **Top-*k* Recommendations**:
   - Once the recommendation scores r_{ui} are computed for all items, the system can recommend the top k items with the highest scores to the user.
6. **Normalization**:
   - It's common to normalize the recommendation scores to ensure that they fall within a specific range. This can be done using a sigmoid function or other normalization techniques.

   Collaborative Filtering: Here, we try to search for lookalike customers and offer products based on what his/her lookalike has chosen. This algorithm is very effective but takes a lot of time and resources.

In Figure 19.3, we see that filtering is being done on similar types of users based on their previously read books, and the new book is recommended to one person by checking whether the book has been read by a similar type of person or not. It is a technique that uses information from a user, such as ratings or purchases made by the user, to recommend items to other users of the site who have similar tastes. This can be done by either matching similarities between users or items.

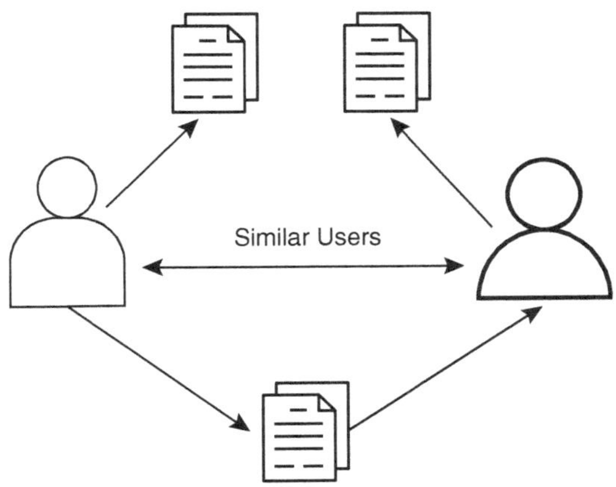

FIGURE 19.3 Collaborative filtering approach.

The two approaches have been given the names User-based collaborative filtering and Item-based collaborative filtering.

There are two main types of collaborative filtering: user-based and item-based. Here's a mathematical explanation:

1. **User-Item Matrix**:
   - Let U be the set of users and I be the set of items. The user-item matrix R is a matrix where R_{ui} represents the rating or interaction that user u has given to item i. This matrix is often sparse since not all users rate all items.
2. **User Similarity**:
   - Define a similarity measure between the users. Cosine similarity is a common choice. The similarity score s_{uv} between user u and user v can be calculated as:

$$s_{uv} = \frac{p_u \cdot p_v}{|p_u| \cdot |p_v|} \tag{19.3}$$

   where p_u and p_v are the preference vectors of users u and v, respectively.
3. **Rating Prediction**:
   - To predict the rating R_{ui} that user u would give to item i, the system takes a weighted average of the ratings of u's similar users for item i. The predicted rating

     R_{ui} can be computed as:

$$R_{ui} = \frac{\Sigma_{v \epsilon N(u)} s_{uv} . R_{vi}}{\Sigma_{v \epsilon N(u)} |s_{uv}|} \tag{19.4}$$

   where $N(u)$ is the set of users similar to user u.

4. **Top-*k* Recommendations**:
   - Once the predicted ratings are computed for all items, the system can recommend the top k items with the highest predicted ratings to the user.

19.2.1 Why Collaborative-Based Filtering?

The dataset we downloaded from Kaggle.com comprises sections such as user_ratings, users, book_titles, and so on. We utilized a collaborative filtering approach because, in this case, we try to search for similar customers and offer products based on what their lookalikes have chosen. Since we don't have sections like "book_type" or "genre of the book," we cannot use a content-based filtering approach, as this algorithm aims to find items that resemble each other. Once we have an item look-alike matrix, we can easily recommend similar items to a customer who has purchased an item from the store.

For example, assume a user 'A' purchases books on 'horror' and 'thriller' on a continuous basis. It can then be predicted from their purchasing pattern that A is more likely to prefer horror and thriller books. Therefore, the next recommended book to A should either be a horror or a thriller-based book.

In summary, collaborative filtering uses the preferences of similar users to predict the preferences of a target user, and recommendations are generated based on these predicted preferences. The similarity between users is a key component, and various similarity measures can be employed depending on the context.

19.2.2 User-Based Method of Collaborative Filtering

Suppose, there is a user A. Now, many people like "Harry Potter" books.

But can we say how much A likes it? The idea is to pick 20–50 people who share similar tastes with A, then how much Alike depends on how much those picked people liked.

It can be said as a whole that "we may like that book because our friends like it".

19.2.3 Item-Based Method of Collaborative Filtering

Suppose A has read many good and bad books. Would we recommend him to read "The Thousand Splendid Suns" book? -

The idea is to pick 20–50 books from the dataset that share a similar audience to that of "The Thousand Splendid Suns", then how much A will like depends on how much A liked those previous books.

In a nutshell, it can be said that "People who have bought these books also liked this book" and likewise the books are recommended.

19.2.4 Drawbacks of Collaborative Filtering

- **Scale**: The data to deal with in Collaborative Filtering are very huge in range. This algorithm is very effective but takes a lot of time and resources.
- **Sparse Data**: There might be users who have rated only one book.

- **Cold-Start**: There might be users and items who are new and do not have any history or previous records.
- **Popularity Bias**: There are books like "Harry Potter" which are being read by almost all the users.

19.3 THE PLATFORM-ANACONDA

19.3.1 What is an Anaconda Navigator?

Anaconda Navigator is a desktop graphical user interface(GUI) incorporated into Anaconda®'s dissemination that enables you to dispatch applications and effectively oversee conda packages, situations, and channels without utilizing direction line directions [4,5]. A pilot can scan for bundles on Anaconda Cloud or in a neighbourhood Anaconda Repository. It is accessible for Windows, macOS, and Linux.

19.3.2 Why Use Navigator?

To run, many scientific packages rely on specific versions of other packages. Data scientists often use multiple versions of many packages and use different environments to isolate these versions. The command-line program conda is both a package manager and an environment manager, helping researchers ensure that each version of every package has all the conditions it requires and works effectively. Navigator is a simple, point-and-click approach to working with packages and environments without needing to type conda commands in a terminal window. You can use it to find the packages you need, install them in an environment, run the packages, and update them, all within Navigator.

19.3.3 How Anaconda is Useful for Data Scientists?

The pulsating heart of Anaconda Enterprise is the collection of core AI/ML tools in Python and R that data scientists use to build models. With Anaconda Enterprise, data scientists have access to 1,500+ packages for data science out of the box, hosted internally to ensure access and availability behind your organization's firewall. When users log in to Anaconda Enterprise, they can browse model projects and preconfigured project templates that address common problems and make getting started easy. Anaconda Enterprise requires zero setup from users—just log in, pick a notebook template, and then start importing packages, exploring data, and building models. Directly from the platform and without involving DevOps, data scientists can quickly develop and deploy AI and machine learning models into production. Anaconda provides the tools needed to easily:

- Collect data from files, databases, and data lakes
- Manage environments with Conda (all package dependencies are taken care of at the time of download)
- Share, collaborate on and reproduce projects
- Deploy projects into production with the single click of a button.

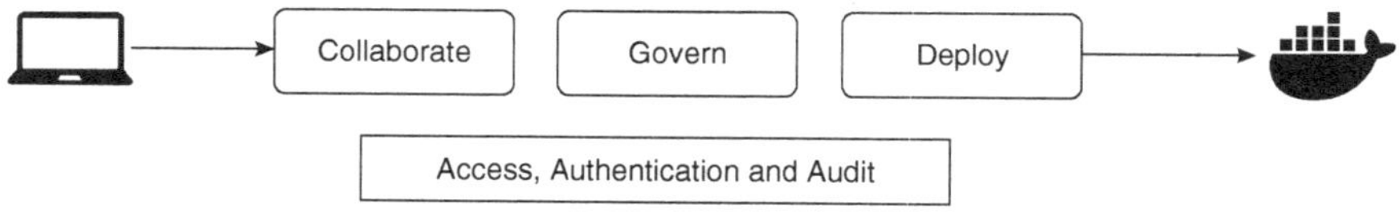

FIGURE 19.4 Anaconda Architecture for AI/ML.

19.4 ARCHITECTURE OF ANACONDA

Figure 19.4 shows the architecture of Anaconda for the projects based on Artificial Intelligence (AI) and Machine Learning (ML). One can use Jupyter Notebook, import the libraries required either in Python or R, import the dataset, and work on the project.

19.4.1 What Applications One Can Access Using Navigator?

The following applications are available by default in Navigator:

- JupyterLab
- Jupyter Notebook
- QTConsole
- Spyder
- VSCode
- Glueviz
- Orange 3 App
- Rodeo
- RStudio

19.4.2 Version Used in the project

Anaconda 3 has been used to implement the project on the Book Recommendation System.

19.5 PROGRAMMING LANGUAGE

The language used for the implementation of the Book Recommendation System in Python of version 3.6.

19.5.1 About Python

The "Python library" contains several different kinds of components.

It contains data types that would normally be considered part of the "core" of a language, such as numbers and lists. For these types, the Python language core defines the form of literals and places some constraints on their semantics, but it does not fully define the semantics. (On the other hand, the language core does define syntactic properties like the spelling and precedence of operators.) The library also contains built-in functions and exceptions—objects that can be used by all Python code without the need for an import statement. Some of these are defined by the core language, but many are not essential for the core semantics and are only described here. The bulk of the library, however, consists of a collection of modules. There are many ways to categorize this collection. Some modules are written in C and built into the Python interpreter; others are written in Python and imported in source form.

19.5.2 Why Python?

Python is widely considered the preferred language for teaching and learning ML (Machine Learning). A few simple reasons are:

It's simple to learn. As compared to C, C++, and Java the syntax is simpler and Python also consists of a lot of code libraries for ease of use.

Though it is slower than some of the other languages, the data handling capacity is great.

- Open Source – Python along with R is gaining momentum and popularity in the Analytics domain since both of these languages are open source.
- Capability of interacting with almost all third-party languages and platforms.
- Python is the tool of choice for people who want to explore and apply machine learning to their research projects, such as data analysis, pattern recognition, etc. Python comes with several highly attractive machine-learning libraries that can be applied out of the box.
- It has numerous packages for machine learning and other computations. Prime examples include NumPy, pandas, seaborn, and matplotlib. These packages are well-documented, which helps in starting new projects and solutions. They also speed up the process of fixing bugs.

19.6 LIBRARIES

Libraries used in Python

19.6.1 NumPy

NumPy is a low-level library written in C (and FORTRAN) for high-level mathematical functions. NumPy cleverly overcomes the problem of running slower algorithms in Python by using multidimensional arrays and functions that operate on arrays. Any algorithm can then be expressed as a function on arrays, allowing the algorithms to run quickly.

- **Applications of NumPy**:
  1. a powerful N-dimensional array object
  2. sophisticated (broadcasting) functions

3. tools for integrating C/C++ and Fortran code
4. useful linear algebra, Fourier transform, and random number capabilities

19.6.2 Pandas

Pandas is a software library written for the Python programming language for data manipulation and analysis. In particular, it offers data structures and operations for manipulating numerical tables and time series.

- **Features of Pandas**:
  1. DataFrame object for data manipulation
  2. Tools for reading and writing data between in-memory data structures and different file formats.
  3. Integrated handling of missing data.
  4. Reshaping and pivoting of data sets.
  5. Data set merging and joining.

19.7 ARCHITECTURE

The architecture of the recommendation system is shown in Figure 19.5.

There are three datasets named 'BX-Books.csv', 'BX-Users.csv', and 'BX-Book-Ratings.csv'. The datasets contain information about books, such as their

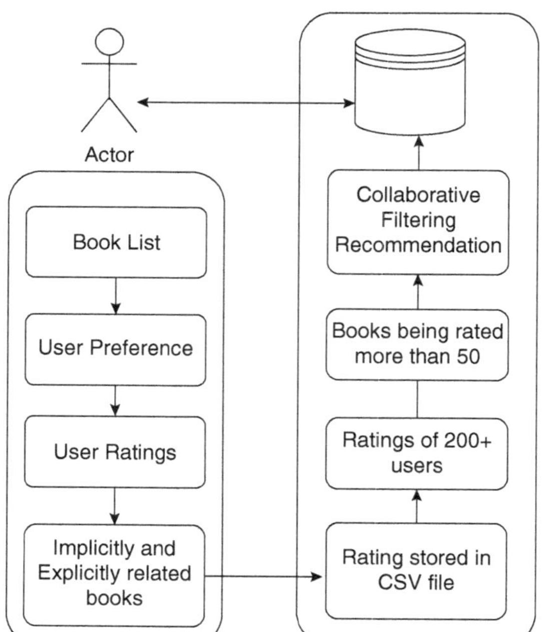

FIGURE 19.5 Architecture of Book Recommendation System.

publisher, title, and ISBN; the user dataset contains information about the user_id, age, and location; and the rating dataset contains data about ISBN, user_id, and their ratings. These datasets are loaded, cleaned, and prepared for the algorithm. To prepare the data, we ensured its accuracy by discarding repetitive data, considering only those users who have rated more than 200 books and books that have been rated more than 50 times. A collaborative filtering approach is then applied in the implementation of the system.

19.8 METHODOLOGY

Step 1: Sample Data Set

- Downloading a sample dataset from Kaggle.com for Book Recommendation System
- Importing the libraries
- Loading

  The libraries used for the implementation of the Book Recommendation system, namely NumPy and Pandas, are imported, and the downloaded dataset with the .csv extension is loaded. In addition to the book dataset, there are two more datasets—users and ratings—which are also loaded.

Step 2: Cleaning After loading the data, the unwanted data is removed because not all data is required for the algorithm. For instance, in the BOOKS dataset, there were columns for images of books, which are unnecessary for our system.

Step 3: Column name changing for the user's interest. This step is optional. The column names are changed according to the user's preference, but they may also remain unchanged.

Step 4: Calculating value counts of the rating where it is greater than 200. In the fourth step, we check the user_id to identify users who have rated more than 200 books.

Step 5: Merging the tables and saving them in a tuple variable.

Step 6: Checking the ratings of each book and changing the index. An index is added to the table, and the number of ratings for each book is counted and saved in "number_rating."

Step 7: Merging the previous table with a new index. Now, a merge operation is performed with the previous table and "number_rating," and the index is added.

Step 8: Keep the books that have more than 50 ratings. After the previous operation, we discarded other book details and kept only those books that have been rated more than 50 times, which are saved in "final_rating."

Step 9: Removing the same user ratings on the same book.

Now, certain users have rated the same book more than once, which is not proper data to work with as it could lead to incorrect results. Therefore, we have removed the duplicate ratings from the same user for the same book.

Step 10: Converting the rating table into a 2D matrix

After removing the unwanted data we changed the final_rating table into a 2D matrix. There are many '0' values in the matrix which need to be removed. In the next step, it is being done.

```
In [72]  for i in range(len(suggestions)):

Index(['Harry Potter and the Chamber of Secrets (Book2)',
'Harry Potter and the Prisoner of Azkaban (Book3)',
'Harry Potter and the Goblet of Fire (Book 4)',
'Harry Potter and the Sorcerer's Stone (Book1)', 'Exclusive',
'The Cradle Will Fall'],
dtype='object', name='title')
```

FIGURE 19.6 The recommendation of books after the pivot point.

Step 11: Removing '0' values in matrix

The '0' values are removed from the matrix to make the table suitable for the algorithm. This is done using csr_matrix imported from scipy. sparse, and it is stored in book_sparse. The type of book_sparse is then displayed.

Step 12: Finding the nearest neighbours:

Finding the distance and suggestion using *k*-neighbours

Step 13: Generating a generalized function

After the suggestions and distances are calculated, i.e., the recommended books are found using Nearest Neighbors, the names of the suggested books are printed, as shown in Figure 19.6. Finally, a generalized function for recommending books is created to find the recommended books concerning a pivot point.

19.9 CONCLUSION

There is a surge of interest in Artificial Intelligence and Machine Learning nowadays due to Big Data. The issue of scalability was addressed to some extent by Mahout, thanks to the presence of the Hadoop system. However, with the increasing size of datasets, we need to explore other Machine Learning frameworks as well. A large number of Machine Learning toolkits are available in the market, but one size does not fit all. Experts sometimes reject them due to the lack of feature availability. For instance, in the case of a recommender system, we need at least one type of filtering mechanism as part of the framework. Keeping these possibilities in mind, we can compare the following Machine Learning frameworks supporting Collaborative Filtering: Distributed-Weka, Flink-ML, and Oryx, with Apache Mahout in terms of scalability, speed, coverage, and so on. Recommender systems are becoming an effective tool embedded into websites to enhance the user experience. The successful implementation of a recommender system can be a boon for websites aiming to integrate recommendations as part of their system. We have learned a lot about recommendation systems, the tools and algorithms used for them, as well as the types of Machine Learning. Additionally, the paradigms of recommendation systems, such as Collaborative Filtering and Content-based Filtering, have been discussed.

REFERENCES

[1] Zhou, Z. H. (2021). *Machine Learning*. Springer Nature, Singapore.

[2] Harrison Jr, J. H., Gilbertson, J. R., Hanna, M. G., Olson, N. H., Seheult, J. N., Sorace, J. M., & Stram, M. N. (2021). Introduction to artificial intelligence and machine learning for pathology. *Archives of Pathology & Laboratory Medicine*, 145(10), 1228–1254.

[3] Singh, A. V., & Sharma, M. (2017, December). Building an effective recommender system using machine learning based framework. In *2017 International Conference on Infocom Technologies and Unmanned Systems (Trends and Future Directions) (ICTUS)* (pp. 215–219). IEEE.

[4] Hukkeria, G. S., Goudarb, R. H., & Kulkarnic, S. (2019). Powering the internet of things and big data through machine learning. *International Journal of Engineering Sciences and Management- A Multidisciplinary Publication of VTU*, 1(1), 56–61.

[5] Anaconda documentation. Accessed April 11 2019, https://docs.anaconda.com/anaconda/install/hashes/win-3-64/.

Index

Kolibri 360 GmbH
Ferdinand-Jühlke-Straße 7
99095 Erfurt, Deutschland
produktsicherheit@kolibri360.de